Ancient India as described by Ptolemy; being a translation of the chapters which describe India and Eastern Asia in the treatise on Geography written by Klaudios Ptolemaios, with introduction, commentary, map of India according to Ptolemy, and index.

Ptolemy, John Watson. Maccrindle

Ancient India as described by Ptolemy; being a translation of the chapters which describe India and Eastern Asia in the treatise on Geography written by Klaudios Ptolemaios ... with introduction, commentary, map of India according to Ptolemy, and ... index, by J. W. McCrindle.
Ptolemy
British Library, Historical Print Editions
British Library
1885
xii. 373 p. ; 8°.
T 4055/4

Works by J. W. McCRINDLE, M.A., M.R.A.S., Late Principal of the Government College, Patna, and Fellow of the University of Calcutta.

ANCIENT INDIA

AS DESCRIBED BY THE CLASSICAL AUTHORS;

BEING a series of copiously annotated Translations of all extant works relating to India, written by The ANCIENT GREEK and ROMAN Authors. The following volumes of the series have appeared :—

I.—ANCIENT INDIA AS DESCRIBED BY MEGASTHENÊS AND ARRIAN : Being a translation of the fragments of the Indika of Megasthenês, and of the first part of the Indika of Arrian: with Introduction, Notes, and Map of Ancient India ....Rs. 2-8

II.—THE COMMERCE AND NAVIGATION OF THE ERYTHRÆAN SEA: Being a translation of the Periplus Maris Erythraei (Circumnavigation of the Red Sea and Indian Ocean), and of Arrian's account of the Voyage of Nearkhos: with Introduction, Commentary and Index ....Rs. 3-0

III.—ANCIENT INDIA AS DESCRIBED BY KTÊSIAS THE KNIDIAN: Being a translation of the fragments of his Indika, with Introduction, Notes and Index ....Rs. 3

IV.—ANCIENT INDIA AS DESCRIBED BY PTOLEMY: Being a translation of the Chapters on India and on Central and Eastern Asia in the Treatise on Geography by Klaudios Ptolemaios, the celebrated Astronomer: with Introduction, Commentary, Map of India according to Ptolemy, and a very copious Index..Rs. 4-4

The 5th and last volume will contain Strabo's Geography of India and the accounts of the Makedonian Invasion of India given by Arrian and Curtius.

BOMBAY: EDUCATION SOCIETY'S PRESS.

CALCUTTA: THACKER, SPINK & Co.

LONDON : TRÜBNER & Co.

33

OPINIONS OF THE PRESS.

In rendering the results of Dr. Schwanbeck's industry accessible to English readers by this translation of the collected fragments of the lost *Indika* of Megasthenês, perhaps the most trustworthy of the Greek writers on India, Mr. McCrindle would have performed a most valuable service even had he not enriched the original by the addition of copious critical notes, and a translation of Arrian's work on the same subject.—*Calcutta Review.*

Mr. McCrindle's translations of the accounts of Ancient India by Megasthenês and Arrian is a most valuable contribution to our knowledge of the subject in the days when Greeks and Romans were ruling the world . . . . Mr. McCrindle has conferred a great boon on society by translating Dr. Schwanbeck's learned work into English, illustrating it by a valuable map of Ancient India, and publishing it at a small price. There is more *bonâ fide* information regarding Ancient India in this unpretending volume than is to be found in the great bulk of Sanskrit Puranas; whilst it forms a most valuable adjunct to the mass of traditions and myths which have been preserved in the Hindu epics of the Mahabharata and Ramayana, &c.—*Pioneer.*

Mr. J. W. McCrindle, of Patna, has given us a readable translation both of Schwanbeck's Megasthenês, and of the first part of Arrian's *Indika.* Mr. McCrindle deserves the thanks of all who take an interest in Ancient India, and, should he be able to fulfil his promise to translate "the entire series of classical works relating to India," he will give an impetus to the study of the early civilization of this country among native as well as European scholars. His work is well printed, and, as far as we have been able to judge, carefully edited.—*The Madras Times.*

Mr. McCrindle, who has already published a portion of the translation of Arrian, reprints these valuable contributions to our scanty knowledge of Ancient India. . . . . . An introduction and notes add value to the translation, a value which happens to be very great in this case, and to centre in one long note on the identification of the old Palibothra or Pataliputra with the modern Patna.—*The Daily Review.*

Mr. McCrindle, who holds a very high position in the Education Department of the Indian Government, has collected into a volume some translations which he has lately contributed to the "Indian Antiquary" from Megasthenês and Arrian. . . . . Strabo and Pliny thought fit to condemn the writings of Megasthenês as absolutely false and incredible, although they were glad to copy into their own works much that he had written. We moderns, however, with our longer experience of travellers' tales, and of the vitality of fabulous statements, and practised in comparing accounts that vary, find much in these fragments that agrees with what we can reasonably conjecture of the past of India. . . . . . We may observe that many of the singularities of the human race, which are depicted on the famous *Mappemonde*, at Hereford, are described by Megasthenês. Mr. McCrindle's volume ends with an excellent translation of the first part of Arrian's Indika. He is to be congratulated on having made a very useful contribution to the popular study of Indian antiquities.—*Westminster Review.*

A good notion of the extent of the knowledge respecting India possessed by the old Greeks and Romans may be formed from the translation of the writings on the subject of Megasthenês and Arrian, presented by Mr. McCrindle, under the title of *Ancient India.* Many of the statements made by the old writers are unmixed fable, although Megasthenês, there can be no doubt, travelled as far as Bengal, but on the whole, as much accurate knowledge was possessed by the Romans in the first century after Christ, as by the European nations in the 16th century. An introduction, notes, and map of India add to the practical utility of Mr. McCrindle's work.—*Scotsman.*

Both of these ancient works are very interesting as illustrating the knowledge possessed by the later Greeks and the Romans respecting the geography of India and the neighbouring regions. Mr. McCrindle's prefaces, each with an informatory introduction, embody the results of the most recent investigations of modern scholarship on the subjects to which they relate.—*Scotsman.*

Mr. J. W. McCrindle, Principal of the Government College, Patna, has set himself the task of publishing, from time to time, translations of the Greek and Latin works which relate to Ancient India, and in pursuance of this intention, some time since he published a work entitled *Ancient India as described by Megasthenês and Arrian.* A second instalment has now appeared under the title of *The Commerce and Navigation of the Erythræan Sea; being a translation of the Periplus Maris Erythraei*, by an anonymous writer, *and of Arrian's account of the Voyage of Nearkhos, from the mouth of the Indus to the head of the Persian Gulf*, with introduction, commentary, notes, and index. The introduction and commentary embody the main substance of Müller's prolegomena and notes to the *Periplus* and of Vincent's *Commerce and Navigation of the Ancients*, so far as it relates specially to that work. The identification of places on the Malabar and Coromandel Coasts is derived from Bishop Caldwell's *Dravidian Grammar.* Other recent works have been resorted to for verification and correction of the contents of the narrative. To those students who have neither the learned work of Dr. Vincent, nor the *Geographi Graeci Minores* of C. Müller, within reach, this handy volume will prove very serviceable.—*The Academy*.

The careful and scholarly translations of ancient texts relating to India, which Mr. McCrindle is preparing in serial order, promises to be of great value. The method which he follows is in accordance with the best traditions of English scholarship . . . . . As to the historical importance of these texts there can be only one opinion. History in Sanskrit literature is conspicuous by its absence, so that external authorities are at once the only ones available, and at least redeem by their unbiassed character their relatively deficient opportunities of information. Those who are best acquainted with the difficulties of English rule in India, are best aware that the problems of Indian administration are, in fact, problems of Indian history. . . . It is thus of vital importance that every possible hint and clue as to the course of the legal, social, and economical history of the country should be made available. It is the special value of Mr. McCrindle's work that it will form a solid, positive basis for the earliest period of authentic Indian history, &c.—*The Civil and Military Gazette*, Lahore.

The fragments of the *Indika* of Megasthenês, collected by Dr. Schwanbeck, with the first part of the *Indika* of Arrian, the *Periplus Maris Erythraei*, and Arrian's account of the voyage of Nearkhos have been translated, in two most useful volumes, by Mr. J. W. McCrindle, M.A. The Indika of Ktesias with the fifteenth book of Strabo is also promised, and the sections referring to India in Ptolemy's Geography would complete a collection of the highest value to Indian history.—*Note under the article India, in the new edition of the Encyclopædia Britannica.*

We are glad to learn that the papers by Mr. J. W. McCrindle (on Ptolemy's Geography of India) which have recently been appearing in the *Indian Antiquary* are to be published separately. The amount of patient and scholarly work which they indicate is of the kind that we are rather accustomed to look for from a German *savant*, and can hardly be properly appreciated by one who does not know by experience the difficulties of such investigations.—*The Scottish Geographical Magazine.*

ANCIENT INDIA

AS DESCRIBED BY

PTOLEMY;

BEING

A TRANSLATION OF THE CHAPTERS WHICH DESCRIBE INDIA AND CENTRAL AND EASTERN ASIA IN THE TREATISE ON GEOGRAPHY WRITTEN BY KLAUDIOS PTOLEMAIOS, THE CELEBRATED ASTRONOMER,

WITH

INTRODUCTION, COMMENTARY, MAP OF INDIA ACCORDING TO PTOLEMY, AND A VERY COPIOUS INDEX,

BY

J. W. McCRINDLE, M.A., M.R.A.S.,

FORMERLY PRINCIPAL OF THE GOVERNMENT COLLEGE, PATNA, AND FELLOW OF THE UNIVERSITY OF CALCUTTA; MEMBER OF THE GENERAL COUNCIL OF THE UNIVERSITY OF EDINBURGH.

Reprinted from the "Indian Antiquary," 1884.

Calcutta:
THACKER, SPINK & Co.

Bombay:
B. E. S. PRESS.

London:
TRÜBNER & Co.

1885.

BOMBAY:

PRINTED AT THE EDUCATION SOCIETY'S PRESS.

PREFACE.

PTOLEMY'S "Treatise on Geography," like his famous work on astronomy to which it formed the sequel, was destined to govern the world's opinion on the subject of which it treated, from the time of its publication until the dawn of the modern era, a period of about 1,300 years. This treatise must have been composed in the interests of chartography rather than of geography, for the author's aim is not so much to describe the earth's surface as to lay down the principles on which maps should be constructed, and to determine the latitude and longitude of places with a view to their being mapped in their proper positions. The principles he here laid down have proved of permanent validity, and are still practically applied in the art of map-construction, but his determinations of the position of places, owing to the paucity and imperfection of the astronomical observations on which, in combination with the existing measurements of terrestrial distances his conclusions were based, are all, with very few exceptions, incorrect. The work lost, of course, much of its old authority as soon as the discoveries of modern times had brought its grave and manifold errors to light. It did not, however, on this account cease to be of high interest and value as an antiquarian record, if we may judge from the multiplicity of the learned disquisitions which have from time to time been published in elucidation of many points of Ptolemaic Geography.

There is perhaps no part of the contents which has received more attention from scholars than the chapters relating to India, where the tables abound to a surprising extent with names which are found nowhere else in classical literature, and which were doubtless obtained directly from Indian sources, rather than from reports of travellers or traders who had visited the country. On glancing over these names one cannot fail to remark how very few of them have any but the most distant resemblance to the indigenous names which they must have been intended to represent. Philologists, however, have made persistent efforts to penetrate the disguise which conceals the original forms of the names so much distorted by Ptolemy, and have succeeded in establishing a great number of satisfactory identifications, as well as in hitting upon others which have a balance of probability in their favour—a similar service has been rendered by the archæological investigations which have now for many years been systematically prosecuted under the auspices of the Indian Government.

The present work has for its main object to show concisely what has been accomplished up to this time in this department of enquiry. It has been compiled from multifarious sources which are not easily accessible, as for instance from foreign publications not yet translated into our own language, and from the Journals and Transactions of various societies at home and abroad which concern themselves with Oriental literature.

I venture therefore to hope that my compendium, which it has taken much time and laborious research to prepare, may meet with recognition and acceptance as a useful contribution to general literature, while proving also serviceable to scholars as a work of reference.

I proceed now to indicate the method which I have followed in the treatment of my subject, and to specify the authorities on which I have principally relied. I have then, in an introductory chapter, attempted to give a succinct account of the general nature of Ptolemy's geographical system, and this is followed by a translation of several chapters of his First Book which serve to exhibit his general mode of procedure in dealing with questions of Geography, and at the same time convey his views of the configuration of the coasts of India, both on this side the Ganges and beyond. In translating the text I have taken it in detachments of convenient length, to each of which I have subjoined a commentary, the main object of which is—1st, to show, as far as has been ascertained, how each place named by Ptolemy in his Indian Tables has been identified; 2nd, to trace the origin or etymology of each name, so far as it is possible to do so; and 3rd, to notice very concisely the most prominent facts in the ancient history of the places of importance mentioned. I have, as a rule, quoted the sources from which my information has been derived, but may here state that I have generally adopted the views of M. Vivien de Saint-Martin and those of Colonel Yule, whose map of ancient India in Smith's well-known historical *Atlas of Ancient Geography* is allowed on all hands to be the best that has yet been produced. These authors have examined the greater part of the Ptolemaic Geography of India, and their conclusions are for the most part coincident. The works of Saint-Martin, which I have consulted, are these: *Étude sur la Géographie Grecque et Latine de l'Inde, et en particulier sur l'Inde de Ptolémée, dans ses rapports avec la Géographie Sanskrite*; *Mémoire Analytique sur la Carte de l'Asie Centrale et de l'Inde*; et *Étude sur la Géographie et les populations primitives du Nord-Ouest de l'Inde d'après les hymnes védiques.* Colonel Yule has expressed his views chiefly in the notes upon the map referred to, but also occasionally in the notes to his edition of *Marco Polo* and in other works from his pen. Frequent reference will be found in my notes to that work of vast erudition, Prof. Lassen's *Indische Alterthumskunde.* Unfortunately the section which he has devoted to a full examination of Ptolemy's India is the least satisfactory portion of his work. His system of identification is based on a wrong principle, and many of the conclusions to which it has led are such as cannot be accepted. His work is notwithstanding, as Yule says, "a precious mine of material for the study of the ancient grography of India." For elucidations of the Ptolemaic geography of particular portions of India I have consulted with great advantage such works as the following:—Wilson's *Ariana Antiqua*, General Cunningham's *Geography of Ancient India*, Vol. I. (all yet published), and his *Reports on the Archæological Survey of India*; Bishop Caldwell's Introduction to his *Dravidian Grammar*, valuable for identification of places in the south of the Peninsula; the *Bombay Gazetteer*, edited by Mr. J. M. Campbell, who has carefully investigated the antiquities of that Presidency; the volumes of *Asiatic Researches*; the *Journals of the Royal Asiatic Society* and of the kindred Societies in India; the *Journals of the Royal Geographical Society*; the articles on India and places in India in Smith's *Dictionary of Classical Geography*, written almost all by Mr.

Vaux; articles in the *Indian Antiquary*; Benfey's *Indien* in the *Encyclopädie* of Ersch and Grüber; the Abbé Halma's *Traité de Géographie de Claude Ptolémée*, Paris, 1828; the Chapters on Marinus and Ptolemy's System of Geography in Bunbury's *History of Ancient Geography*; Priaulx's *Indian Travels of Apollonius of Tyana*, &c.; Stephanos of Byzantium *On Cities*; Sir Emerson Tennent's *Ceylon*; Sir H. Rawlinson's articles on Central Asia which have appeared in various publications, and other works which need not here be specified.

There has recently been issued from the press of Firmin-Didot, Paris, the first volume of a new and most elaborate edition of Ptolemy's Geography, prepared by C. Müller, the learned editor of the *Geographi Graeci Minores*, but the work unfortunately has not advanced so far as to include the chapters which contain the geography of India.

I would here take the opportunity of expressing my obligations to Dr. Burgess, the late editor of the *Indian Antiquary*, for his careful revision of the proofs, and for sundry valuable suggestions.

Having thought it advisable to extend the scope of the work beyond the limits originally contemplated, I have included in it those chapters of the geography in which China, Central Asia, and all the provinces adjacent to India are described. The reader is thus presented with the Ptolemaic Geography of the whole of Asia, with the exception only of those countries which from propinquity and frequency of intercourse were well known to the nations of the West.

In a short Appendix will be found some additional notes.

The present volume forms the fourth of the Series of Annotated Translations of the Works of the Classical Writers which relate to India. Another volume, containing Strabo's Indian Geography and the Accounts given by Arrian and Curtius of the Makedonian Invasion of India, will complete the series.

3, ABBOTSFORD PARK, EDINBURGH,
June, 1885.

CONTENTS.

PTOLEMY'S GEOGRAPHY OF INDIA AND SOUTHERN ASIA.

Introduction.

Ptolemy and his System of Geography.

Klaudios Ptolemaios, or as he is commonly called, Ptolemy, was distinguished alike as a Mathematician, a Musician, an Astronomer and a Geographer, and was altogether one of the most accomplished men of science that antiquity produced. His works were considered as of paramount authority from the time of their publication until the discoveries of modern times had begun to show their imperfections and errors. It is surprising that with all his fame, which had even in his own lifetime become pre-eminent, that the particulars of his personal history should be shrouded in all but total darkness. Nothing in fact is known for certain regarding him further than that he flourished in Alexandria about the middle of the 2nd century of our æra, in the reign of Antoninus Pius, whom he appears to have survived.

His work on Geography formed a sequel to his great work on Astronomy, commonly called the *Almagest.* From its title Γεωγραφικὴ Ὑφήγησις, *an Outline of Geography*, we might be led to infer that it was a general treatise on the subject, like the comprehensive work of Strabo, but in reality it treats almost exclusively of Mathematical, or what may be called Cosmical, Geography. Ptolemy's object in composing it was not like that of the ordinary Geographer to describe places, but to correct and reform the map of the world in accordance with the increased knowledge which had been acquired of distant countries and with the improved state of science. He therefore limits his argument to an exposition of the geometrical principles on which Geography should be based, and to a determination of the position of places on the surface of the earth by their latitudes and longitudes. What he considered to be the proper method of determining geographical positions he states very clearly in the following passage: "The proper course," he says, "in drawing up a map of the world is to lay down as the basis of it those points that were determined by the most correct (astronomical) observations, and to fit into it those derived from other sources, so that their positions may suit as well as possible with the principal points thus laid down in the first instance."[1]

Unfortunately, as Bunbury remarks, it was impossible for him to carry out in practice—even approximately—the scheme that he had so well laid down in theory. The astronomical observations to which he could refer were but few—and they were withal either so defective or so inaccurate that he could not use them with con-

[1] Book I. cap. 4. The translation is Bunbury's.

fidence. At the same time his information concerning many parts of the earth, whether owing to their remoteness or the conflicting accounts of travellers regarding them, was imperfect in the extreme. The extent, however, of his geographical knowledge was far greater than that possessed by any of his predecessors, and he had access to sources of information which enabled him to correct many of the errors into which they had fallen.

He was induced to undertake the composition of his *Geography* through his being dissatisfied more or less with all the existing systems. There was however one work—that of his immediate precursor, Marinos of Tyre—which approximated somewhat closely to his ideal, and which he therefore made the basis of his own treatise. Marinos, he tell us, had collected his materials with the most praiseworthy diligence, and had moreover sifted them both with care and judgment. He points out, however, that his system required correction both as to the method of delineating the sphere on a plane surface, and as to the computation of distances, which he generally exaggerated. He censures him likewise for having assigned to the known world too great a length from west to east, and too great a breadth from north to south.

Of Ptolemy's own system, the more prominent characteristics may now be noted: He assumed the earth to be a sphere, and adopting the estimate of Poseidônios fixed its circumference at 180,000 stadia, thus making the length of a degree at the equator to be only 500 stadia, instead of 600, which is its real length.[2] To this fundamental miscalculation may be referred not a few of the most serious errors to be found in his work. With regard to the question of the length and the breadth of the inhabited part of the earth, a question of first importance in those days, he estimated its length as measured along the parallel of Rhodes[3] which divided the then known world into two nearly equal portions at 72,000 stadia, and its breadth at 10,000. The meridian in the west from which he calculated his longitudes was that which passed through the Islands of the Blest (Μακάρων Νῆσοι) probably the Canary Islands,[4] and his most

---

[2] The Olympic *stadium*, which was in general use throughout Greece, contained 600 Greek feet, which were equal to 625 Roman feet, or 606¾ English feet. The Roman mile contained 8 stadia, or about half a stadium less than an English mile. A stadium of 600 Greek feet was very nearly the 600th part of a degree, and 10 stadia are therefore just about equal to a Nautical or Geographical mile. According to Eratosthenes, a degree at the Equator was equal to 700 stadia, but according to Poseidônios it was equal to only 500. The truth lay between, but Ptolemy unfortunately followed Poseidônios in his error.

[3] "The equinoctial line was of course perfectly fixed and definite in Ptolemy's mind, *as an astronomical line*; but he had no means of assigning its position on the Map of the World, except with reference to other parallels, such as the tropic at Syene, or the parallels of Alexandria and Rhodes, which had been determined by direct observation."—Bunbury, *Hist. of Anc. Geog.*, vol. II, p. 560, n. 2.

[4] The Island of Ferro—the westernmost of the Group of the Canaries, which was long taken as the prime meridian, and is still so taken in Germany—is really situated 18° 20′ west of Greenwich, while Cape St. Vincent (called anciently *the Sacred Cape*) is just about 9°, so that the real difference between the two amounted to 9° 20′ instead of only 2½°. Two corrections must therefore be applied to Ptolemy's longitudes—one-sixth must be deducted because of his under-estimate of the length

eastern meridian was that which passed through the Metropolis of the Sinai, which he calls Sinai or Thinai, and places in 180° 40′ E. Long. and 3° S. Lat. The distance of this meridian from that of Alexandria he estimated at 119½ degrees, and the distance of the first meridian from the same at 60½ degrees, making together 180 degrees, or exactly one-half of the circumference of the earth. His estimate of the breadth he obtained by fixing the southern limit of the inhabited parts in the parallel of 16¼ degrees of South Latitude, which passes through a point as far south of the Equator as Meroë is north of it. And by fixing the northern limit in the parallel of 63 degrees North Latitude, which passes through Thoulê (probably the Shetland Islands), a space of nearly 80 degrees was thus included between the two parallels, and this was equivalent in Ptolemy's mode of reckoning to 40,000 stadia.

Having made these determinations he had next to consider in what mode the surface of the earth with its meridians of longitude and parallels of latitude should be represented on a sphere and on a plane surface—of the two modes of delineation that on the sphere is the much easier to make, as it involves no method of projection, but a map drawn on a plane is far more convenient for use, as it presents simultaneously to the eye a far greater extent of surface. Marinos had drawn his map of the world on a plane, but his method

---

of a degree along the Equator, and 6° 50′ must be added because Ferro was so much further west than he supposed. Subject to these corrections his longitudes would be fairly accurate, provided his calculations of distances were otherwise free from error.

of projection was altogether unsatisfactory. It is thus described by Ptolemy: Marinos, he says, on account of the importance of the countries around the Mediterranean, kept as his base the line fixed on of old by Eratosthenes, *viz.*, the parallel through Rhodes in the 36th degree of north latitude. He then calculated the length of a degree along this parallel, and found it to contain 400 stadia, the equatorial degree being taken at 500. Having divided this parallel into degrees he drew perpendiculars through the points of division for the meridians, and his parallels of latitude were straight lines parallel to that which passed through Rhodes. The imperfections of such a projection are obvious. It represented the parts of the earth north of the parallel of Rhodes much beyond, and those south of it much below, their proper length. Places again to the north of the line stood too far apart from each other, and those to the south of it too close together. The projection, moreover, is an erroneous representation, since the parallels of latitude ought to be circular arcs and not straight lines

Ptolemy having pointed out these objections to the system of Marinos proceeds to explain the methods which he himself employed. We need say nothing more regarding them than that they were such as presented a near approximation to some of those which are still in use among modern Geographers.

Ptolemy's treatise is divided into 8 books. In the 1st or introductory book he treats first of Geography generally—he then explains and criticizes the system of Marinos, and concludes by describing the methods of projection which may be employed in the construction of maps. The next 6 books and the first 4 chapters of the 7th book consist of tables which give distinctly in degrees and parts of a degree the latitudes and longitudes of all the places in his map. These places are arranged together in sections according to the country or tribe to which they belong, and each section has prefixed to it a brief description of the boundaries and divisions of the part about to be noticed. Descriptive notices are also occasionally interspersed among the lists, but the number of such is by no means considerable. The remainder of the 7th book and the whole of the 8th are occupied with a description of a series of maps which, it would appear, had been prepared to accompany the publication of the work, and which are still extant. The number of the maps is twenty-six, viz. 10 for Europe, 4 for Libya, and 12 for Asia. They are drawn to different scales, larger or smaller, according as the division represented was more or less known. He gives for each map the latitudes and longitudes of a certain number of the most important cities contained in it, but these positions were not given in the same manner as in the tables, for the latitudes are now denoted by the length of the longest day and the longitudes according to the difference of time from Alexandria. It might be supposed that the positions in question were such as had been determined by actual astronomical observations, as distinguished from those in the Tables, which were for the most part derived from itineraries, or from records of voyages and travels. This supposition is however untenable, for we find that while the statements as to the length of the longest days at the selected places are always correct for the latitudes assigned them, they are often glaringly wrong for their real positions. Ptolemy, it is evident, first mapped out in the best way he could the places, and then calculated for the more important of these places the astronomical phenomena incident to them as so situated. I conclude by presenting the reader with a translation of some chapters of the Introductory Book,[5] where Ptolemy in reviewing the estimate made by Marinos of the length of the known world from west to east, has frequent occasion to mention India and the Provinces beyond the Ganges, which together constitute what is now called Indo-China.

Book I., Cap. 11.

§ 1. What has now been stated will suffice to show us what extent in *breadth* it would be fair to assign to the inhabited world. Its *length* is given by Marinos at 15 hours, this being the distance comprised between his two extreme meridians—but in our opinion he has unduly extended the distance towards the east. In fact, if the estimate be properly reduced in this direction the entire length must be fixed at less than 12 hours, the Islands of the Blest being taken as the limit towards

---

[5] The edition used is that of C. F. A. Noble, Leipsic, 1843.

the west, and the remotest parts of Sêra and the Sinai[6] and Kattigara[7] as the limit towards

[6] "China for nearly 1,000 years has been known to the nations of Inner Asia, and to those whose acquaintance with it was got by that channel, under the name of Khitai, Khata, or Cathay, *e.g.*, the Russians still call it Khitai. The pair of names, Khitai and Machin, or Cathay and China, is analogous to the other pair Seres and Sinai. Seres was the name of the great nation in the far east as known by land, Sinai as known by sea; and they were often supposed to be diverse, just as Cathay and China were afterwards." Yule's *Marco Polo*, 2nd ed., Introd., p. 11 and note.

[7] The locality of Kattigara has been fixed very variously. Richthofen identified it with Kian-chi in Tong-king, and Colonel Yule has adopted this view. "To myself," he says, "the arguments adduced by Richthofen in favour of the location of Kattigara in the Gulf of Tong-king, are absolutely convincing. This position seems to satisfy every condition. For 1st, Tong-king was for some centuries at that period (B. C. 111 to A.D. 263), only incorporated as part of the Chinese Empire. 2nd, the only part mentioned in the Chinese annals as at that period open to foreign traffic was Kian-chi, substantially identical with the modern capital of Tong-king, Kesho or Hanoi. Whilst there are no notices of foreign arrivals by any other approach, there are repeated notices of such arrivals by this province, including that famous embassy from Antun, King of Ta-t'sin, *i.e.*, M. Aurelius Antoninus (A.D. 161-180) in A.D. 166. The province in question was then known as Ji-nan (or Zhi-nan, French); whence possibly the name Sinai, which has travelled so far and spread over such libraries of literature. The Chinese Annalist who mentions the Roman Embassy adds: 'The people of that kingdom (Ta-t'sin or the Roman Empire) came in numbers for trading purposes to Fu-nan, Ji-nan, and Kian-chi.' Fu-nan we have seen, was Champa, or Zabai. In Ji-nan with its chief port Kian-chi, we may recognize with assurance Kattigara, Portus Sinarum. Richthofen's solution has the advantages of preserving the true meaning of Sinai as the Chinese, and of locating the Portus Sinarum in what was then politically a part of China, whilst the remote Metropolis Thinae remains unequivocally the capital of the Empire, whether Si-gnan-fu in Chen-si, or Lo-yang in Ho-nan be meant. I will only add that though we find Katighora in Edrisi's *Geography*, I apprehend this to be a mere adoption from the *Geogra-*

the east. § 2. Now the entire distance from the Islands of the Blest to the passage of

---

phy of Ptolemy, founded on no recent authority. It must have kept its place also on the later mediæval maps; for Pigafetta, in that part of the circumnavigation where the crew of the *Victoria* began to look out for the Asiatic coast, says that Magellan 'changed the course . . . until in 13° of N. Lat. in order to approach the land of Cape Gaticara, which Cape (under correction of those who have made cosmography their study, for they have never seen it), is not placed where they think, but is towards the north in 12° or thereabouts.' [The Cape looked for was evidently the extreme S. E. point of Asia, actually represented by Cape Varela or Cape St. James on the coast of Cochin-China.] It is probable that, as Richthofen points out, Kattigara, or at any rate Kianchi, was the Lukin or Al-Wâkin of the early Arab Geographers. But the terminus of the Arab voyagers of the 9th century was no longer in Tong-King, it was Khân-fû, apparently the Kan-pu of the Chinese, the haven of the great city which we know as Hang-chow, and which then lay on or near a delta-arm of the great Yang-tse.'' These arguments may be accepted as conclusively settling the vexed question as to the position of Kattigara. In a paper, however, recently read before the R. Asiatic Society, Mr. Holt, an eminent Chinese scholar, expressed a different view. He "showed that there was good evidence of a very early communication from some port on the Chinese coast to near Martaban, or along the valley of the Irâwaḍî to the north-west capital of China, then at Si-gnan-fu or Ho-nan-fu. He then showed that the name of China had been derived from the Indians, who first knew China, and was not due to the Tsin Dynasty, but more probably came from the name of the Compass, specimens of which were supplied to the early envoys, the Chinese being thus known in India as the 'Compass-people,' just as the Seres, another Chinese population, derived their western name from 'Silk.' That the knowledge of this fact was lost to both Indians and Chinese is clear from the use by Hiuen-Tsiang and later writers of two symbols (see Morrison's *Dictionary*, syllabic part, No. 8,033) to designate the country, as these, while giving the sound 'Che-ha,' indicate that they are substitutes for original words of like sounds, the true sense of which cannot now be recovered. Having shown that M. Reinaud's view of an intercourse between China and Egypt in the first century A.D. has no real foundation, Mr. Holt the Euphrates at Hierapolis, as measured along the parallel of Rhodes, is accurately determined by summing together the several intervening distances as estimated in stadia by Marinos, for not only were the distances well ascertained from being frequently traversed, but Marinos seems moreover in his computation of the greater distances, to have taken into account the necessary corrections for irregularities and deviations.[8] He understood, besides, that while the length of a single degree of the 360 degrees into which the equatorial circle is divided measures, as in the commonly accepted estimate, 500 stadia, the parallel circle which passes through Rhodes in 36 degrees of N. latitude, measures about 400 stadia. § 3. It measures, in fact, a little over that number if we go by the exact proportion of the parallels, but the excess is so trifling as in the case of the equatorial degree, that it may be neglected. But

further stated that there was no evidence of an embassy from M. Aurelius having gone by sea to China in A.D. 166. In conclusion, he urged, that in his judgment, there was no proof whatever of any knowledge of a maritime way to China before the 4th century A.D., the voyage even of Fa-hian, at that period being open to serious criticism. He believes therefore with M. Gosselin that the Kattigara of Ptolemy was probably not far from the present Martaban, and that India for a considerable period up to the 7th century A.D. dominated over Cambodia."

[8] Deviations from the straight line by which the route would be represented in the map. The irregularities refer to the occasional shortening of the daily march by obstacles of various kinds, bad roads, hostile attacks, fatigue, &c.

his estimates of the distances beyond Hierapolis require correction. § 4. He computes the distance from the passage of the Euphrates already mentioned to the Stone Tower[9] at 876

[9] "One of the circumstances of the route that Ptolemy has reproduced from Marinos is that on leaving Baktra the traveller directed his course for a long enough time towards the North. Assuredly the caravans touched at Samarkand (the Marakanda of Greek authors) which was then, as now, one of the important centres of the region beyond the Oxus. For passing from Sogdiana to the east of the snowy range, which covers the sources of the Jaxartes and the Oxus, three main routes have existed at all times: that of the south, which ascends the high valleys of the Oxus through Badakshân; that in the centre, which goes directly to Kâshgar by the high valleys of the Syr-Darya or Jaxartes; and lastly that of the north, which goes down a part of the middle valley of the Jaxartes before turning to the east towards Chinese Tartary. Of these three routes, the itinerary of the Greek merchants could only apply to the 2nd or the 3rd; and if, as has been for a long time supposed with much probability, the Stone Tower of the Itinerary is found in an important place belonging to the valley of the Jaxartes, of which the name Tâshkand has precisely the same meaning in the language of the Turkomâns, it would be the northern route that the caravan of Maës would have followed. The march of seven months in advancing constantly towards the east leads necessarily towards the north of China (Saint-Martin, *Étude*, pp. 428-9.) Sir H. Rawlinson however assigns it a more southern position, placing it at Tash-kurghan, an ancient city which was of old the capital of the Sarik-kul territory, a district lying between Yarkand and Badakshan, and known to the Chinese as Ko-panto. The walls of Tash-kurghan are built of unusually large blocks of stone. It was no doubt, Sir Henry remarks, owing to the massive materials of which it was built, that it received the name of Tash-kurghan or the 'Stone Fort,' and it seems to have every claim to represent the λίθινος πύργος of Ptolemy, where the caravans rendezvoused before entering China, in preference to Tashkand or Ush, which have been selected as the site of the Stone Tower by other geographers."—*Jour. R. Geog. Soc.* vol. XLII, p. 327.

schœni[10] or 26,280 stadia, and from the Stone Tower to Sêra, the metropolis of the Sêres, at a 7 months' journey or 36,200 stadia as reckoned along the same parallel. Now in neither case has he made the proper deductions for the excess caused by deviations ; and for the second route he falls into the same absurdity as when he estimated the distance from the Garamantes to Agisymba.[11] § 5. Where he had to deduct above half of the stadia in the march of the 3 months and 14 days, since such a march could not possibly have been accomplished without halting.

---

[10] According to Herodotos (lib. II, c. vi), the *schoinos* was equal to two Persian *parasangs* or 60 stadia, but it was a very vague and uncertain measure, varying as Strabo informs us (lib. XVII, c. i, 24) from 30 to 120 stadia. In the case before us, it was taken as equivalent to the *parasang* of 30 stadia and afforded with correction some approximation to the truth.

[11] "The Roman arms had been carried during the reign of Augustus (B. C. 19) as far as the land of the Garamantes, the modern Fezzan, and though the Roman Emperors never attempted to establish their dominion over the country, they appear to have permanently maintained friendly relations with its rulers, which enabled their officers to make use of the oasis of the Garamantes as their point of departure from which to penetrate further into the interior. Setting out from thence, a General named Septimius Plancus 'arrived at the land of the Ethiopians, after a march of 3 months towards the south.' Another Commander named Julius Maternus, apparently at a later date, setting out from Leptis Magna, proceeded from thence to Garama, where he united his forces with those of the king of the Garamantes, who was himself undertaking a hostile expedition against the Ethiopians, and their combined armies 'after marching for four months towards the south,' arrived at a country inhabited by Ethiopians, called Agisymba, in which rhinoceroses abounded."—Bunbury, *Hist. of Anc. Geog.*, vol. II, pp. 522-3.

The necessity for halting would be still more urgent when the march was one which occupied 7 months. § 6. But the former march was accomplished even by the king of the country himself, who would naturally use every precaution, and the weather besides was all throughout most propitious. But the route from the Stone Tower to Sêra is exposed to violent storms, for as he himself assumes, it lies under the parallels of the Hellespont and Byzantium,[12] so that the progress of travellers would be frequently interrupted. § 7. Now it was by means of commerce this became known, for Marinos tells us that one Maës, a Makedonian, called also Titianus, who was a merchant by hereditary profession, had written a book giving the measurement in question, which he had obtained not by visiting the Sêres in person, but from the agents whom he had sent to them. But Marinos seems to have distrusted accounts borrowed from traders. § 8. In giving, for instance, on the authority of Philêmon, the length of Ivernia (Ireland) at a 20 days' journey, he refuses to accept this estimate, which was got, he tells us, from merchants, whom he reprobates as a class of men too much engrossed with their own proper business to care about ascertaining the truth, and who also from mere vanity frequently exaggerated distances. So

---

[12] Lat. 40° 1′—Lat. of Tâsh-kurghân.

too, in the case before us, it is manifest that nothing in the course of the 7 months' journey was thought worthy either of record or remembrance by the travellers except the prodigious time taken to perform it.

CAP. 12.

§ 1. Taking all this into consideration, together with the fact that the route does not lie along one and the same parallel (the Stone Tower being situated near the parallel of Byzantium, and Sêra lying farther south than the parallel through the Hellespont) it would appear but reasonable in this case also to diminish by not less than a half the distance altogether traversed in the 7 months' journey, computed at 36,200 stadia, and so let us reduce the number of stadia which these represent at the equator by one-half only, and we thus obtain (22,625) stadia or $45\frac{1}{4}$ degrees.[13] § 2. For it would be absurd, and show a want of proper judgment, if, when reason enjoins us to curtail the length of both routes, we should follow the injunction with respect to the African route, to the length of which there is the obvious objection, *viz.*, the species of animals in the neighbourhood of Agisymba,

[13] 36,200 stadia along the parallel of Rhodes are equivalent, according to Ptolemy's system, to 45,250 stadia along the equator, and this sum reduced by a half gives the figures in the text.

which cannot bear to be transplanted from their own climate to another, while we refuse to follow the injunction with regard to the route from the Stone Tower, because there is not a similar objection to its length, seeing that the temperature all along this route is uniform, quite independently of its being longer or shorter. Just as if one who reasons according to the principles of philosophy, could not, unless the case were otherwise clear, arrive at a sound conclusion.[14]

§ 3. With regard again to the first of the two Asiatic routes, that, I mean which leads from the Euphrates to the Stone Tower, the estimate of 870 *schœni* must be reduced to 800 only, or 24,000 stadia, on account of deviations. § 4. We

---

[14] Marinos was aware that Agisymba lay in a hot climate, from the fact that its neighbourhood was reported to be a favourite resort for rhinoceroses, and he was thus compelled to reduce his first estimate of its distance, which would have placed it in far too cold a latitude for these animals, which are found only in hot regions. But no such palpable necessity compelled him to reduce his estimate of the distance from the Stone Tower to the Metropolis of the Sêres, for here the route had an equable temperature, as it did not recede from the equator but lay almost uniformly along the same parallel of latitude A little reflexion, however, might have shown Marinos that his enormous estimate of the distance to the Seric Metropolis required reduction as much as the distance to Agisymba, though such a cogent argument as that which was based on the habitat of the rhinoceros was not in this instance available. It is on the very face of it absurd to suppose that a caravan could have marched through a difficult and unknown country for 7 months consecutively at an average progress of 170 stadia (about 20 miles) daily.

may accept as correct his figures for the entire distance as the several stages had been frequently traversed and had therefore been measured with accuracy. But that there were numerous deviations is evident from what Marinos himself tells us. § 5. For the route from the passage of the Euphrates at Hierapolis through Mesopotamia to the Tigris, and the route thence through the Garamaioi of Assyria, and through Media to Ekbatana and the Kaspian Gates, and through Parthia to Hekatompylos Marinos considers to lie along the parallel which passes through Rhodes, for he traces (*in his map*) this parallel as passing through these regions. § 6. But the route from Hekatompylos to the capital city of Hyrkania must, of necessity, diverge to the north, because that city lies somewhere between the parallel of Smyrna and that of the Hellespont, since the parallel of Smyrna is traced as passing below Hyrkania and that of the Hellespont through the southern parts of the Hyrkanian Sea from the city bearing the same name, which lies a little farther north. § 7. But, again, the route herefrom to Antiokheia (Merv) of Margiana through Areia, at first bends towards the south, since Areia lies under the same parallel as the Kaspian Gates, and then afterwards turns towards the north, Antiokheia being situated under the parallel of the Hellespont.[15] The route after this runs in an eastward direction to Baktra whence it turns towards the north in ascending the mountains of the Kômêdoi, and then in passing through these mountains it pursues a southern course as far as the ravine that opens into the plain country. § 8. For the northern parts of the mountain region and those furthest to the west where the ascent begins, are placed by him under the parallel of Byzantium, and those in the south and the east under the parallel of the Hellespont. For this reason, he says, that this route makes a detour of equal length in opposite directions, that in advancing to the east it bends towards the south, and thereafter probably runs up towards the north for 50 *schœni*, till it reaches the Stone Tower. § 9. For to quote his own

---

[15] The actual latitudes of the places here mentioned may be compared with those of Ptolemy :—

	Real Lat.	Ptolemy's Lat.
Byzantium	41°	43° 5′
Hellespont	40°	41° 15′
Smyrna	38° 28′	38° 35′
Issus	37°	36° 35′
Rhodes	36° 24′	36° 25′
Hierapolis	36° 28′	36° 15′
Ekbatana	34° 50′	37° 45′
Kaspian Gates	35° 30′	37°
Hekatompylos	35° 40′	37° 50′
Antiokheia (Merv)	37° 35′	40° 20′
Baktra (Balkh)	36° 40′	41°
Stone Tower (Tâshkand)	42° 58′	43°
Sêra Metropolis (Ho-nan)	38° 35′	33° 58′

words, "When the traveller has ascended the ravine he arrives at the Stone Tower, after which the mountains that trend to the east unite with Imaus, the range that runs up to the north from Palimbothra." § 10. If, then, to the 60 degrees made up of the 24,000 stadia, we add the 45¼ degrees which represent the distance from the Stone Tower to Sêra, we get 105¼ degrees as the distance between the Euphrates and Sêra as measured along the parallel of Rhodes.[16] § 11. But, further, we

[16] Saint-Martin identifies Sêra, the Metropolis of the Sêres, with a site near Ho-nan-fu. He says, (*Etudes*,' p. 432) "At the time when the caravan journey reported by Maës was made (in the first half of the first century of our era), the Han surnamed Eastern held the reins of government, and their residence was at Lo-yang near the present City of Ho-nan-fou, not far from the southern bank of the lower Hoang-ho. It is there then we should look to find the place which in their ignorance of the language of the country, and in their disdain for barbarous names, the Greek traders designated merely as the Metropolis of the Sêres." The road these traders took appears to have been the same by which Hiuen-Tsiang travelled towards India.

We may here insert for comparison with Ptolemy's distances two itineraries, one by Strabo and the other by Pliny. Strabo (lib. XI, c. viii, 9) says: "These are the distances which he (Eratosthenes) gives:—

	Stadia.
From the Kaspian Sea to the Kyros about ...	1,800
Thence to the Kaspian Gates	5,600
Thence to Alexandreia of the Areioi (Herat)..	6,400
Thence to Baktra, called also Zariaspa (Balkh)	3,870
Thence to the Jaxartes, which Alexander reached, about	5,000
Making a total of	22,670."

He also assigns the following distances from the Kaspian Gates to India:—

	Stadia.
"To Hekatompylos	1,960
To Alexandreia of the Areioi (Herat)............	4,530

can infer from the number of stadia which he gives as the distance between successive places lying along the same parallel, that the distance from the Islands of the Blest to the sacred Promontory in Spain (*Cape St. Vincent*), is 2½ degrees, and the distance thence to the mouth of the Bætis (*Guadalquivir*), the same.

	Stadia.
Thence to Prophthasia in Dranga (a little north of lake Zarah	1,600
Thence to the City Arakhotos (Ulan Robût)...	4,120
Then to Ortospana (Kâbul) on the 3 roads from Baktra	2,000
Thence to the confines of India	1,000
Which together amount to	15,300."
The sum total however is only.....................	15,210

Pliny (lib. VI, c. xxi) says: " Diognetus and Baeton, his (Alexander's) measurers, have recorded that from the Kaspian Gates to Hekatompylos of the Parthians there were as many miles as we have stated, thence to Alexandria Arion a city built by that king, 575 miles, to Prophthasia of the Drangae 198 miles, to the town of the Arakhosii 565 miles, to Hortospanum 175 miles, thence to Alexander's town (Opianê) 50 miles. In some copies numbers differing from these are found. They state that the last-named city lay at the foot of Caucasus; from that the distance to the Cophes and Peucolatis, a town of the Indians, was 237 miles, and thence to the river Indus and town of Taxila 60 miles, to the Hydaspes, a famous river, 120 miles, to the Hypasis, no mean river [IXXXIXI] 390—which was the limit of Alexander's progress, although he crossed the river and dedicated altars on the far-off bank, as the letters of the king himself agree in stating." The Kaspian Gates formed a point of great importance in ancient Geography, and many of the meridians were measured from it. The pass has been clearly identified with that now known as the Sirdar Pass between Verâmîn and Kishlak in Khowar. Arrian states that the distance from the city of Rhagai to the entrance of the Gates was a one day's march. This was, however, a forced march, as the ruins of Rhagai (now Rai, about 5 miles from Tehran) are somewhere about 30 miles distant from the Pass.

From the Bætis to Kalpê, and the entrance of the Straits, 2½ degrees. From the Straits to Karallis in Sardinia, 25 degrees. From Karallis to Lilybaion, in Sicily, 4½ degrees. From this Cape to Pakhynos, 3 degrees. Then again, from Pakhynos to Tainaros, in Lakonia, 10 degrees. Thence to Rhodes, 8¼ degrees. From Rhodes to Issus, 11¼ degrees, and finally from Issos to the Euphrates, 2½ degrees.[17] § 12. The

[17] I may present here the tabular form in which Mr. Bunbury (vol. II, p. 638) exhibits the longitudes of the principal points in the Mediterranean as given by Ptolemy, and the actual longitudes of the same points computed from Ferro:

	Longitude in Ptolemy.	Real longitude E. of Ferro.
Sacred Promontory	2° 30′	9° 20′
Mouth of Bætis	5° 20′	12°
Calpe (at mouth of Straits).	7° 30′	13°
Caralis in Sardinia...........	32° 30′	27° 30′
Lilybæum in Sicily	37°	30° 45′
Pachynus (Prom.) in Sicily.	40°	33° 25′
Tænarus (Prom.)...............	50°	40° 50′
Rhodes	58° 20′	46° 45′
Issus	69° 20′	54° 30′

The same authority observes (vol. II, p. 564) "Ptolemy thus made the whole interval from the Sacred Cape to Issus, which really comprises only about 45° 15′ to extend over not less than 67 degrees of longitude, and the length of the Mediterranean itself from Calpe to Issus, to amount to 62 degrees: rather more than 20 degrees beyond the truth. It is easy to detect one principal source of this enormous error. Though the distances above given are reported by Ptolemy in degrees of longitude, they were computed by Marinos himself from what he calls *stadiasmi*, that is from distances given in maritime itineraries and reported in stadia. In other words, he took the statements and estimates of preceding authorities and converted them into degrees of longitude, according to his own calculation that a degree on the equator was equal to 500 stadia, and

sum of these particular distances gives a total of 72 degrees, consequently the entire length of the known world between the meridian of the Islands of the Blest and that of the Sêres is 177¼ degrees, as has been already shown.[15]

Cap. 13.

§ 1. That such is the length of the inhabited world may also be inferred from his estimate of the distances in a voyage from India to the Gulf of the Sinai and Kattigara, if the sinuosities of the coast and irregularity of the navigation be taken into account, together with the positions as drawn into nearer proximity in the projections; for, he says, that beyond the Cape called Kôry where the Kolkhic Gulf terminates, the Argaric Gulf begins, and that the distance thence to the City of Kouroula, which is situated to the north-east of Kôry is 3,400 stadia. § 2. The

---

consequently a degree of longitude in latitude 36° would be equal (approximately) to 400 stadia." The total length of the Mediterranean computed from the stadiasmoi must have been 24,800. This was an improvement on the estimate of Eratosthenes, but was still excessive. In the ancient mode of reckoning sea distances the tendency was almost uniformly towards exaggeration.

[15] The different corrections to be applied to Ptolemy's eastern longitudes have been calculated by Sir Henry Rawlinson to amount to *three-tenths*, which is within one-seventieth part of the empirical correction used by M. Gossellin. [If we take *one-fifth* from Ptolemy's longitude of a place, and deduct 17° 43′ for the W. longitude of Ferro, we obtain very approximately the modern English longitude. Thus, for Barygaza, Ptolemy's longitude is 113°15′ and 113°15′—22°39′—17°43′=72°53′, or only 5′ less than the true longitude W. of Greenwich. —J. B.]

distance right across may, therefore, be estimated at about 2,030 stadia, since we have to deduct a third because of the navigation having followed the curvature of the Gulf, and have also to make allowances for irregularities in the length of the courses run. § 3. If now we further reduce this amount by a third, because the sailing, *though subject to interruption*, was taken as continuous, there remain 1,350 stadia, determining the position of Kouroula as situated north-east from Kôry. § 4. If now this distance be referred to a line running parallel to the equator and towards the East, and we reduce its length by half in accordance with the intercepted angle, we shall have as the distance between the meridian of Kouroula and that of Kôry, 675 stadia, or $1\frac{1}{3}$ degree, since the parallels of these places do not differ materially from the great circle.[19]

§ 5. But to proceed: the course of the voyage from Kouroura lies, he says, to the south-east as far as Paloura, the distance being 9,450 stadia. Here, if we deduct as before one-third for the irregularities in the length of the courses, we shall have the distance on account of the navigation having been continuous to

---

[19] By the intercepted angle is meant the angle contained by two straight lines drawn from Kôry, one running north-east to Kouroula and the other parallel to the Equator. In Ptolemy's map Kouroula is so placed that its distance in a straight line from Kôry is about double the distance between the meridians of those two places.

the south-east about 6,300 stadia. § 6. And if we deduct from this in like manner as before one-sixth, in order to find the distance parallel to the equator, we shall make the interval between the meridians of these two places 5,250 stadia, or $10\frac{1}{2}$ degrees.

§ 7. At this place the Gangetic Gulf begins, which he estimates to be in circuit 19,000 stadia. The passage across it from P a l o u r a to S a d a in a direct line from west to east is 1,300 stadia. Here, then, we have but one deduction to make, *viz.*, one-third on account of the irregularity of the navigation, leaving as the distance between the meridians of Paloura and Sada 8,670 stadia, or $17\frac{1}{3}$ degrees. § 8. The voyage is continued onward from Sada to the City of T a m a l a, a distance of 3,500 stadia, in a south-eastward direction. If a third be here again deducted on account of irregularities, we find the length of the continuous passage to be 2,330 stadia, but we must further take into account the divergence towards the south-east, and deduct one-sixth, so we find the distance between the meridians in question to be 1,940 stadia, or 3° 50′ nearly. § 9. He next sets down the passage from T a m a l a to the Golden Khersonese at 1,600 stadia, the direction being still towards the south-east, so that after making the usual deductions there remain as the distance between the two meridians 900 stadia, or 1° 48′. The sum of these particulars makes the distance from Cape Kôry to the Golden Khersonese to be 34° 48′.

Cap. 14.

§ 1. Marinos does not state the number of stadia in the passage from the Golden Khersonese to Kattigara, but says that one Alexander had written that the land thereafter faced the south, and that those sailing along this coast reached the city of Zaba in 20 days, and by continuing the voyage from Zaba southward, but keeping more to the left, they arrived after some days at Kattigara. § 2. He then makes this distance very great by taking the expression "some days" to mean "many days," assigning as his reason that the days occupied by the voyage were too many to be counted,—a most absurd reason, it strikes me. § 3. For would even the number of days it takes to go round the whole world be past counting? And was there anything to prevent Alexander writing "many" instead of "some," especially when we find him saying that Dioskoros had reported that the voyage from Rhapta to Cape Prasum took "many days." One might in fact with far more reason take "some" to mean "a few," for we have been wont to censure this style (*of expression*).[20] § 4. So now lest we

[20] To account for the seeming caprice which led Marinos to take the expression *some days* as equivalent to *ever so many days* it has been supposed that he had

should appear to fall ourselves into the same error, that of adapting conjectures about distances to some number already fixed on, let us compare the voyage from the Golden Khersonese to

---

adopted the theory that Kattigara, the furthest point eastward that had been reached by sea, was situated nearly under the same meridian as Sêra, the furthest point in the same direction that had been reached by land. Unfortunately the expression used by Alexander *some days* did not square with this theory, and it was all the worse in consequence for that expression. "The result," says Mr. Bunbury (vol. II, p. 537), "derived by Marinos from these calculations was to place Kattigara at a distance of not less than 100 degrees of longitude, or nearly 50,000 stadia, east of Cape Kôry; and as he placed that promontory in 125½° of longitude east of the Fortunate Islands, he arrived at the conclusion that the total length of the inhabited world was, in round numbers, 225°, equivalent, according to his calculation to 112,500 stadia. As he adopted the system of Poseidônios, which gave only 180,000 stadia for the circumference of the globe, he thus made the portion of it which he supposed to be known, to extend over nearly two-thirds of the whole circumference. This position of Cape Kôry, which was adopted by Ptolemy as a position well established, was already nearly 34° too far to the east; but it was by giving the enormous extension we have pointed out to the coast of Asia beyond that promontory, that he fell into this stupendous error, which though partly corrected by Ptolemy, was destined to exercise so great an influence upon the future progress of geography." Columbus by accepting Ptolemy's estimate of the circumference of the globe greatly under-estimated the distance between the western shores of the Atlantic and the eastern shores of Asia, and hence was led to undertake his memorable enterprise with all the greater hope and courage.

With reference to the position of Cape Kôry as given by Ptolemy, Bunbury says (Vol. II, p. 537, note): "Cape Kôry is placed by Ptolemy, who on this point apparently follows Marinos, in 125° E. Longitude. It is really situated 80° E. of Greenwich and 98° E. of Ferro; but as Ptolemy made a fundamental error in the position of his primary meridian of nearly 7° this must be added to the amount of his error in this instance. He himself states that Cape Kôry was 120° E. of the mouth of the Bætis, the real difference of longitude being only 86°20′."

Kattigara, consisting of the 20 days to Zaba and the "some days" thence to Kattigara with the voyage from Arômata to Cape Prasum, and we find that the voyage from Arômata to Rhapta took also 20 days as reported by Theophilos, and the voyage from Rhapta to Prasum "many more days" as reported by Dioskoros, so that we may set side by side the "some days" with the "many days" and like Marinos take them to be equivalent. § 5. Since then, we have shown both by reasoning and by stating ascertained facts, that Prasum is under the parallel of 16° 25′ in South latitude, while the parallel through Cape Arômata is 4° 15′ in North latitude, making the distance between the two capes 20° 40′, we might with good reason make the distance from the Golden Khersonese to Zaba and thence to Kattigara just about the same. § 6. It is not necessary to curtail the distance from the Golden Khersonese to Zaba, since as the coast faces the south it must run parallel with the equator. We must reduce, however, the distance from Zaba to Kattigara, since the course of the navigation is towards the south and the east, in order that we may find the position parallel to the equator. § 7. If again, in our uncertainty as to the real excess of the distances, we allot say one-half of the degrees to each of these distances, and from the 13° 20′ between Zaba and Kattigara we deduct a third on account of the divergence, we shall have the distance from the Golden Khersonese to Kattigara along a line parallel to the equator of about 17° 10′. § 8. But it has been shown that the distance from Cape Kôry to the Golden Khersonese is 34° 48′, and so the entire distance from Kôry to Kattigara will be about 52°.

§ 9. But again, the meridian which passes through the source of the River Indus is a little further west than the Northern Promontory of Taprobanê, which according to Marinos is opposite to Kôry, from which the meridian which passes through the mouths of the River Bætis is a distance of 8 hours or 120°. Now as this meridian is 5° from that of the Islands of the Blest, the meridian of Cape Kôry is more than 125° from the meridian of the Islands of the Blest. But the meridian through Kattigara is distant from that through the Islands of the Blest a little more than 177° in the latitude of Kôry, each of which contains about the same number of stadia as a degree reckoned along the parallel of Rhodes. § 10. The entire length then of the world to the Metropolis of the Sinai may be taken at 180 degrees or an interval of 12 hours, since it is agreed on all hands that this Metropolis lies further east than Kattigara, so that the length along the parallel of Rhodes will be 72,000 stadia.

CAP. 17, (part).

§ 3. For all who have crossed the seas to those places agree in assuring me that the district of

Sakhalitês in Arabia, and the Gulf of the same name, lie to the east of S y a g r o s, and not to the west of it as stated by Marinos, who also makes S i m y l l a, the emporium in India, to be further west not only than Cape K o m a r i, but also than the Indus. § 4. But according to the unanimous testimony both of those who have sailed from us to those places and have for a long time frequented them, and also of those who have come from thence to us, S i m y l l a, which by the people of the country is called T i m o u l a, lies only to the south of the mouths of the river, and not also to west of them. § 5. From the same informants we have also learned other particulars regarding India and its different provinces, and its remote parts as far as the Golden Khersonese and onward thence to Kattigara. In sailing thither, the voyage, they said, was towards the east, and in returning towards the west, but at the same time they acknowledged that the period which was occupied in making the voyages was neither fixed nor regular. The country of the Sêres and their Metropolis was situated to the north of the Sinai, but the regions to the eastward of both those people were unknown, abounding it would appear, in swamps, wherein grew reeds that were of a large size and so close together that the inhabitants by means of them could go right across from one end of a swamp to the other. In travelling from these parts there was not only the road that led to Baktrianê by way of the Stone Tower, but also a road that led into India through Palimbothra. The road again that led from the Metropolis of the Sinai to the Haven at Kattigara runs in a south-west direction, and hence this road does not coincide with the meridian which passes through Sêra and Kattigara, but, from what Marinos tell us, with some one or other of those meridians that are further east.

I may conclude this prefatory matter by quoting from Mr. Bunbury his general estimate of the value of Ptolemy's Indian Geography as set forth in his criticism of Ptolemy's Map of India.

His strictures, though well grounded, may perhaps be considered to incline to the side of severity. He says (vol. II, pp. 642-3), "Some excellent remarks on the portion of Ptolemy's work devoted to India, the nature of the different materials of which he made use, and the manner in which he employed them, will be found in Colonel Yule's introduction to his Map of India, in Dr. Smith's *Atlas of Ancient Geography* (pp. 22-24). These remarks are indeed in great measure applicable to the mode of proceeding of the Alexandrian Geographer in many other cases also, though the result is particularly conspicuous in India from the fulness of the information—crude and undigested as it was—which he had managed to bring together. The result, as presented to us in the tables of Ptolemy, is a map of utter confusion, out of which it is very difficult to extract in a few instances any definite conclusions." The attempt of Lassen to identify the various places mentioned by Ptolemy, is based throughout upon the fundamental error of supposing that the geographer possessed a Map of India similar to our own, and that we have only to compare the ancient and modern names in order to connect the two. As Col. Yule justly observes: " Practically, he (Lassen) deals with Ptolemy's compilation as if that Geographer had possessed a collection of real Indian surveys, with the data systematically co-ordinated. The fact is, that if we should take one of the rude maps of India that appeared in the 16th century (*e.g.* in Mercator or in Lindschoten), draw lines of latitude and longitude, and then *more Ptolemaico* construct tables registering the co-ordinates of cities, sources and confluences as they appeared in that map, this would be the sort of material we have to deal with in Ptolemy's India." But, in fact, the case is much stronger than Col. Yule puts it. For such a map as he refers to, of the 16th century, however rude, would give a generally correct idea of the form and configuration of the Indian Peninsula. But this, as we have seen, was utterly misconceived by Ptolemy. Hence he had to fit his data, derived from various sources, such as maritime and land itineraries, based upon real experience, into a framework to which they were wholly unsuited, and this could only be effected by some Procrustean process, or rather by a repetition of such processes, concerning which we are left wholly in the dark.

Col. Yule's map of Ancient India is undoubtedly by far the best that has yet been produced: it is indeed the only attempt to interpret Ptolemy data, upon which such a map must mainly be founded upon anything like sound critical principles. But it must be confessed that the result is far from encouraging. So small a proportion of Ptolemy's names can find a place at all, and so many of those even that appear on the map are admitted by its author to rest upon very dubious authority; that we remain almost wholly in the dark as to the greater part of his voluminous catalogues; and are equally unable to identify the localities which he meant to designate, and to pronounce an opinion upon the real value of his materials."

Book VII.

Contents.

Description of the furthest parts of Greater Asia, according to the existing provinces and Satrapies.

1. [*Tenth Map*]
   of India within the River Ganges.
2. [*Eleventh Map*]
   of India beyond the Ganges.
   of the Sinai.
3. [*Twelfth Map*]
   of the Island of Taprobanê and the islands surrounding it.
4. *Outline Sketch of the Map of the Inhabited World.*
   Delineation of the Armillary Sphere with the Inhabited World.
   Sketch of the World in Projection.

[5. *There are 400 Provinces and 30 Maps.*]

Cap. I.

Description of India within the Ganges.

§ 1. India within the river Ganges is bounded on the west by the Paropanisadai and Arakhôsia and Gedrôsia along their eastern sides already indicated; on the north by Mount Imaös along the Sogdiaioi and the Sakai lying above it; on the east by the river Ganges; and on the south and again on the west by a portion of the Indian Ocean. The circuit of the coast of this ocean is thus described:—

2. In Syrastrênê, on the Gulf called Kanthi,

a roadstead and harbour	109° 30′	20°
The most western mouth of the River Indus called Sagapa	110° 20′	19° 50′
The next mouth called Sinthôn	110° 40′	19° 50′
The 3rd mouth called Khrysoun (the Golden)	111° 20′	19° 50′
The 4th called Kariphron	111° 40′	19° 50′
The 5th called Sapara	112° 30′	19° 50′
The 6th called Sabalaessa	113°	20° 15′
The 7th called Lônibarê	113° 30′	20° 15′
3. Bardaxêma, a town	113° 40′	19° 40′
Syrastra, a village	114°	19° 30′
Monoglôsson, a mart	114° 10′	18° 40′

Comment.—Strabo, following Eratosthenes, regarded the Indus as the boundary of India on the west, and this is the view which has been generally prevalent. Ptolemy, however, included within India the regions which lay immediately to the west of that river, comprehending considerable portions of the countries now known as Balûchistân and Afghânistân. He was fully justified in this determination, since many places beyond the Indus, as the sequel will show, bore names of Sanskrit origin, and such parts were ruled from the earliest times down to the Muḥammadan conquests by princes of Indian descent. The western boundary as given by Ptolemy would be roughly represented by a line drawn from the mouth of the Indus and passing through the parts adjacent to Kandahâr, Ghaznî, Kâbul, Balkh, and even places beyond. The Paropanisadai inhabited the regions lying south of the mountain range called Paropanisos, now known as the Central Hindû-Kûsh. One of these towns was Ortospana, which has been identified with the city of Kâbul, the Karoura of our author. He gives as the eastern boundary of the Paropanisadai a line drawn south from the sources of the river Oxus through the Kaukasian Mountains (the eastern portion of the Hindû-Kûsh) to a point lying in long. 119° 30′ and lat. 39°. Arakhôsia lay to the south of the Paropanisadai—its chief city was Arakhôtos, whose name, according to Rennell, is preserved in Arokhaj. There is a river of the same name which has been identified with the Helmand (the Etymander or Erymanthos of the ancients) but also and more probably with the Urghand-âb or Arkand-âb, which passes by Kandahâr. Gedrôsia, the modern Balûchistân, had for its eastern boundary the River Indus. The boundary of India on the north was formed by Mount Imaös (Sansk. *hima*, cold), a name which was at first applied by the Greeks to the Hindû-Kûsh and the chain of the Himâlayas running parallel to the equator, but which was gradually in the course of time transferred to the Bolor range which runs from north to south and intersects them. Ptolemy, however, places Imaös further east than the Bolor, and in the maps which accompany his *Geography*, this meridian chain, as he calls it, is prolonged up to the most northernly plains of the Irtish and Obi.

Sogdiana lay to the north of Baktria and abutted on Skythia, both towards the north and towards the west. The name has been preserved in that of Soghd, by which the country along the Kohik from Bokhârâ to Samarkand has always been known. Our author places the Sogdian Mountains (the Pâmîr range) at the sources of the Oxus, and the mountains of the Kômêdai between the sources of that river and the Jaxartes.

The Sakai were located to the east of the Sogdians—Ptolemy describes them as nomadic, as without towns and as living in woods and caves. He specifies as their tribes the Karatai (probably connected with the Kirâtai of India), the Komaroi, the Kômêdai, the Massagetai, the Grynaioi Skythai, the Toörnai and the Byltai. The Sakai it would appear therefore were the Mountaineers of Kâfiristân, Badakshân, Shignân, Roshan, Baltistân or Little Tibet, &c.

Syrastrênê and Larikê.

Syrastrênê:—The name is formed from the Sanskṛit Surâshṭra (now Soraṭh) the ancient name of the Peninsula of Gujarât. It is mentioned in the *Periplûs of the Erythraean Sea* as the sea-board of Abêria, and is there praised for the great fertility of its soil, for its cotton fabrics, and for the superior stature of its inhabitants.

Kanthi:—The Gulf of this name is now called the Gulf of Kachh. It separates Kachh, the south coast of which is still called Kantha, from the Peninsula of Gujarât. In the *Periplûs* the gulf is called Barakê and is described as of very dangerous navigation. In Ptolemy, Barakê is the name of an island in the Gulf.

Two mouths only of the Indus are mentioned by the followers of Alexander and by Strabo. The *Periplûs* gives the same number (7) as Ptolemy. There are now 11, but changes are continually taking place. Sagapa, the western mouth, was explored by Alexander. It separates from the main stream below Ṭhaṭha. In the chronicles of Sindh it is called Sâgâra, from which perhaps its present name Ghâra, may be derived. It has long ceased to be navigable.

Sinthôn:—This has been identified with the Piti branch of the Indus, one of the mouths of the Baghâr River. This branch is otherwise called the Sindhi Khrysoun. This is the Kediwârî mouth.

Khariphron:—Cunningham identifies this with the Kyâr river of the present day which, he says, leads right up to the point where the southern branch of the Ghâra joins the main river near Lâri-bandar.

Sapara:—this is the Wârî mouth.

Sabalaessa is now the Sir mouth.

Lonibarê in Sanskṛit is Lôṇavâri (or Lôṇavaḍâ, or Lavaṇavâri or Lâvaṇavâṭâ.[21] It is now the Korî, but is called also the Launî which preserves the old name.

Bardaxêma:—This, according to Yule, is now Pur-bandar, but Dr. Burgess prefers Srînagar, a much older place in the same district, having near it a small village called Bardiyâ, which, as he thinks, may possibly be a reminiscence of the Greek name.

Syrastra:—This in the Prakritized form is Soraṭh. It has been identified by Lassen with Junâgaḍh, a place of great antiquity and historical interest in the interior of the Peninsula, about 40 miles eastward from the coast at Navi-bandar. The meaning of the name is *the old fort.* The place was anciently called Girnagara, from its vicinity to the sacred mountain of Girnâr, near which is the famous rock inscribed with the edicts of Aśôka, Skandagupta and Rudra Dâma. Yule identifies Syrastra with Navi-bandar, a port at the mouth of the Bhâdar, the largest river of the Peninsula, said to be fed by 99 tributaries. Junâgaḍh was visited by Hiuen Tsiang, who states that after leaving the kingdom of Valabhî (near Bhaunagar) he went about 100 miles to the west and reached the country of Su-la-ch'a (Saurâshṭra) that was subject to the kingdom of Valabhî See *Tarîkh-i-Soraṭh,* edited by Dr. Burgess, pp. 33-199.

Monoglôsson:—This is now represented by Mangrol, a port on the S. W. coast of the Penin-

---

[21] *Lavaṇa* is the Sanskṛit word for *salt.*

sula below Navi-bandar. It is a very populous place, with a considerable traffic, and is tributary to Junâgaḍh.

4. In L a r i k ê.

Mouth of the River Môphis...	114°	18° 20′
Pakidarê, a village	113°	17° 50′
Cape Maleô	111°	17° 30′

5. In the Gulf of B a r y g a z a.

Kamanê	112°	17°
Mouth of the River Namados	112°	17° 45′
Nausaripa	112° 30′	16° 30′
Poulipoula	112° 30′	16°

L a r i k ê, according to Lassen, represents the Sansk. R â s h ṭ r i k a in its Prakrit form L a ṭ i k a. Lâr-deśa, however, the country of L â r (Sansk. Lâṭa) was the ancient name of the territory of Gujarât, and the northern parts of Koṅkaṇ, and L a r i k ê may therefore be a formation from Lâr with the Greek termination *ikê* appended. The two great cities of Barygaza (Bharoch) and Ozênê (Ujjain) were in Larikê, which appears to have been a political rather than a geographical division.

M a l e ô must have been a projection of the land somewhere between the mouth of the Mahî and that of Narmadâ—but nearer to the former if Ptolemy's indication be correct.

The Gulf of B a r y g a z a, now the Gulf of Khambhat, was so called from the great commercial emporium of the same name (now Bharoch) on the estuary of the Narmadâ at a distance of about 300 stadia from the Gulf. This river is called the Namados or Namadês by Ptolemy and the Namnadios by the Author of the *Periplûs*, who gives a vivid account of the difficulties attending the navigation of the gulf and of the estuary which was subject to bores of great frequency and violence.

Kamanê is mentioned as Kammônê in the *Periplûs*, where it is located to the south of the Narmadâ estuary. Ptolemy probably errs in placing it to northward of it.

Nausaripa has been identified with Nausâri, a place near the coast, about 18 miles south from Sûrat.

Poulipoula is in Yule's map located at Sanjan, which is on the coast south from Nausâri. It was perhaps nearer Balsâr.

6. Ariakê Sadinôn.

Soupara	112° 30′	15° 30′
Mouth of the River Goaris ...	112° 15′	15° 10′
Dounga	111° 30′	15°
Mouth of the River Bênda ...	110° 30′	15°
Simylla, a mart and a cape...	110°	14° 45′
Hippokoura	111° 45′	14° 10′
Baltipatna	110° 30′	14° 20′

Âriakê corresponds nearly to Mahârâshṭra—the country of the Marâṭhâs. It may have been so called, because its inhabitants being chiefly Aryans and ruled by Indian princes were thereby distinguished from their neighbours, who were either of different descent or subject to foreign domination. The territory was in Ptolemy's time divided among three potentates, one of whom belonged to the dynasty of the Sadineis and ruled the prosperous trading communities that occupied the seaboard. This dynasty is mentioned in the *Periplûs* (cap. 52) whence we learn that Sandanes after having made himself master of Kalliena (now Kalyâṇa), which had formerly belonged to the house of Saraganes the elder, subjected its trade to the severest restrictions, so that if Greek vessels entered its port even accidentally, they were seized and sent under guard to Barygaza, the seat evidently of the paramount authority. Sadanes, according to Lassen, corresponds to the Sanskṛit word Sâdhana, which means *completion* or *a perfecter*, and also an agent or representative. By Saraganes is probably indicated one of the great Śâtakarṇi or Andhra dynasty. The *Periplûs* makes Ariâkê to be the beginning of the kingdom of Mambares and of all India.

Soupara has been satisfactorily identified by Dr. Burgess with Supârâ, a place about 6 miles to the north of Vasai (Bassein). It appears to have been from very early times an important centre of trade, and it was perhaps the capital of the district that lay around it. Among its ruins have been preserved some monuments, which are of historical interest, and which also attest its high antiquity. These are a fragment of a block of basalt like the rocks of Girnâr, inscribed with edicts of Aśôka, and an old Buddhist Stûpa. The name of Supârâ figures conspicuously in the many learned and elaborate treatises which were evoked in the course of the famous controversy regarding the situation of Ophir to which Solomon despatched the ships he had hired from the Tyrians. There can now be little doubt that if Ophir did not mean India itself it designated some place in India, and probably Supârâ, which lay on that part of the coast to which the traders of the west, who took advantage of the monsoon to cross the ocean, would naturally direct their course. The name moreover of Supârâ is almost identical with that of Ophir when it assumes, as it often does, an initial S, becoming Sôphara as in the *Septuagint* form of the name, and Sofir which is the Coptic name for India, not to mention other similar forms. (See Benfey's *Indien*, pp. 30-32).

The mouths of the Goaris and Bênda Yule takes to be the mouths of the Strait that isolates Salsette and Bombay. The *names* represent, as he thinks, those of the Gôdâvarî and Bhîma respectively, though these rivers flow in a direction different from that which Ptolemy assigns to them, the former discharging into the Bay of Bengal and the latter into the Kṛishṇâ, of which it is the most considerable tributary. Ptolemy's rivers, especially those of the Peninsula, are in many instances so dislocated, that it is difficult to identify them satisfactorily. It appears to have been his practice to connect the river-mouths which he found mentioned in records of coasting voyages with rivers in the interior concerning which he had information from other sources, and whose courses he had only partially traced. But, as Yule remarks, with his erroneous outline of the Peninsula this process was too hazardous and the result often wrong. Mr. J. M. Campbell, Bo.C.S., would identify the Goaris with the Vaitarna River, as Gore is situated upon it and was probably the highest point reached by ships sailing up its stream. The sources of the Vaitarna and the Gôdâvarî are in close propinquity. The Bênda he would identify with the Bhîwandî River, and the close similarity of the names favours this view.

Dounga is placed in Yule's map to the S. E. of Supârâ on the Strait which separates Salsette from the mainland. Ptolemy, however, through his misconception of the configuration of this part of the coast, places it a whole degree to the west of Supârâ. Mr. Campbell, from some similarity in the names, suggests its identity with Dugâḍ—a place about 10 miles N. of Bhîwandî and near the Vajrabâî hot springs. Dugâḍ, however, is too far inland to have been here mentioned by Ptolemy, and moreover, it lies to the north of Supârâ, whereas in Ptolemy's enumeration, which is from north to south, it is placed after it.

Simylla:—Yule identifies this with Chaul and remarks: "Chaul was still a chief port of Western India when the Portuguese arrived. Its position seems to correspond precisely both with Simylla and with the Ṣaimûr or Jaimûr (*i.e.* Chaimur, the Arabs having no *ch*) of the Arabian geographers. In Al-Bîrûnî the coast cities run: Kambâyat, Bahruj, Sindân (Sanjân), Sufâra (Supârâ), Tana (near Bombay). "There you enter the country of Lârân, where is Jaimûr." Istakhri inverts the position of Sindân and Sufâra, but Saimûr is still furthest south." In a note he adds: "Ptolemy mentions that Simylla was called by the natives Timula (probably Tiamula); and putting together all these forms, Timula, Simylla, Ṣaimûr, Chaimûr, the real name must have been something like Chaimul or Châmul, which would modernize into Chaul, as Chamari and Prâmara into Chauri and Pawâr." Chaul or Chêṅwal lies 23 miles S. of Bombay. Paṇḍit Bhagvânlâl Indraji, Ph.D., suggested as a better identification Chimûla in Trombay Island, this being supported by one of the Kaṇhêri inscriptions in which Chimûla is mentioned. apparently as a large city, like Supârâ and Kalyâṇa in the neighbourhood. Mr. Campbell thus discusses the merits of these competing identifications:—" Simylla has a special interest, as Ptolemy states that he learned some of his Geography of Western India from people who traded to Simylla and had been familiar with it for many years, and had come from there to him—Ptolemy speaks of Simylla as a point and emporium, and the author of the *Periplûs* speaks of it as one of the Koṅkaṇ local marts. Simylla till lately was identified with Chaul. But the discovery of a village Chembur on Trombay Island in Bombay Harbour, has made it doubtful whether the old trade centre was there or at Chaul. In spite of the closer resemblance of the names, the following reasons seem to favour the view that Chaul, not Chimûla, was the Greek Simylla. First, it is somewhat unlikely that two places so close, and so completely on the same line of traffic as Kalyân (the Kalliena of the *Periplûs*) and Chimûla should have flourished at the same time. Second, the expression in the *Periplûs* 'below (μετα) Kalliena other local marts are Semulla' points to some place down the coast rather than to a town in the same Harbour as Kalliena, which according to the Author's order north to south should have been named before it. Third, Ptolemy's point (promontorium) of Simylla has no meaning if the town was Chembur in Trombay. But it fits well with Chaul, as the headland would then be the south shore of Bombay Harbour, one of the chief capes in this part of the coast, the south head of the gulf or bay whose north head is at Bassein. This explanation of the Simylla point is borne out by Fryer (1675) *New Account* (pp. 77-82), who talked of Bombay 'facing Chaul' and notices the gulf or hollow in the shore stretching from Bassein to Chaul Point. The old (1540) Portuguese name Chaul Island' for the isle of Kennery of the south point of Bombay, further supports this view." Ptolemy's map gives great prominence to the projection of land at Simylla, which (through a strange misconception on his part, for which it is impossible to account) is therein represented as the great south-west point of India, whence the coast bends at once sharply to the east instead of pursuing its course continuously to the south.

H i p p o k o u r a :—This word may be a Greek translation (in whole or in part) of the native name of the place. Hence Paṇdit Bhagvânlâl Indraji was led to identify it with Ghoḍabandar (Horse-port) a town on the Ṭhaṇa Strait, whose position however is not in accordance with Ptolemy's data. Mr. Campbell again has suggested an identification free from this objection. Ghoṛegâoṅ (Horse-village) in Kolâba, a place at the head of a navigable river, which was once a seat of trade. Yule takes it, though doubtingly, as being now represented by Kuḍâ near Râjapûr. Hippokourios was one of the Greek epithets of Poseidôn. Ptolemy mentions another Hippokoura, which also belonged to Ariâkê and was the Capital of Baleokouros. Its situation was inland.

Baltipatna:—This place is mentioned in the *Periplûs* under the somewhat altered form Palaipatmai. Yule locates it, but doubtingly, at Daibal. Fra Paolino identified it with Balaerpatam (the Baleopatam of Rennell) where the king of Cananor resided, but it lies much too far south to make the identification probable. Mr. Campbell has suggested Pali, which he describes as "a very old holy town at the top of the Nagôtna river." Its position, however, being too far north and too far from the sea, does not seem to suit the requirements.

7. (Ariakê) of the Pirates.

Mandagara	113°	14°
Byzanteion	113° 40′	14° 40′
Khersonêsos	114° 20′	14° 30′
Armagara	114° 20′	14° 20′
Mouth of the River Nanagouna	114° 30′	13° 50′
Nitra, a mart	115° 30′	14° 40′

Ariakê.

Piracy, which from very early times seems to have infested, like a pernicious parasite, the commerce of the Eastern Seas, flourished nowhere so vigorously as on the Koṅkaṇ Coast, along which richly freighted merchantmen were continually plying. Here bands of pirates, formed into regularly organized communities like those of the Ṭhags in the interior of the country, had established themselves in strongholds contiguous to the creeks and bays, which were numerous on the coast, and which afforded secure harbourage to their cruisers. The part of the coast which was subject to their domination and which was in consequence called the Pirate Coast, extended from the neighbourhood of Simylla to an emporium called Nitra, the Mangaruth of Kosmas and the Mangalûr of the present day. Whether the native traders took any precautions to protect their ships from these highwaymen of the ocean is not known, but we learn from Pliny, that the merchantmen which left the Egyptian ports heading for India carried troops on board well-armed for their defence. Mr. Campbell has ingeniously suggested that by 'Ανδρῶν Πειρατῶν Ptolemy did not mean pirates, but the powerful dynasty of the Ândhrabhṛitya that ruled over the Koṅkaṇ and some other parts of the Dekhan. He says (*Bombay Gazetteer*, Thâna, vol. II., p. 415 n. 2nd), "Perhaps because of Pliny's account of the Koṅkaṇ pirates, Ptolemy's phrase *Ariâkê Andrōn Peiratōn* has been taken to mean Pirate Ariâkê. But Ptolemy has no mention of pirates on the Koṅkaṇ Coast, and, though this does not carry much weight in the case of Ptolemy, the phrase *Andrōn Peiratōn* is not correct Greek for pirates. This and the close resemblance of the words suggest that *Andrōn Peiratōn* may originally have been Andhrabhrityon." On this it may be remarked, that though Ptolemy has no mention of pirates on the Koṅkaṇ Coast this is not in the least surprising, since his work is almost exclusively geographical, and whatever information on points of history we obtain from it is more from inference than direct statement. Further, I do not see why the expression ἀνδρῶν Πειρατῶν if taken to mean pirates should be called incorrect Greek, since in later Attic it was quite a common usage to join ἀνήρ with titles, professions and the like.

Mandagara:—This may be a transliteration, somewhat inexact, of Madangaṛh (House of Love) the name of a fort about 12 miles inland from Bankûṭ. More likely the place is Mândlâ on the north bank of the Sautrî river, opposite Bankûṭ, and now known as Kolmândlâ, and Bâg and Bâgmândlâ. Mangalûr, to which as far as the name goes it might be referred, is too far south for the identification.

Byzanteion:—The close correspondence of this name with that of the famous capital on the Bosporos has led to the surmise that a colony of Greeks had established themselves on this coast for commercial purposes, notwithstanding the danger to be apprehended from attacks by the pirates in their neighbourhood. It appears however quite unlikely that Greeks should have formed a settlement where few, if any, of the advantages could be enjoyed which generally determined their choice of a locality in which to plant a colony. The name may perhaps be a transliteration of Vijayanta, now Vijayadurga, the south entrance of the Vâghotan river in Ratnagiri. The word means the Fort of Victory.

Khersonêsos:—This seems to be the peninsula which is in the neighbourhood of Goa. It is mentioned in the *Periplûs* as one of the haunts of the pirates, and as being near the island of the Kaineitai, that is, St. George's Island.

Armagara:—This is placed near the mouth of the Nanagouna river, which may be taken to mean here the river on which Sadâśivagaṛh stands. The Nanagouna however must be identified with the Tâptî, whose embouchure is about 6° farther north. Its name is Sanskṛit, meaning 'possessed of many virtues.' To account for this extraordinary dislocation, Yule supposes that Ptolemy, having got from his Indian lists a river Nânâguna rising in the Vindhyas, assigns to it three discharges into the sea by what he took for so many delta branches, which he calls respectively Goaris, Benda, and Nanaguna. This, he adds, looked possible to Ptolemy on his map, with its excessive distortion of the western coast, and his entire displacement of the Western Ghâts. Mr. Campbell suggests that Ptolemy may have mistaken the Nânâ Pass for a river.

Nitra is the most southern of the pirate ports, and is mentioned by Pliny in a passage where he remarks that ships frequenting the great emporium of Mouziris ran the risk of being attacked by pirates who infested the neighbourhood, and possessed a place called Nitra. Yule refers it as has been already stated to Mangalur.

8. Limyrikê.

Tyndis, a city	116°	14° 30′
Bramagara	116° 45′	14° 20′
Kalaikarias	116° 40′	14°
Mouziris, an emporium	117°	14°

Mouth of the River Pseudostomos	117° 20′	14°
Podoperoura	117° 40′	14° 15′
Semnê	118°	14° 20′
Koreoura	118° 40′	14° 20′
Bakarei	119° 30′	14° 30′
Mouth of the River Baris	120°	14° 20′

Limyrikê:—Lassen was unable to trace this name to any Indian source, but Caldwell has satisfactorily explained its origin. In the introduction to his *Dravidian Grammar* he states (page 14), that in the Indian segment of the Roman maps called the *Peutinger Tables* the portion of India to which this name is applied is called Damirike, and that we can scarcely err in identifying this name with the Tamiḻ country, since Damirike evidently means *Damir-ikê*. In the map referred to there is moreover a district called Scytia Dymirice, and it appears to have been this word which by a mistake of Δ for Λ Ptolemy wrote Lymirike. The D, he adds, retains its place in the *Cosmography* of the Geographer of Ravenna, who repeatedly mentions Dimirica as one of the 3 divisions of India. Ptolemy and the author of the *Periplûs* are at one in making Tyndis one of the first or most northern ports in Limyrikê. The latter gives its distance from Barygaza at 7,000 stadia, or nearly 12 degrees of latitude, if we reckon 600 stadia to the degree. Notwithstanding this authoritative indication, which makes Limyrikê begin somewhere near Kalikat (11° 15′ N. lat.) its frontier has generally been placed nearly 3 degrees further north, Tyndis having been located at Barcelôr. This error has been rectified by Yule, whose adherence to the data of the *Periplûs* has been completely justified by the satisfactory identification of Mouziris (the southern rival in commercial prosperity of Barygaza) with Kranganur, instead of with Mangalur as previously accepted. The capital of Limyrikê was Karûr, on the Kâvêrî, where resided Kêrobothros, *i.e.*, Kêralaputra, the Chêra king.

Tyndis is described in the *Periplûs* as a place of great note pertaining to the kingdom of Kêprobotras, and situate near the sea at a distance of 500 stadia from Mouziris. This distance north from Kranganur with which, as has been stated, Mouziris has been identified, brings us to Tanûr. "Tanûr itself," says Yule, "may be Tyndis; it was an ancient city, the seat of a principality, and in the beginning of the 16th century had still much shipping and trade. Perhaps, however, a more probable site is a few miles further north, Kaḍaluṇḍi, *i. e.* Kaḍal-tuṇḍi, 'the raised ground by the sea,' standing on an inlet 3 or 4 miles south of Bêpur. It is not now a port, but persons on the spot seem to think that it must formerly have been one, and in communication with the Backwater." He adds in a note supplied by Dr. Burnell, "The composition of *Kaḍal* and *Tuṇḍi* makes *Kaḍaluṇḍi* by Tamiḻ rules." The pepper country called Kottonarike was immediately adjacent to Tyndis, which no doubt exported great quantities of that spice.

Bramagara is placed in the table half a degree to the east of Tyndis, *i.e.*, really to the south of it, since Ptolemy makes the Malabar

Coast run east instead of south. The name may be a transliteration of the Sanskrit *Brahmâgâra*, which means 'the abode of the Brâhmaṇs.' The Brâhmaṇs of the south of India appear in those days to have consisted of a number of isolated communities that were settled in separate parts of the country, and that were independent each of the other. This, as Lassen remarks (*Ind. Alt.*, vol. III, p. 193) is in harmony with the tradition according to which the Ârya Brâhmaṇs were represented as having been settled by Paraśurâma in 61 villages, and as having at first lived under a republican constitution. In section 74 Ptolemy mentions a town called Brâhmê belonging to the Brâhmaṇoi Magoi, *i.e.*, 'sons of the Brâhmaṇs.'

Kalaikarias:—The last half of this word (*Karias*) is doubtless the Tamiḷ word for "coast," *karei*, which appears also in another of Ptolemy's names, Peringkarei, mentioned as one of the inland towns Kandionoi (sec. 89). I find in Arrowsmith's large Map of India a place called 'Chalacoory' to the N. E. of Kranganur, and at about the same distance from it as our author makes Kalaikarias distant from Mouziris.

Mouziris may unhesitatingly be taken to, represent the Muyiri of Muyiri-Koḍu, which says Yule, appears in one of the most ancient of Malabar inscriptions as the residence of the King of Koḍangalur or Kranganur, and is admitted to be practically identical with that now extinct city. It is to Kranganur he adds that all the Malabar traditions point as their oldest seaport of renown; to the Christians it was the landing-place of St. Thomas the Apostle.

Mouth of the river Pseudostomos, or 'false-mouth.' According to the table the river enters the sea at the distance of $\frac{1}{3}$ of a degree below Mouziris. It must have been one of the streams that discharge into the Backwater.

Podoperoura must be the Poudopatana of Indikopleustês—a word which means 'new town,' and is a more correct form than Ptolemy's Podoperoura.

Semnê:—The Sanskrit name for Buddhist Ascetics was *Śramaṇa*, in Tamil *Śamaṇa*, and as we find that this is rendered as *Semnoi* by Clemens Alexandrinus, we may infer that Semnê was a town inhabited by Buddhists, having perhaps a Buddhist temple of noted sanctity. For a different explanation see Lassen's *Ind. Alt.* vol. III, p. 194.

Bakarei is mentioned by Pliny as Becare, and as Bakarê by the Author of the *Periplûs*, who places it at the mouth of the river on which, at a distance of 120 stadia from the sea was situated the great mart called Nelkynda, or Melkynda as Ptolemy writes it. The river is described as difficult of navigation on account of shallows and sunken reefs, so that ships despatched from Nelkynda were obliged to sail down empty to Bakarê and there take in their cargoes. The distance of Nelkynda from Mouziris is given at about 500 stadia, and this whether the journey was made by sea or by river or by land. Upon this Yule thus remarks: "At this distance south from Kranganur we are not able to point to a quite satisfactory Nelkynda. The site which has been selected as the most probable is nearly 800 stadia south of Mouziris. This is Kallaḍa, on a river of the same name entering the Backwater, the only navigable river on this south-west coast except the Perri-âr near Kranganur. The Kallaḍa river is believed to be the *Kanĕṭṭi* mentioned in the Kêralotatti legendary history of Malabar, and the town of Kallaḍa to be the town of Kanĕṭṭi. It is now a great entrepôt of Travankor pepper, which is sent from this to ports on the coast for shipment. That Nelkynda cannot have been far from this is clear from the vicinity of the Πυῤῥὸν ὄρος or *Red-Hill* of the *Periplûs* (sec. 58). There can be little doubt that this is the bar of red laterite which, a short distance south of Quilon, cuts short the Backwater navigation, and is thence called the Warkallê barrier. It forms abrupt cliffs on the sea, without beach, and these cliffs are still known to seamen as the *Red Cliffs*. This is the only thing like a sea cliff from Mount d'Ely to Cape Comorin." The word Bakarei may represent the Sanskrit *dvâraka*, 'a door.'

Mouth of the river B a r i s :—The Baris must be a stream that enters the Backwater in the neighbourhood of Quilon.

9. Country of the A ï o i.

Melkynda	120° 20′	14° 20′
Elangkôn (or Elangkôr), a mart	120° 40′	14°
Kottiara, the metropolis	121°	14°
Bammala	121° 20′	14° 15′
Komaria, a cape and town	121° 45	13° 30′

Limirike and country of the Aïoi.

The Aïoi:—This people occupied the southern parts of Travankor. Their name is perhaps a transliteration of the Sanskrit *ahi*, 'a snake,' and if so, this would indicate the prevalence among them of serpent worship. Cunningham, in his *Geography of Ancient India* (p. 552), states that in the Chino-Japanese Map of India the alternative name of Malyakûṭa is Hai-an-men, which suggests a connection with Ptolemy's Aïoi. I note that the entrance to the Backwater at Kalikoulan is called the Great Ayibicca Bar, and an entrance farther south the Little Ayibicca Bar. The first part of this name may also be similarly connected.

Melkynda, as already stated is the Nelkynda of the *Periplûs*, which places it, however, in Limyrikô. Pliny speaks of it as *portus gentis Neacyndon* (v. ll. Neacrindon, Neachyndon, Nelcyndon.) The name, according to Caldwell, probably means West Kynda, that is Kannetri, the south boundary of Kêrala Proper. When Mangalur was taken as the representative of Mouziris, Nelkynda was generally identified with Nelisuram, which besides the partial resemblance of its name, answered closely in other respects to the description of Nelkynda in the *Periplûs*—*Cff.* C. Müller, *not. ad Peripl., Sec.* 54. Lassen, *Ind. Alt.*, vol. III, p. 190. Bunbury, *Hist. of Anc. Geog.* vol. I, pp. 467-8.

Elangkôn or Elangkôr is now Quilon, otherwise written Kulam.

"Kottiara," says Caldwell, "is the name of a place in the country of the Aïoi of Ptolemy in the Paralia of the Author of the *Periplûs*, identical in part with South Travankor. Apparently it is the Cottara of Pliny, and I have no doubt it is the Cottara of the *Peutinger Tables.* It is called by Ptolemy the Metropolis, and must have been a place of considerable importance. The town referred to is probably Kôttâra, or as it is ordinarily written by Europeans 'Kotaur,' the principal town in South Travankor, and now as in the time of the Greeks distinguished for its commerce." *Dravid. Gram.*, Introd. p. 98. The name is derived from *kôḍ* 'a foot,' and *âr-û* 'a river.'

Bammala:—Mannert would identify this with Bulita, a place a little to the north of Anjenga, but this is too far north. It may perhaps be the Balita of the *Periplûs.*

Komaria, a cape and a town:—We have no difficulty in recognizing here Cape Comorin, which is called in the *Periplûs* Komar and Komarei. The name is derived from the Sanskrit *kumâri*, 'a virgin,' one of the names of the Goddess Dûrgâ who presided over the place, which was one of peculiar sanctity. The Author of the *Periplûs* has made the mistake of extending the Peninsula southward beyond Comorin.

We may here compare Ptolemy's enumeration of places on the west coast with that of the *Periplûs* from Barygaza to Cape Comorin.

Ptolemy.	*Periplûs.*
Barygaza	Barygaza
Nousaripa	Akabarou
Poulipoula	
Soupara	Souppara
Dounga	Kalliena

Ptolemy.	*Periplûs.*
Simylla	Semylla
Island of Milizêgyris	Mandagora
Hippokoura	
Baltipatna	Palaipatmai
Mandagora	Melizeigara
Is. of Heptanêsia	
Byzanteion	Byzantion
	Toparon
	Tyrannosboas
	3 separate groups of islands
Khersonêsos	Khersonêsos
Armagara	Is. of Leukê
Is. of Peperine	
Nitra	Naoura
Tyndis	Tyndis
Trinêsia Islands	
Bramagara	
Kalaikarias	
Mouziris	Mouziris
Podoperoura	
Semnê	
Is. Leukê	
Koreoura	
Melkynda	Nelkynda
Bakarei	Bakarê
Elangkôn	Mons Pyrrhos
Kottiara	
Bammola	Balita
Komaria	Komar.

There is a striking agreement between the two lists, especially with respect to the order in which the places enumerated succeed each other. There are but three exceptions to the coincidence and these are unimportant. They are, Milizegyris, Mandagora and the Island Leukê, i.e. 'white island,' if the name be Greek. The Melizeigara of the *Periplûs*, Vincent identifies with Jayagaḍh or Sidi, perhaps the Sigerus of Pliny (lib. VI, c. xxvi, 100). Ptolemy makes Milizêgyris to be an island about 20 miles south of Simylla. There is one important place which he has failed to notice, Kalliena now Kalyâṇa, a well-known town not far from Bombay.

10. Country of the Kareoi.

In the Kolkhic Gulf, where there is the Pearl Fishery:—

Sôsikourai	122°	14° 30′
Kolkhoi, an emporium	123°	15°
Mouth of the river Sôlên......	124°	14° 40′

The country of the Kareoi corresponds to South Tinneveli. The word *karei*, as already stated is Tamiḻ, and means 'coast.' The Kolkhic Gulf is now known as the Gulf of Manâr. The pearl fishery is noticed in the *Periplûs*.

Sôsikourai:—By the change of S into T we find the modern representative of this place to be Tutikorin (Tuttukuḍi) a harbour in Tinneveli, where there are pearl banks, about 10 miles south of Kolkhoi. This mart lay on the Sôlên or Tâmraparṇî river. Tutikorin in the *Peutinger Tables* is called *Colcis Indorum*. The Tamiḻ name is Kolkei, almost the same as the Greek. Yule in his work on Marco Polo (vol. II, pp. 360-61) gives the following account of this place, based on information supplied by Dr. Caldwell:—

"Kolkhoi, described by Ptolemy and the Author of the *Periplūs* as an emporium of the pearl trade, as situated on the sea-coast to the east of Cape Comorin, and as giving its name to the Kolkhic Gulf or Gulf of Manâr has been identified with Korkai, the mother-city of Kayal (the Coël of Marco Polo). Korkai, properly Kolkai (the *l* being changed into *r* by a modern refinement, it is still called Kolka in Malayalam), holds an important place in Tamiḻ traditions, being regarded as the birth-place of the Pâṇḍya dynasty, the place where the princes of that race ruled previously to their removal to Madurâ. One of the titles of the Pâṇḍya kings is 'Ruler of Korkai.' Korkai is situated two or three miles inland from Kayal, higher up the river. It is not marked in the G. Trig. Surv. map, but a village in the immediate neighbourhood of it, called Mâramangalam 'the good fortune of the Pâṇḍyas' will be found in the map. This place, together with several others in the neighbourhood, on both sides of the river, is proved by inscriptions and relics to have been formerly included in Korkai, and the whole intervening space between Korkai and Kayal exhibits traces of ancient dwellings. The people of Kayal maintain that their city was originally so large as to include Korkai, but there is much more probability in the tradition of the people of Korkai, which is to the effect that Korkai itself was originally a seaport; that as the sea retired it became less and less suitable for trade, that Kayal rose as Korkai fell, and that at length, as the sea continued to retire, Kayal also was abandoned. They add that the trade for which the place was famous in ancient times was the trade in pearls."

Mouth of the River Sôlên:—This river is identified by Lassen with the Sylaur, which he says is the largest northern tributary of the Tâmraparṇî. On this identification Yule remarks:—"The 'Syllâr' of the maps, which Lassen identifies with Sôlên, originates, as Dr. Caldwell tells me, in a mistake. The true name is 'Sitt-âr,' 'Little River,' and it is insignificant." The Tâmraparṇî is the chief river of Tinneveli. It entered the sea south of Kolkhoi. In Tamiḻ poetry it is called Porunei. Its Pâli form is Tambapanni. How it came to be called the Sôlên remains as yet unexplained. *Śôḷa* is an element in several South Indian geographical names, meaning Chôḷa. The word Tâmraparṇî itself means 'red-leaved' or 'copper-coloured sand.' Taprobane, the classical name for Ceylon, is this word in an altered form.

11. Land of Pandion.

In the Orgalic Gulf, Cape Kôry, called also Kalligikon..	125° 40′	12° 20′
Argeirou, a town	125° 15′	14° 30′
Salour, a mart	125° 20′	15° 30′

The land of Pandion included the greater portion of the Province of Tinneveli, and extended as far north as to the highlands in the neighbourhood of the Koimbatur gap. Its western boundary was formed by the southern range of the Ghâts, called by Ptolemy Mount Bêttigô, and it had a sea-board on the east, which extended for some distance along the Sinus Orgalicus, or what is now called Palk's Passage.

The Author of the *Periplûs* however, assigns it wider limits, as he mentions that Nelkynda, which lay on the Malabar Coast, as well as the pearl-fishery at Kolkhoi, both belonged to the Kingdom of Pandion. The kingdom was so called from the heroic family of the Pâṇḍya, which obtained sovereign power in many different parts of India. The Capital, called Madurâ, both by Pliny and by our author, was situated in the interior. Madurâ is but the Tamiḻ manner of pronouncing the Sanskrit *Mathurâ*, which also designated the sacred city on the Jamnâ famous as the birthplace and the scene of the exploits of Kṛishṇa, who assisted the Pâṇḍus in their war with the Kurus. The city to this day retains its ancient name, and thus bears, so to speak, living testimony to the fact that the Âryans of Northern India had in early times under Pâṇḍya leaders established their power in the most southern parts of the Peninsula.

The Orgalic Gulf lay beyond the Kolkhic Gulf, from which it was separated by the Island of Râmêsvaram and the string of shoals and small islands which almost connect Ceylon with the mainland. It derived its name from Argalou, a place mentioned in the *Periplûs* as lying inland and celebrated for a manufacture of muslin adorned with small pearls. The northern termination of the gulf was formed by Cape Kalimîr.

Cape Kôry:—Ptolemy makes Kôry and Kalligikon to be one and the same cape. They are however distinct, Kôry being the headland which bounded the Orgalic Gulf on the south, and Kalligikon being Point Kalimîr, which bounded it on the north. The curvature of this Gulf was called by the Hindûs Râmadhanuḥ, or '*Râma's bow*,' and each end of the bow Dhanuḥ-kôṭi or simply Kôṭi. The Sanskrit word *kôṭi* (which means '*end, tip* or *corner*') becomes in Tamiḻ *kôḍi*, and this naturally takes the form of Kôṛi or Kôry. The southern Kôṭi, which was very famous in Indian story, was formed by the long spit of land in which the Island of Râmêśvaram terminates. It is remarkable, as Caldwell remarks, that the Portuguese, without knowing anything of the Κῶρυ of the Greeks, called the same spit of land Cape Ramancoru. Ptolemy's identification of Cape Kôry with Kalligikon or Point Kalimîr is readily explained by the fact just stated that each of these projections was called Kôṭi.

This word Kôṭi takes another form in Greek and Latin besides that of Kôry, viz., Kôlis, the name by which Pomponius Mela and Dionysios Periêgêtês (v. 1148) designate Southern India. The promontory is called Coliacum by Pliny, who describes it as the projection of India nearest Ceylon, from which it was separated by a narrow coral sea. Strabo (lib. XV, c. i, 14) quoting Onêsikritos, speaks of Taprobane as distant from the most southern parts of India, which are opposite the Kôniakoi, 7 days' sail towards the south. For Kôniakoi the reading Kôlîakoi has been with reason suggested.

Ptolemy, like the author of the *Periplûs* and other writers, regarded Cape Kôry as the most important projection of India towards the south, and as a well-established point from which the distances of other places might conveniently be calculated. He placed it in 125 degrees of E. longitude from Ferro, and at 120 degrees east of the mouth of the River Bætis in Spain from which, however, its distance is only 86⅓ degrees. Its latitude is 9° 20′ N. and that of Cape Comorin 8° 5′, but Ptolemy makes the difference in latitude to be only 10′.

The identity of Kalligikon with Point Kalimîr has already been pointed out. *Calimere* is a corrupt form of the Tamil̤ compound Kallimedu, *Euphorbia eminence*, and so the first part of the Greek name exactly coincides with the Tamil̤ Kal̤l̤i, which means the *Euphorbia* plant, or perhaps a kind of cactus. Pliny mentions a projection on the side of India we are now considering which he calls Calingon, and which the similarity of name has led some to identify with Kalligikon, and therefore with Point Kalimîr. It seems better, however, taking into account other considerations which we need not here specify, to identify this projection with Point Gôdâvarî.

Before concluding this notice we may point out how Ptolemy has represented the general configuration of the eastern coast beyond the Orgalic Gulf. His views here are almost as erroneous as those he entertained concerning the west coast, which, it will be remembered, he did not carry southward to Cape Comorin, but made to terminate at the point of Simylla, thus effacing from the Map of India the whole of the Peninsula.

The actual direction of the east coast from point Kalimîr is first due north as far as the mouths of the Kṛishṇâ, and thereafter north-east up to the very head of the Bay of Bengal. Ptolemy, however, makes this coast run first towards the south-east, and this for a distance of upwards of 600 miles as far as Paloura, a place of which the site has been fixed with certainty as lying near the southern border of Kaṭak, about 5 or 6 miles above Ganjâm. Ptolemy places it at the extremity of a vast peninsula, having for one of its sides the long stretch of coast just mentioned, and he regards it also as marking the point from which the Gangetic Gulf begins. The coast of this gulf is made to run at first with an inclination to westward, so that it forms at its outlet the other side of the peninsula. Its curvature is then to the north-east, as far as to the most eastern mouth of the Ganges, and thence its direction is to the south-east till it terminates at the cape near Têmala, now called Cape Negrais, the south-west projection of Pegu.

12. Country of the Batoi.

Nikama, the Metropolis	126°	16°
Thelkheir	127°	16° 10′
Kouroula, a town	128°	16°

13. In Paralia specially so called: the country of the Tôringoi.

Mouth of the River Khabêros	129°	15° 15′
Khabêris, an emporium	128° 30′	15° 40′
Sabouras, an emporium	130°	14° 30′

The Batoi occupied the district extending from the neighbourhood of Point Kâlimîr to the southern mouth of the River Kâvêrî and corresponding roughly with the Province of Tanjore.

Nikama, the capital, has been identified with Nagapatam (Nâgapaṭṭanam) by Yule, who also identifies (but doubtingly) Thelkyr with Nagor and Kouroula with Karikal.

Paralia, as a Greek word, designated generally any maritime district, but as applied in India it designated exclusively (ἰδίως) the seaboard of the Tôringoi. Our author is here at variance with the *Periplûs*, which has a Paralia extending from the Red Cliffs near Quilon to the Pearl-Fishery at the Kolkhoi, and comprising therefrom the coast-lines of the Aïoi and the Kareoi. "This Paralia," says Yule, "is no doubt Purali, an old name of Travankor, from which the Râja has a title *Puraliśan*, 'Lord of Purali.' But the "instinctive striving after meaning" which so often modifies the form of words, converted this into the Greek Παραλία, 'the coast.' Dr. Caldwell however inclines rather to think that *Paralia* may possibly have corresponded to the native word meaning *coast*, viz. *karei*.

In sec. 91, where Ptolemy gives the list of the inland towns of the Tôringoi, he calls them the Sôrêtai, mentioning that their capital was Orthoura, where the king, whose name was Sôrnagos, resided. In sec. 68 again he mentions the Sôrai as a race of nomads whose capital was Sôra where their king, called Arkatos, resided. Caldwell has pointed out the identity of the different names used to designate this people. Σῶρα, he says, "which we meet alone and in various combinations in these (Ptolemy's) notices represents the name of the northern portion of the Tamilian nation. This name is Chôla in Sanskrit, Chôḷa in Telugu, but in Tamiḻ Sôṛa or Chôṛa. The accuracy with regard to the name of the people is remarkable, for in Tamiḻ they appear not only as Sôṛas, but also as Sôṛagas and Sôṛiyas, and even as Sôṛingas. Their country also is called Sôṛagam. The *r* of the Tamiḻ word Sôṛa is a peculiar sound not contained in Telugu, in which it is generally represented by *ḍ* or *ḷ*. The transliteration of this letter as *r* seems to show that then, as now, the use of this peculiar *ṛ* was a dialectic peculiarity of Tamiḻ."

The River Khabêros is the Kâvêrî. *Kâvêra* is the Sanskrit word for *saffron.* Kâvêrî, according to a legend in the *Harivaṅśa*, was changed by her father's curse from one-half of the Gaṅgâ into the river which bears her name, and which was therefore also called Ardha-gaṅgâ, *i.e.*, half-gaṅgâ. Karoura, the residence of the Chera king, was upon this river.

Dr. Burnell identified Khabêris with Kâvêrîpaṭṭam (*Ind. Ant.*, vol. VII, p. 40) which lies a little to the north of Tranquebar (Tallangambadi) at the mouth of the Pudu-Kâvêrî (New Kâvêrî).

Sabouras:—This mart Yule refers doubtingly to Gudalur (Cuddalore) near the mouth of the S. Penn-âr River.

14. The Arouarnoi (Arvarnoi).

Pôdoukê, an emporium	130° 15′	14° 30′
Melangê, an emporium.........	131°	14° 20′
Mouth of the River Tyna.......	131° 40′	12° 45′

Kottis ....132° 20′ 12° 10′
Manarpha (or Manaliarpha,
a mart) .... 133° 10′ 12°

15. Maisôlia.

Mouth of the River Maisôlos 134° 11° 40′
Kontakossyla, a mart ....134° 30′ 11° 40′
Koddoura ....135° 11° 30′
Allosygnê, a mart ....135° 40′ 11° 20′
The point of departure (*aphe-
têrion) for ships bound for
Khrysê ....136° 20′—11°

The territory of the Arouarnoi (Arvarnoi) was permeated by the River Tyna, and extended northward to Maisôlia, the region watered by the River Maisôlos in the lower parts of its course. Opinions differ with regard to the identification of these two rivers, and consequently also of the places mentioned in connection with them. Some of the older commentators, followed by Yule, take the Tyna to be the Pinâka or Penn-âr River, and the Maisôlos the Kṛishṇâ. Lassen again, and recent writers generally, identify the Tyna with the Kṛishṇâ and the Maisôlos with the Gôdâvarî. To the former theory there is the objection that if the Gôdâvarî be not the Maisôlos, that most important of all the rivers on this coast is left unnoticed, and Lassen accordingly asks why should the small Penn-âr appear and the great Gôdâvarî be omitted. To this Yule rejoins, "We cannot say why; but it is a curious fact that in many maps of the 16th and 17th and even of the 18th century the Gôdâvarî continues to be omitted altogether. A beautiful map in Valentijn (vol. V), shows Gôdâvarî only as a river of small moment, under a local name." He argues further that the name Tynna if applied to the Kṛishṇâ is unaccounted for. As identified with the Penn-âr or Pinâka, TYNNA is an easy error for ΠΥΝΝΑ.

Pôdoukê:—This mart is mentioned in the *Periplûs* along with Kamara and Sôpatma as ports to which merchants from Limyrikê and the north were wont to resort. According to Böhlen, Ritter and Benfey, it is Puduchchêri (Pondicherry). Lassen and Yule agree, however, in placing it at Pulikât, which is nearly two degrees further north.

In Yule's map Melangê is placed at Kṛishṇapatam, a little to the south of the North Penn-âr River, which as we have seen, he identifies with the Tyna. Its name closely approximates to that of the capital Malanga, and hence Cunningham, who takes the Maisôlos to be the Gôdâvarî, and who locates Malanga in the neighbourhood of Elûr, identifies Melangê with Bandar Malanka (near one of the Gôdâvarî mouths) which he assumes to have been so called from its being the port (*bandar*) with which the capital that lay in the interior communicated with the sea. See *Geog. of Anc. Ind.*, pp. 539-40.

Manarpha (or Manaliarpha):—This mart lay at the mouth of a river which still preserves traces of its name, being called the Manâra. Kottis lay not very far to the north of it.

Maisôlia is the name of the coast between the Kṛishṇâ and the Gôdâvarî, and onward thence to the neighbourhood of Paloura. It is the Masalia of the *Periplûs* which describes it as the sea-board of a country extending far inland, and noted for the manufacture, in immense quantities, of the finer kinds of cotton fabrics. The name is preserved in Masulipaṭṭam, which has been corrupted for the sake of a meaning into Machhlipatam, which means *fish-town.* The Metropolis called Pityndra was seated in the interior.

Kontakossyla transliterates, though not quite correctly, the Sanskrit Kaṇṭakasthala, '*place of thorns.*' In Yule's map it is placed inland near the Kṛishṇâ, in the neighbourhood of Koṇḍapalle, in which its name seems to be partly preserved.

Koddoura has been identified with Gûdrû, a town near Masulipatam.

Allosygnê may perhaps be now represented by Koringa (Koranja) a port situated a little beyond Point Gôdâvarî. Its distance from the point next mentioned in the Tables may be roughly estimated at about 230 miles, but Ptolemy makes it to be only ⅔ of a degree, and thus leaves undescribed an extensive section of the coast comprising the greater part of the sea-board of the Kalingai. A clue to the explanation of this error and omission is supplied by a passage in the *Periplûs*, which runs to the effect that ships proceeding beyond Maisôlia stood out from the shore and sailing right across a bay made a direct passage to the ports of Dêsarênê, *i.e.* Orissa. It may hence be inferred that navigators who came from a distance to trade in those seas would know little or nothing of a coast which they were careful to avoid, and that Ptolemy in consequence was not even so much as aware of its existence.

The point whence ships took their departure for Khrysê Yule places at the mouth of a little river called the Baroua (the Puacotta of Lindschoten) lying under Mt. Mahendra in lat. 18° 54′ N. This *aphetêrion*, he points out, was not a harbour as Lassen supposed, from which voyages to Khrysê were made, but the point of departure from which vessels bound thither struck off from the coast of India, while those bound for the marts of the Ganges renewed their coasting. The course of navigation here described continued to be followed till modern times, as Yule shows by a quotation from Valentijn's book on the Dutch East Indies (1727) under a notice of Bimlipatam :—" In the beginning of February, there used to ply . . . to Pegu, a little ship with such goods as were in demand, and which were taken on board at Masulipatam. . . . From that place it used to run along the coast up to 18° N. Lat., and then crossed sea-wards, so as to hit the land on the other side about 16°, and then, on an offshore wind, sailed very easily to the Peguan River of Syriang." (Syriam below Rangun).

16. In the Gangetic Gulf.

Paloura or Pakoura, a town..	136° 40′	11° 20′
Nanigaina	136° 20′	12°
Katikardama........................	136° 20′	12° 40′
Kannagara	136° 30′	13° 30′
Mouth of the River Manada. .	137°	14°
Kottobara	137° 15′	14° 40′

Sippara	137° 40′	15° 30′
Mouth of the River Tyndis...	138° 30′	16°
17. Mapoura	139°	16° 30′
Minagara	140°	17° 15′
Mouth of the Dôsarôn.........	141°	17° 40′
Kôkala	142°	18°
Mouth of the River Adamas	142° 40′	18°
Kôsamba or Kôsaba	143° 30′	18° 15′

Paloura :—Ptolemy, as we have seen, placed this town at the extremity of a great peninsula projecting to the south-east, which had no existence however, except in his own imagination. The following passage, quoted by Yule from Lindschoten, shows that the name of Paloura survived till modern times, and indicates at the same time where its site is to be looked for :— " From the river of Puacota to another called Paluor or Palura, a distance of 12 leagues, you run along the coast with a course from S. W. to E. Above this last river is a high mountain called Serra de Palura, the highest mountain on the coast. This river is in 19½°." The Palura River must be the river of Ganjâm, the latitude of which is at its mouth 19° 23′. Ptolemy fixes at Paloura the beginning of the Gangetic Gulf.

Nanigaina may perhaps be placed at Purî, famous for the temple of Jagannâtha Katikardama.

The first part of the name points to the identification of this place with Kaṭak, the capital of Orissa.

Kannagara :—There can be little doubt that we have here the Kanarak of modern times, called also the Black Pagoda.

Mouth of the Manada:—Ptolemy enumerates four rivers which enter the Gulf between Kannagara and the western mouth of the Ganges, the Manada, the Tyndis, the Dôsarôn and the Adamas. These would seem to be identical respectively with the four great rivers belonging to this part of the coast which succeed each other in the following order:—The Mahânadî, the Brâhmaṇî, the Vaitaraṇî and the Suvarṇarêkhâ, and this is the mode of identification which Lassen has adopted. With regard to the Manada there can be no doubt that it is the Mahânadî, the great river of Orissa at the bifurcation of which Kaṭak the capital is situated. The name is a Sanskrit compound, meaning 'great river.' Yule differs from Lassen with regard to the other identifications, making the Tyndis one of the branches of the Mahânadî, the Dôsarôn,—the Brâhmaṇi, the Adamas,—the Vaitaraṇî, and the Kambyson (which is Ptolemy's western mouth of the Ganges)—the Suvarṇarêkhâ.

The Dôsarôn is the river of the region inhabited by the Daśârnas, a people mentioned in the *Vishṇu Purâna* as belonging to the south-east of Madhya-dêśa in juxta-position to the Sabaras, or Suars. The word is supposed to be from *daśan* 'ten,' and ṛiṇa 'a fort,' and so to mean 'the ten forts.'

Adamas is a Greek word meaning *diamond.* The true Adamas, Yule observes, was in all probability the Sank branch of the Brâhmaṇî, from which diamonds were got in the days of Mogul splendour.

Sippara:—The name is taken by Yule as

representing the Sanskrit *Śûrpâraka*. *Pâra* in Sanskrit means 'the further shore or opposite bank of a river.'

Minagara:—The same authority identifies this with Jajhpûr. In Arrowsmith's map I find, however, a small place marked, having a name almost identical with the Greek, Mungrapûr, situated at some distance from Jajhpûr and nearer the sea.

Kôsamba is placed by Yule at Balasôr, but by Lassen at the mouth of the Subanrêkhâ which, as we have seen, he identifies with the Adamas. There was a famous city of the same name, Kauśâmbî, in the north-west of India, on the River Jamnâ, which became the Pâṇḍû capital after Hastinâpura had been swept away by the Ganges, and which was noted as the shrine of the most sacred of all the statues of Buddha. It is mentioned in the *Râmâyaṇa*, the *Mahâvaṁśa*, and the *Mêghadûta* of Kâlidâsa. It may thus be reasonably concluded that the Kôsamba of Ptolemy was a seat of Buddhism established by propagandists of that faith who came from Kauśâmbî.

18. Mouths of the Ganges.

The Kambyson mouth, the most western	144° 30′	18° 15′
Poloura, a town	145°	18° 30′
The second mouth, called Mega	145° 45′	18° 30′
The third called Kambêrikhon	146° 30′	18° 40′
Tilogrammon, a town	147° 20′	18°

The fourth mouth, Pseudostomon..........................147° 40′ 18° 30′
The fifth mouth, Antibolê ...148° 30′ 18° 15′

Ptolemy appears to have been the first writer who gave to the western world any definite information concerning that part of the Bengal Coast which receives the waters of the Ganges. His predecessors had indeed excelled him in the fulness and accuracy with which they had described the general course of the river, but they did not know, except in the very vaguest way, either where or how it entered the sea. Strabo, for instance, was not even aware that it had more than a single mouth. Ptolemy, on the other hand, mentions by name five of its mouths, and his estimate of the distance between the most western and the most eastern of these (4 degrees of latitude) is not very wide of the mark. Some traces also of his nomenclature are still to be found. It is difficult, however, to identify the mouths he has named with those now existing, as the Ganges, like the Indus, has shifted some of its channels, and otherwise altered the hydrography of its delta. Opinions differ regarding the western mouth, called the Kambyson. One would naturally take it to be the Hughlî river, on which Calcutta stands, and V. de Saint-Martin accordingly adopts this identification. It is impossible to doubt, he says, that the Kambysum is the Hughlî river, which must have been at all times one of the principal outlets, as is proved historically by the mention of Tâmraliptâ, 600 years before our æra, as one of the most frequented ports of Eastern India. It would be possible enough, he continues, that below Diamond Point, the principal channel, instead of passing as now in front of Kalpî remounted to the west in front of Tamluk (the ancient Tâmraliptâ) by the mouth of Tingorcally, and came thus to touch at a locality of which the actual name Nungabusan recalls that of Kambysum or Kambusum. Wilford and Yule, on the other hand, agree in identifying the Kambyson with the Subanrêkhâ river, which was formerly but erroneously supposed to be a branch of the Ganges, and they are thus free to take the Hughlî river as representing the second mouth called by Ptolemy the Mega, the Greek word for 'great.' Saint-Martin identifies this estuary with the River Matlâ to which in recent years an attempt was made to divert the commerce of Calcutta, in consequence of the dangers attending the navigation of the Hughlî. With regard to the Kambêrikhon, or third mouth, there is no difference of opinion. "It answers," says Saint-Martin, "to the Barabaṅgâ, a still important estuary, which receives the river of Kobbadak (or rather Kobbarak), which traverses the whole extent of the delta. The *Kshêtra Samâsa*, a modern treatise of Sanskrit Geography, which Wilford has often quoted in his Memoir on the Ancient Geography of the Gangetic basin, calls this river Koumâraka. Here the Kambêrikhon of the Greek navigators is easily recognized." The fourth mouth was called Pseudostomon, that is, 'false mouth,' because it lay concealed behind numerous islands, and was often mistaken for the easternmost mouth of the Ganges. This Ptolemy calls Antibolê, a name which has not yet been explained. It is the Ḍhakka or old Ganges river, and seems to have been the limit of India and the point from which measurements and distances relating to countries in India were frequently made.

In connexion with the river-mouths Ptolemy mentions two towns, Poloura and Tilogrammon. The former is placed in Yule's map at Jelasur, near the Subanrêkhâ, and the latter at Jesor. Its name seems to be compounded of the two Sanskrit words *tîla, 'sesamum,'* and *grâma, 'a village or township.'*

Ptolemy having thus described the whole sea-coast of India, from the mouths of the Indus to those of the Ganges, gives next a list of its mountain ranges, together with figures of Latitude and Longitude, showing the limits of the length of each range as well as the direction.

19. The mountains belonging to Intra-gangetic India are named as follows :—

The Apokopa, called *Poinai Theôn*, which extend from long. 116° to 124° and from lat. 23° at their western limit to 26° at the eastern.

20. Mount Sardônyx, in which is found the precious stone of the same name, and whose middle point is in long. 117° and lat. 21°.

21. Mount Ouïndion (Vindion) which extends from 126° to 135°, and preserves from its western to its eastern limit a uniform latitude of 27°.

Ptolemy enumerates seven of these, probably following some native list framed in accordance with the native idea that seven principal mountains existed in each division of a continent. A

Paurâṅik list gives us the names of the seven which pertained to India, Mahêndra, Malaya, Sahya, Śuktimat, Riksha, Vindhya and Pâripâtra or Pâriyâtra. This can hardly be the list which Ptolemy used, as only two of his names appear in it, Ouxenton (—) Ṛiksha, and Ouindion (—) Vindhya. As his views of the configuration of India were so wide of the mark, his mountain ranges are of course hopelessly out of position, and the latitudes and longitudes assigned to them in the tables afford no clue to their identification. Some help however towards this, as Yule points out, lies in the river-sources ascribed to each, which were almost certainly copied from native lists, in which notices of that particular are often to be found.

The Apokopa, or '*punishment*' of the '*gods*':—There is a consensus of the authorities in referring the range thus named to the Aravali mountains. Mount Arbuda (Abu) which is by far the most conspicuous summit, is one of the sacred hills of India. It was mentioned by Megasthenes in a passage which has been preserved by Pliny (N.H. lib.VI,c. xxi) who calls it Mons Capitalia,*i.e.* the 'Mount of Capital Punishment,' a name which has an obvious relation to the by-name which Ptolemy gives it, 'the punishment of the gods.' The word *apokopa* is of Greek origin, and means primarily 'what has been cut off,' and is therefore used to denote 'a cleft,' 'a cliff,' 'a steep hill.' It occurs in the *Periplûs* (sec. 15) where it designates a range of precipitous hills running along the coast of Azania, *i.e.* of Ajan in Africa. Its Sanskrit equivalent may have been given as a name to Mount Arbuda because of its having been at some time rent by an earthquake. In point of fact the *Mahâbhârata* has preserved a tradition to the effect that a cleft (*chhidra*) had here been made in the earth. Such an alarming phenomenon as the cleaving of a mountain by an earthquake would naturally in superstitious times be ascribed to the anger of the gods, bent on punishing thereby some heinous crime. (See Lassen's *Ind. Alt.* vol. III, pp. 121-2).

Mount Sardônyx is a short range, a branch of the Vindhya, now called Sâtpura, lying between the Narmadâ and the Tâptî: it is mentioned by Ktêsias (frag. 8) under the name of Mount Sardous. It has mines of the carnelian stone, of which the sardian is a species. The *Periplûs* (sec. 49) notices that onyx-stones were imported into Barygaza from the interior of the country, and that they were also among the articles which it exported.

Mount Ouindion:—This is a correct transliteration of *Vindhya*, the native name of the extensive range which connects the northern extremities of the Western and Eastern Ghâṭs, and which separates Hindûstân proper—the Madhya-dêśa or middle region, regarded as the sacred land of the Hindûs—from the Dekhan. Ptolemy, as Lassen remarks (*Ind. Alt.* vol. III, p. 120), is the only geographer of classical antiquity in whose writings the indigenous name of this far-spread range is to be found. His Vindion however does not embrace the whole of the Vindhya system, but only the portion which lies to the west of the sources of the Sôṇ. Sanskrit writers speak of the Vindhyas as a family of mountains. They extended from Baroda to Mirzapur, and were continued thence to Chunar.

22. Bêttigô, which extends from 123° to 130°, and whose western limit is in lat. 21° and its eastern in 20°.

23. Adeisathron, whose middle point is in long. 132° and in lat. 23°.

24. Ouxenton, which extends from 136° to 143°, and whose western limit is in lat. 22° and its eastern in 24°.

25. The Oroudian Mountains, which extend from 138° to 133°, and whose eastern limit is in 18° lat. and its western 16°.

Mount Bêttigô:—As the rivers which have their sources in this range—the Pseudostomos, the Baris, and the Sôlên or Tâmraparṇî, all belong to South Malabar, there can be no doubt that Bêttigô denotes the southern portion of the Western Ghâṭs extending from the Koimbatur gap to Cape Comorin—called Malaya in the Paurâṇik list already quoted. One of the summits of this range, famous in Indian mythology as the abode of the Ṛishi Agastya, bears the name in Tamiḷ of Podigei, or as it is pronounced Pothigei. It is visible from the mouth of the Tâmraparṇî, which has its sources in it, and from Kolkhoi, and the Greeks who visited those parts, and had the mountain pointed out to them would no doubt apply the name by which they heard it called to the whole range connected with it. (See Caldwell's *Dravid. Gram.* Introd. p. 101.)

A d e i s a t h r o n:—If we take Ptolemy's figures as our guide here, we must identify this range with the chain of hills which Lassen describes in the following passage:—" Of the mountain system of the Dekhan Ptolemy had formed an erroneous conception, since he represented the chain of the Western Ghâṭs as protruded into the interior of the country, instead of lying near to the western coast with which it runs parallel, and he was misled thereby into shortening the courses of the rivers which rise in the Western Ghâṭs. The chain which he calls Adeisathron begins in the neighbourhood of Nâgpur and stretches southward to the east of the rivers Wain + Gaṅgâ and Praṇîtâ, separates the Gôdâvarî from the Kṛishṇâ, and comes to an end at the sources of the Kâvêrî. This view of his meaning is confirmed by the fact that he locates the two cities Baithana or Pratishṭhâna which lies to the east of the Western Ghâṭs, on the Gôdâvarî, and Tagara both to the west of Adeisathron. He was led into this misrepresentation partly through the incompleteness and insufficiency of the accounts which he used, and partly through the circumstance that the Eastern Ghâṭ does not consist of a single chain, but of several parallel chains, and that to the south of the sources of the Kâvêrî the Eastern Ghâṭ is connected with the Western Ghâṭ through the Nilgiri Mountains. The name Adeisathron, one sees, can only refer to the West Ghâṭ in which the Kâvêrî rises." (*Ind. Alt.* vol. III, pp. 162-3). Yule explains the source of Ptolemy's error thus: " No doubt his Indian lists showed him Kâvêrî rising in Sahyâdri (as does Wilford's list from the

Brahmanda Purâṇa, As. Res. vol. VIII, p. 335f.). He had no real clue to the locality of the Sahyâdri, but found what he took for the same name (*Adisathra*) applied to a city in the heart of India, and there he located the range." Adeisathron must therefore be taken to denote properly that section of the Western Ghâṭs which is immediately to the north of the Koimbatur gap, as it is there the Kâvêrî rises. The origin of the name Adeisathron will be afterwards pointed out.

Ouxenton designates the Eastern continuation of the Vindhyas. All the authorities are at one in referring it to the mountainous regions south of the Sôṇ, included in Chhutia Nâgpûr, Râmgarh, Sirgujâ, &c. Ptolemy places its western extremity at the distance of one degree from the eastern extremity of the Vindhyas. The rivers which have their sources in the range are the Tyndis, the Dôsarôn, the Adamas and an unnamed tributary of the Ganges. The name itself represents the Sanskrit Ṛikshavant, which however did not designate the Eastern Vindhyas, but a large district of the central. This difference in the application of the names need not invalidate the supposition of their identity. The authors whom Ptolemy consulted may have misled him by some inaccuracy in their statements, or the Hindûs themselves may have intended the name of Ṛikshavant to include localities further eastward than those which it primarily denoted. *Ṛiksha* means 'a bear,' and is no doubt connected with the Greek word of the same meaning, *arktos*.

The Oroudian Mountains:—"This we take,"

says Yule, "to be the Vaidûrya just mentioned, as the northern section of the Western Ghâṭs, though Ptolemy has entirely misconceived its position. We conceive that he found in the Indian lists that the great rivers of the eastern or Maesolian Coast rose in the Vaidûrya, and having no other clue he places the Orûdia (which seems to be a mere metathesis of Odûrya for Vaidûrya) near and parallel to that coast. Hence Lassen and others (all, as far as is known) identify these Oroudian Mountains with those that actually exist above Kaliṅga. This corresponds better, no doubt, with the position which Ptolemy has assigned. But it is not our business to map Ptolemy's errors; he has done that for himself; we have to show the real meaning and application of the names which he used, whatever false views he may have had about them."

26. The rivers which flow from Mount Imaös into the Indus are arranged as follows:—

Sources of the River Kôa ...	120°	37°
Sources of the River Souastos..	122° 30′	36°
Sources of the River Indus ..	125°	37°
Sources of the River Bidaspês	127° 30′	36° 40′
Sources of the River Sandabal	129°	36°
Sources of the River Adris or Rouadis..................	130°	37°
Sources of the River Bidasis..	131°	35° 30′

Regarding the origin and meaning of the name Indus, Max Müller (*India, what it can teach us*) says: "In the *Vêdas* we have a number of names of the rivers of India as they were known to one single poet, say about 1000 B.C. We then hear nothing of India till we come to the days of Alexander, and when we look at the names of the Indian rivers represented by Alexander's companions in India, we recognize without much difficulty nearly all of the old Vedic names. In this respect the names of rivers have a great advantage over the names of towns in India. I do not wonder so much at the names of the Indus and the Ganges being the same. The Indus was known to early traders, whether by sea or land. Skylax sailed from the country of the Paktys, *i.e.* the Pushtus, as the Afghans still call themselves, down to the mouth of the Indus. That was under Darius Hystaspês (B.C. 521-486). Even before that time India and the Indians were known by their name, which was derived from Sindhu, the name of their frontier river. The neighbouring tribes who spoke Iranic languages all pronounced, like the Persian, the *s* as an *h* (Pliny, lib. VI, c.xx, 7) '*Indus incolis Sindus appellatus.*' Thus *Sindhu* became *Hindhu* (*Hidhu*) and as *h's* were dropped, even at that early time, *Hindhu* became *Indu.* Thus the river was called Indus, the people Indoi by the Greeks, who first heard of India from the Persians. *Sindhu* probably meant originally the *divider*, *keeper* and *defender*, from *sidh* to *keep off*. No more telling name could have been given to a broad river, which guarded peaceful settlers both against the inroads of hostile tribes and the attacks of wild animals. . . . Though Sindhu was used as an appellative noun for river in general, it remained throughout the whole history of India, the name of its powerful guardian river, the Indus." For a full discussion of the origin of the name I may refer the reader to Benfey's *Indien*, pp. 1—2, in the *Encyclopædia* of Ersch and Grüber.

The Indus being subject to periodic inundations, more or less violent, has from time to time undergone considerable changes. As has been already indicated it not unfrequently shifts the channels by which it enters the sea, and in the upper part of its course it would seem to be scarcely less capricious. Thus while at the time of the Makedonian invasion it bifurcated above Arôr, the capital of the Sogdi, to run for about the distance of 2 degrees in two beds which enclosed between them the large island called by Pliny (lib. VI, c. xx, 23) Prasiakê, the Prârjuna of the inscription on the Allâhâbâd column, it now runs at that part in a single stream, having forsaken the eastern bed, and left thereby the once flourishing country through which it flowed a complete desert.

In his description of the Indus, Ptolemy has fallen into error on some important points. In the first place, he represents it as rising among the mountains of the country of the Daradrae to the east of the Paropanisos, and as flowing from its sources in a southward direction. Its true birthplace is, however, in a much more southern latitude, viz., in Tibet, near the sources of the Satlaj, on the north side of Mount Kailâsa, famous in Indian mythology as the dwelling-place of Kuvêra and as the paradise of Śiva, and its initial direction is towards the north-west, till it approaches the frontiers of Badakshân, where it turns sharply southward. Ptolemy does not stand alone in making this mistake, for Arrian places the sources in the lower spurs of the Paropanisos, and he is here at one with Mela (lib. III, c. vii, 6), Strabo (lib. XV, c. ii, 8), Curtius (lib. VIII, c. ix, 3) and other ancient writers. In fact, it was not ascertained until modern times whence the Indus actually came. His next error has reference to the length of the Indus valley as measured from the mouth of the Indus to its point of junction with the Kâbul river. This he makes to be 11 degrees, while in point of fact it is somewhat less than 10. This error is, however, trivial as compared with the next by which the junction of the Indus with the united stream of the Panjâb rivers is made to take place at the distance of only one degree below its junction with the Kâbul river, instead of at the distance of six degrees or halfway between the upper junction and the sea. This egregious error not only vitiates the whole of his delineation of the river system of the Panjâb, but as it exaggerates by more than 300 miles the distance between the lower junction and the sea, it obscures and confuses all his geography of the Indus valley, and so dislocates the positions named in his tables, that they can only in a few exceptional cases be identified.[22]

---

[22] "It is hard enough," says Major-General Haig, "to have to contend with the vagueness, inconsistencies and contradictions of the old writers; but these are as nothing compared with the obstacles which the physical characteristics of the country itself oppose to the enquirer. For ages the Indus has been pushing its bed across the valley from east to west, generally by the gradual process of erosion, which effectually wipes out every trace of town and village on its banks; but at times also by a more or less sudden shifting of its waters into

All the large tributaries of the Indus, with the exception of the Kâbul river, join it on its left or eastern side. Their number is stated by Strabo (lib. XV, c. i, 33) and by Arrian (lib. V, c. vi) to be 15, but by Pliny (lib. VI, c. xx, 23) to be 19. The most of them are mentioned in one of the hymns of the *Ṛig Veda* (X, 75) of which the following passages are the most pertinent to our subject:—

1. "Each set of seven [*streams*] has followed a threefold course. The Sindhu surpasses the other rivers in impetuosity.

2. Varuṇa hollowed out the channels of thy course, O Sindhu, when thou didst rush to thy contests. Thou flowest from [*the heights of*] the earth, over a downward slope, when thou leadest the van of those streams.

4. To thee, O Sindhu, the [*other streams*] rush . . . Like a warrior king [*in the centre of his army*] thou leadest the two wings of thy host when thou strugglest forward to the van of these torrents.

5. Receive favourably this my hymn, O Gaṅgâ, Yamunâ, Sarasvatî, Śutudrî, Parashṇi; hear, O Marudvṛidhâ, with the Asiknî, and Vitastâ, and thou Arjîkîyâ with the Sushômâ.

---

entirely new channels, leaving large tracts of country to go to waste, and forcing the inhabitants of many a populous place to abandon their old homes, and follow the river in search of new settlements. . . . Perhaps the retiring stream will leave behind it vast quantities of drift-sand which is swept by the high winds over the surrounding country . . . where the explorer may search in vain for any record of the past. I have had, as an enquirer, experience of the difficulties here described." (*J. R. A. S. N. S.* vol. XVI, p. 281).

6. Unite first in thy course with the Trishṭâmâ, the Sasartû, the Rasâ and the Śvêtî; thou meetest the Gomatî, and the Krumu, with the Kubhâ, and the Mehatnû, and with them are borne onward as on the same car." (See *Journ. R. A. S.*, N. S., Vol. XV, pp. 359-60).

As Ptolemy makes the Kôa join the Indus, it must be identified with the Kâbul river, the only large affluent which the Indus receives from the west. Other classical writers call it the Kôphên or Kôphês, in accordance with its Sanskrit name the Kubhâ. Ptolemy's name, it must however be noted, is not applicable to the Kâbul river throughout its whole course, but only after it has been joined by the River Kâmah, otherwise called the Kunâr. This river, which is inferior neither in size nor in length to the arm which comes from Kâbul, is regarded as the main stream by the natives of the country, who call the course of the united streams either the Kâmah or the Kunâr indifferently, as far as the entrance into the plain of Peshâwar. The Kâmah has its sources high up in the north at the foot of the plateau of Pâmîr, not far from the sources of the Oxus, and this suits Ptolemy's description of the Kôa as a river which has its sources in the eastern extremity of Paropanisos, and which joins the Indus after receiving the Souastos or the river of Swât. Kôa is very probably a curtailed form of the name. The Persians appear to have called it the Khoaspês, that being the name of the river on which Susa, their capital city, stood. Under this name it is mentioned by Aristotle (*Meteorolog.* lib. I, c. xiii) who lived long enough to enter in his later writings some of the new knowledge which the expedition of his illustrious pupil had opened up regarding Eastern Countries. It is mentioned also by Strabo (lib. XV, c. i, 26) who followed here the authority of Aristoboulos, one of the companions and one of the historians of the expedition of Alexander, and by Curtius (lib. VIII, c. x), Strabo *l. c.* states that it joins the Kôphês near Plemyrion, after passing by another city, Gorys, in its course through Bandobênê and Gandaritis. The Kôa of Ptolemy is not to be confounded with the Khôês of Arrian (lib. IV, c. xxiii, 2), which must be identified with a river joining the Kôphês higher up its course, viz. that which is formed by the junction of the Alishang and the Alingar. The Euaspla of the latter writer (lib. IV, c. xxiv, 1) is probably only an altered form of Khoaspês.

The identification of the Kôphês and its numerous affluents has been a subject that has much exercised the pens of the learned. They are now unanimous in taking the Kôphês to be the Kâbul river[23] but there are still some important points on which they differ. In the foregoing notice I have adopted as preferable the views of Saint-Martin (*Etude*, pp. 26—34): *Conf.* Lassen, *Ind. Alt.* vol. III, pp. 127-8; Wilson, *Ariana Antiqua*, pp. 138—188. Benfey's *Indien*, pp. 44—46, Cunningham, *Geog. of Anc. India*, pp. 37, 38.

Souastos:—All the authorities are at one in identifying the Souastos with the Swât river—the principal tributary of the Landai or river of Pañjkora (the Gaurî of Sanskrit), which is the

---

[23] Rennell identified it with the Gomul and D'Anville with the Argandâb.

last of the great affluents that the Kâbul river receives from the east before it falls into the Indus. The Souastos, though a small stream, is yet of old renown, being the Śvêtî of the Vedic hymn already quoted, and the Suvâstu of the *Mahâbhârata* (VI, ix, 333), where it is mentioned in conjunction with the Gaurî. Its name figures also in the list of Indian rivers which Arrian (*Indika*, sec. 4) has preserved from the lost work of Megasthenês. Here it is mentioned in conjunction with the Malamantos and the Garoia, which latter is of course the Gaurî. Arrian thus makes the Souastos and the Gouraios to be different rivers, but in another passage of his works (*Anab.* lib. IV, c. xxv) he seems to have fallen into the mistake of making them identical. It is surprising, as Lassen has remarked, that Ptolemy should notice the Souastos, and yet say nothing about the Garoia, especially as he mentions the district of Goryaia, which is called after it, and as he must have known of its existence from the historians of Alexander. He has also, it may be noted, placed the sources of the Souastos too far north.

The five great rivers which watered the region of the Panjâb bear the following names in Ptolemy: Bidaspês, Sandabal, Adris or Rhonadis, Bibasis and Zaradros. This region in early times was called the country of the seven rivers—Sapta Sindhu, a name which, as Sir H. Rawlinson has pointed out, belonged primarily to the seven head streams of the Oxus. As there were only five large streams in the locality in India to which the name was applied, the number was made up to seven by adding smaller affluents or lower branches of combined streams, to which new names were given. The Vedic Âryans, however, as Mr. Thomas remarks, could never satisfactorily make up the sacred seven without the aid of the comparatively insignificant Sarasvatî, a river which no longer exists. These rivers are notably erratic, having more than once changed their bed since Vedic times.

Bidaspês:—This is now the Jhelam or river of Behat, the most western of the five rivers. It drains the whole of the valley of Kaśmîr, and empties into the Akesinês or Chenâb. Ptolemy, however, calls their united stream the Bidaspês. By the natives of Kaśmîr it is called the Bedasta, which is but a slightly altered form of its Sanskrit name the Vitastâ, meaning 'wide-spread.' The classical writers, with the sole exception of our author, call it the Hydaspês, which is not so close to the original as his Bidaspês. It was on the left bank of this river that Alexander defeated Pôros and built (on the battle-field) the city of Nikaia in commemoration of his victory.

Sandabal is an evident mistake of the copyist for Sandabaga. The word in this corrected form is a close transliteration of Chandrabhâgâ (*lunae portio*), one of the Sanskrit names of the River Chenâb. In the Vedic hymn which has been quoted it is called the Asiknî, 'dark-coloured,' whence the name given to it by the Greeks in Alexander's time, the Akesinês. It is said that the followers of the great conqueror discerned an evil omen in the name of Chandrabhâgâ on account of its near similarity to their own word *Androphagos* or *Alexandrophagos*, 'devourer of Alexander' and hence preferred calling it by the more ancient of its two names. It is the largest of all the streams of the Pañchanada. Vigne says that Chandrabhâgâ is the name of a small lake from which the river issues. Pliny has distorted the form Chandabaga into Cantabra or Cantaba (lib. VI, c. xx). According to the historians of Alexander the confluence of this river with the Hydaspês produced dangerous rapids, with prodigious eddies and loud roaring waves, but according to Burnes their accounts are greatly exaggerated. In Alexander's time the Akesinês joined the Indus near Uchh, but the point of junction is now much lower down.

The Adris or Rhouadis is the Râvî, a confluent of the Akesinês, but according to Ptolemy of the Bidaspês. The name Râvî is an abridged form of the Sanskrit Airâvatî. It is called by Arrian (*Anab.* lib. VI, c. viii), the Hydraôtês, and by Strabo (lib. XV, c. i, 21) the Hyarôtis. Arrian (*Indik.* sec. 4) assigns to it three tributaries—the Hyphasis, the Saranges and Neudros. This is not quite correct, as the Hyphasis joins the Akesinês below the junction of the Hydraôtês.

The Bibasis is the river now called the Beiäs, the Vipâśâ of Sanskrit. This word "Vipâśâ" means 'uncorded,' and the river is said to have been so called because it *destroyed the cord* with which the sage Vasishṭha had intended to hang himself. It is called the Hyphasis by Arrian (*Anab.* lib. VI, c. viii), and Diodôros (lib. XVII, c. xciii), the Hypasis by Pliny (lib. VII, c. xvii, 20) and Curtius (lib. IX, c. i), and the Hypanis by Strabo (lib. XV, c. i, 17) and some other writers.

It falls into the Śatadru. It was the river which marked the limit of Alexander's advance into India.

Place	Long.	Lat.
27. Sources of the River Zaradros	132°	36°
Confluence of the Kôa and Indus	124°	31°
Confluence of the Kôa and Souastos	122° 30′	31° 40′
Confluence of the Zaradros and Indus	124°	30°
Confluence of the Zaradros and Bidaspês	125°	30°
Confluence of the Zaradros and Bibasis	131°	34°
Confluence of the Bidaspês and Adris	126° 30′	31° 30′
Confluence of the Bidaspês and Sandabal	126° 40′	32° 40′

The Zaradros is the Satlaj, the most easterly of the five rivers. It is called in Sanskrit the Śatadru, *i.e.*, *flowing in a hundred* (*branches*). Pliny (lib. VI, c. xvii) calls it the Hesydrus, Zadrades is another reading of the name in Ptolemy. The Satlaj, before joining the Indus, receives the Chenâb, and so all the waters of the Pañchanada.

With regard to the nomenclature and relative importance of the rivers of the Panjâb the following remarks of V. de Saint-Martin may be cited:—

"As regards the Hyphasis, or more correctly the Hypasis, the extended application of this name till the stream approaches the Indus, is contrary to the notions which we draw from Sanskrit sources, according to which the Vipâśâ loses its name in the Śatadru (Satlaj), a river which is otherwise of greater importance than the Vipâśâ. Nevertheless the assertion of our author by itself points to a local notion which is confirmed by a passage in the chronicles of Sindh, where the name of the Beiah which is the form of the Sanskrit Vipâśâ in Musalmân authors and in actual use, is equally applied to the lower course of the Satlaj till it unites with the Chenâb not far from the Indus. Arrian, more exact here, or at least more circumstantial than Strabo and the other geographers, informs us that of all the group of the Indus affluents the Akesinês was the most considerable. It was the Akesinês which carried to the Indus the combined waters of the Hydaspês of the Hydraôtês and of the Hyphasis, and each of these streams lost its name in uniting with the Akesinês (Arr. *Anab.* lib. VI, c. v). This view of the general hydrography of the Panjâb is in entire agreement with facts, and with the actual nomenclature. It is correctly recognized that the Chenâb is in effect the most considerable stream of the Panjâb, and its name successively absorbs the names of the Jhelam, the Râvi, and the Gharra or lower Satlaj, before its junction with the Indus opposite Mittankôṭ. Ptolemy here differs from Arrian and the current ideas on the subject. With him it is not the Akesinês (or, as he calls it, the Sandabala for Sandabaga) which carries to the Indus the waters of the Panjâb. It is the Bidaspês (Vitastâ). Ptolemy departs again in another point from the nomenclature of the historians who preceded him in applying to the Gharra or lower Satlaj the name of Zaradros, and not, as did Arrian that of Hypasis. Zadadros is the Śutudrî or Śatadru of the Sanskrit nomenclature, a name which common usage since the Musalmân ascendancy has strangely disfigured into Satlaj. No mention is made of this river in the memoirs relating to the expedition of Alexander, and Megasthenês, it would appear, was the first who made its existence known. The application moreover of the two names of Zadadros and Bibasis to the united current of the Śatadru and the Vipâśâ is justified by the usage equally variable of the natives along the banks, while in the ancient Sanskrit writings the Śatadru goes, as in Ptolemy, to join the Indus. It may be added that certain particularities in the texts of Arrian and Ptolemy suggest the idea that formerly several arms of the Hyphasis existed which went to join, it may be, the Hydraôtês, or, it may be, the lower Akesinês above the principal confluent of the Hyphasis, an idea which the actual examination of the locality appears to confirm. This point merits attention because the obscurities or apparent contradictions in the text of the two authors would here find an easy explanation" (pp. 129-131, also pp. 396-402).

Junction of the Kôa and Indus.—Ptolemy fixes the point of junction in latitude 31°, but the real latitude is 33° 54′. Here the Indus is 872 miles distant from its source, and 942 miles from the sea. The confluence takes place amidst numerous rocks and is therefore turbulent and attended with great noise.

Junction of the Zaradros and Indus:—Ptolemy fixes this great junction in latitude 30°, the real latitude being however 28° 55′. It takes place about 3 miles below Miṭankôt, at a distance of about 490 miles below the junction with the Kâbul River.

Divarication of the Indus towards Mt. Vindion:—The Indus below its junction with the Kâbul river frequently throws out branches (*e.g.* the Nara) which join it again before reaching the sea, and to such branches Ptolemy gives the name of *ἐκτροπαί*. "It is doubtful," Saint-Martin observes, "whether Ptolemy had formed quite a clear idea of this configuration of the valley, and had always distinguished properly the affluents from the branches. Thus one does not quite precisely see what he means by the expression which he frequently employs *ἡ πηγὴ τῆς ἐκτροπῆς*. What he designates thereby must be undoubtedly the streams or currents which descend from the lateral region, and which come to lose themselves in the branches of the river. But the expression, which is familiar to him, is not the less ambiguous and altogether improper"—(p. 235 n.) The branch here mentioned, Lassen (*Ind. Alt.* vol. III, pp. 121, 129) takes to be the Lavaṇî river. "Ptolemy," he says, "in contradiction to fact makes a tributary flow to it from the Vindhya Mountains. His error is without doubt occasioned by this, that the Lavaṇî river, which has its source in the Arâvalî chain falls into the salt lake, the Rin or Iriṇa, into which also the eastern arm of the Indus discharges."

Divarication of the Indus into Arakhôsia:—

Lassen (vol. III, p. 128), takes this to be the Gomal rather than the Korum river. These rivers are both mentioned in the Vedic hymn, where the former appears as the Gômati and the latter as the Krumu.

Branch of the Kôa towards the Paropanisadai:—This is probably the upper Kôphên, which joins the Kôa (Kunâr river) from Kâbul.

Divarication of the Indus towards the Arbita mountains:—Between the Lower Indus and the river called anciently the Arabis or Arbis, was located a tribe of Indian origin called variously the Arabii, the Arbies, the Arabitae, the Ambritae and the Arbiti. There can be no doubt therefore that by the Arbita Mountains Ptolemy designates the range of hills in the territory of that tribe, now called the Hâla Mountains. Towards the northern extremity of this range the Indus receives a tributary called the Gandava, and this we may take to be what Ptolemy calls the divarication of the Indus towards the range. It may perhaps, however, be the Western Nara that is indicated.

Divarication of the Indus into the Paropanisadai:—To judge from the figures in the table this would appear to be a tributary of the Indus joining it from the west a little above its junction with the Kôa or Kâbul river. There is, however, no stream, even of the least note, answering to the description.

28. Divarication (ἐκτροπή) from the Indus running towards Mt. Ouindion 123° 29° 30′
The source of (tributary joining) the Divarication ......127° 27°

Divarication of the Indus towards Arakhôsia	121° 30′	27° 30′
Divarication of the Kôa towards the Paropanisadai ...	121° 30′	33°
The source of (tributary joining) the Divarication	115°	24° 30′
Divarication of the Indus towards the Arbita Mountains	117°	25° 10′
Divarication of the Indus towards the Paropanisadai.	124° 30′	31° 20′
Divarication of the Indus into the Sagapa mouth	113° 40′	23° 15′
From the Sagapa into the Indus	111°	21° 30′
Divarication of the Indus into the Khrysoun (or Golden) mouth	112° 30′	22°
Divarication of the Indus into the Khariphon mouth	113° 30′	22° 20′
From the Khariphon to the Sapara	112° 30′	21° 45′
Divarication of the same River Khariphon into the Sabalaessa mouth...........	113°	21° 20′
Divarication from the River Khariphon into the Lônibare mouth	113° 20′	21° 40′

29. Of the streams which join the Gangês the order is this :—

Sources of the River Diamouna........................	134° 30′	36°

Sources of the Ganges itself...	136°	37°
Sources of the River Sarabos	140°	36°
Junction of the Diamouna and Ganges	136°	34°
Junction of the Sarabos and Ganges	136° 30′	32° 30′

Ptolemy's description of the Ganges is very meagre as compared with his description of the Indus. He mentions by name only 3 of its affluents, although Arrian (quoting from Megasthenês) enumerates no fewer than 17, and Pliny 19. The latitude of its source, Gangotri, which is in the territory of Garhwal, is 30° 54′, or more than 6 degrees further south than its position as given in the table. The name of the river, the Gaṅgâ, is supposed to be from a root *gam*, 'to go,' reduplicated, and therefore to mean the 'Go—go.' The tributaries mentioned by Arrian are these: the Kaïnas, Erannoboas, Kossoanos, Sônos, Sittokatis, Solomatis, Kondokhates Sambos, Magon, Agoranis, Omalis, Kommenases, Kakouthis, Andomatis, Amystis, Oxymagis and the Errhenysis. The two added by Pliny are the Prinas and Jomanes. Regarding these names the following remarks may be quoted from Yule:—"Among rivers, some of the most difficult names are in the list which Pliny and Arrian have taken from Megasthenês, of affluents of the Gangês. This list was got apparently at Palibothra (Patna), and if streams in the vicinity of that city occupy an undue space in the list, this is natural. Thus Magona and Errhenysis,—Mohana and Nirañjana, join to form the river flowing past Gayâ, famous in Buddhist legend under the second name. The navigable Prinas or Pinnas is perhaps Puṇyâ, now Pûnpûn, one of the same cluster. Sonus instead of being a duplicate of Erannoboas, may be a branch of the Gayâ river, still called Soṇâ. Andomatis flowing from the Madiandini, *i.e.*, "Meridionales" is perhaps the Andhela, one of the names of the Chandan river of Bhâgalpûr. Kainas, navigable, is not likely to be the Ken of Bundêlkhand, the old form of which is Karṇavatî, but more probably the Kayâna or Kohâna of Gorakhpûr. It is now a tributary of the lower Ghâgrâ, but the lower course of that river has shifted much, and the map suggests that both the Rapti (Solomatis of Lassen) and Kayâna may have entered the Ganges directly." For the identification of the other rivers in the list see my article in the *Indian Antiquary*, vol. V, p. 331.

Diamouna:—In this it is easy to recognize the Yamunâ, the river which after passing Dehli, Maṭhurâ, Âgrâ, and other places, joins the Ganges, of which it is the largest affluent at Allâhâbâd. It rises from hot springs amid Himâlayan snows, not far westward from the sources of the Ganges. Arrian singularly enough has omitted it from his list of the Ganges affluents, but it is no doubt the river which he subsequently mentions as the Jobares and which flows, he says, through the country of the Sourasenoi, an Indian tribe possessing two large cities, Methora and Kleisobara (Kṛishṇapura?) Pliny (lib. VI, c. xix) calls it the Jomanes, and states that it flows into the Ganges through the Palibothri, between the towns of Methora and Chrysobara (Kṛishṇapura?) The

Ganges at its junction with the Jamnâ and a third but imaginary river called the Sarasvatî, which is supposed to join it underground is called the T r i v ê ṇ î, *i.e.*, 'triple plait' from the intermingling of the three streams.

S a r a b o s :—This is the great river of Kôśala, that is now called the Sarayu or Sarju, and also the Gharghara or Ghogra. It rises in the Himâlayas, a little to the north-east of the sources of the Ganges, and joins that river on its left side in latitude 25° 46′, a little above the junction of the Sôṇ with their united stream. Cunningham regards the Solomatis mentioned in Arrian's list of the tributaries of the Ganges as being the Sarayu under a different name, but Lassen takes it to be the Rapti, a large affluent of the same river from Gôrakhpur. The name, he thinks, is a transliteration or rather abbreviation of Śarâvatî, the name of a city of Kôśala mentioned by Kalidâsa. The river on which the city stood is nowhere mentioned, but its name was in all probability the same as that of the city (*Ind. Alt.*, vol. II, p. 671).

Mouth of the River S ô a :—This river can be no other than the Sôṇ (the Sônos of Arrian's list) which falls into the Ganges about 16 miles above Patna in lat. 25° 37′. It rises in Gôndwana in the territory of Nâgpur, on the elevated tableland of Amarakaṅṭaka, about 4 or 5 miles east of the source of the Narmadâ. It would appear that in former times it joined the Ganges in the immediate neighbourhood of Patna, the modern representative of the Palibothra or Palimbothra of the classical writers. The lat. of the source is 22° 41′; in Ptolemy 28°.

30. Divarication from the Ganges towards the Ouindion range to the mouth of the River Sôa ....136° 10′ 31° 30′

The sources of the river ...	131°	28°
Divarication of the Ganges towards the Ouxenton range	142°	28°
The sources of the divarication	137°	23°
Divarication from the Ganges into the Kambyson Mouth	146°	22°
Divarication from the Ganges into the Pseudostomos	146° 30′	20°
Divarication from the Ganges into the Antibolê Mouth	146° 30′	21°
Divarication from the Kambyson River into the Mega Mouth	145°	20°
Divarication from the Mega Mouth into the Kambêrikhon Mouth	145° 30′	19° 30′

The divarication towards the Ouxenton range:—By this unnamed river, as Lassen has pointed out (*Ind. Alt.*, vol. III, pp. 130, 131) Ptolemy must have meant the Dharmôdaya of the Hindus, although he has assigned far too high a latitude for its junction with the Ganges, 28° instead of only 22° 13′. It is, however, the only considerable stream which flows to the Ganges from the Bear Mountains. It passes Ramgaṛh and Bardhwân, and joins the Hughlî not far from the sea, a little to the east of Tamluk. It is commonly called the Damuda River.

The mouths of the Ganges:—In addition to

the remarks already made regarding these mouths I may here quote a passage from Wilford on this topic: "Ptolemy's description," he says (*Asiat. Researches*, vol. XIV, pp. 464-6) "of the Delta of the Ganges is by no means a bad one, if we reject the latitudes and longitudes, which I always do, and adhere solely to his narrative, which is plain enough. He begins with the western branch of the Ganges or Bhâgîrathî, and says that it sends one branch to the right or towards the west, and another towards the east, or to the left. This takes place at Trivênî, so called from three rivers parting, in three different directions, and it is a most sacred place. The branch which goes towards the right is the famous Sarasvatî; and Ptolemy says that it flows into the Kambyson mouth, or the mouth of the Jelasor river, called in Sanskrit Śaktimatî, synonymous with Kambu or Kambuj, or the river of shells. This communication does not exist, but it was believed to exist, till the country was surveyed. This branch sends another arm, says our author, which affords a passage into the great mouth, or that of the Bhâgîrathî or Ganges. This supposed branch is the Rûpanârâyaṇa, which, if the Sarasvatî ever flowed into the Kambyson mouth, must of course have sprung from it, and it was then natural to suppose that it did so. M. D'Anville has brought the Sarasvatî into the Jelasor river in his maps, and supposed that the communication took place a little above a village called Danton, and if we look into the *Bengal Atlas*, we shall perceive that during the rains, at least, it is possible to go by water, from Hughlî, through the Sarasvatî, and many other rivers, to within a few miles of Danton, and the Jelasor river. The river, which according to Ptolemy branches out towards the east, or to the left, and goes into the Kambarikan mouth is the Jumnâ, called in Bengal Jubunâ. For the Ganges, the Jumnâ and the Sarasvatî unite at the Northern Trivêṇî or Allâhâbâd, and part afterwards at this Trivêṇî near Hughlî . . . called in the spoken dialects Terboni. Though the Jumnâ falls into the Kambarikan mouth, it does by no means form it; for it obviously derives its name from the Kambâdâṛâ or Kambâraka river, as I observed before. Ptolemy says that the Ganges sends an arm towards the east or to the left, directly to the false mouth or Hariṇaghaṭṭâ. From this springs another branch to Antibolê, which of course is the Ḍhâkkâ branch called the Padmâ or Puddâgangâ. This is a mistake, but of no great consequence, as the outlines remain the same. It is the Paddâ or Ḍhâkkâ branch, which sends an arm into the Hariṇaghaṭṭâ. The branching out is near Kasti and Komarkalli, and under various appellations it goes into the Hariṇaghaṭṭâ mouth."

Besides the tributaries of the Ganges already mentioned, Ptolemy refers to two others which it receives from the range of Bêpyrrhos. These are not named, but one is certainly the Kauśikî and the other ought to be either the Gaṇḍakî or the Tîstâ.

31. And of the other rivers the positions are thus:

The sources of the River Namados in the Ouindion range	127°	26° 30′

The bend of the river at Séripala	116° 30′	22°
Its confluence with the River Môphis	115°	18° 30′
32. Sources of the River Nanagouna from the Ouindion range	132°	26° 30′
Where it bifurcates into the Goaris and Binda	114°	16°
33. Sources of the Pseudostomos from the Bêttigô range	123°	21°
The point where it turns	118° 30′	17° 15′
34. Sources of the River Baris in the Bêttigô range	127°	26° 30′
Sources of the River Sôlên in the Bêttigô range	127°	20° 30′
The point where it turns	124°	18°
35. Sources of the River Khabêros in the Adeisathros range	132°	22°
36. Sources of the River Tyna in the Oroudian (or Arouëdan) Mountains	133°	17°
37. Sources of the River Maisôlos in the same mountains	134° 30′	17° 30′
38. Sources of the River Manda in the same mountains	136° 30′	16° 30′
39. Sources of the River Toundis in the Ouxenton range	137°	22° 30′

40. Sources of the River
Dôsarôn in the same range ...140° 24°
41. Sources of the River
Adamas in the same range ...142° 24°

These rivers have been all already noticed, with the exception of the Môphis. This is now the Mahî, a considerable river which flows into the Gulf of Khambât at its northern extremity at a distance of about 35 miles north from the estuary of the Narmâda. Ptolemy is in error in making the two rivers join each other. The Môphis is mentioned in the *Periplûs* as the Maïs. In this list the spelling of the names of two of the rivers of Orissa has been slightly changed, the Manada into Manda and Tyndis into Toundis.

Ptolemy proceeds now (following as much as possible the order already observed) to give a list of the different territories and peoples of India classified according to the river-basins, together with the towns belonging to each territory and each people (§§42—93), *and closes the chapter by mentioning the small islands that lay adjacent to the coast. He begins with the basin of the Kôphês, part of which he had already described in the 6th Book.*

42. The order of the territories in this division (India intra Gangem) and of their cities or villages is as follows:—

Below the sources of the Kôa are located the Lambatai, and their mountain region extends upwards to that of the Kômêdai.

Below the sources of the Souastos is Souastênê.

Below those of the Indus are the Daradrai, in whose country the mountains are of surpassing height.

Below the sources of the Bidaspês and of the Sandabal and of the Adris is Kaspeiria.

Below the sources of the Bibasis and of the Zaradros and of the Diamouna and of the Ganges is Kylindrinê, and below the Lambatai and Souastênê is Gôryaia.

Ptolemy's description of the regions watered by the K ô p h ê n and its tributaries given here and in the preceding book may well strike us with surprise, whether we consider the great copiousness of its details, or the way in which its parts have been connected and arranged. It is evident that he was indebted for his materials here chiefly to native sources of information and itineraries of merchants or caravans, and that he did not much consult the records, whether historical or geographical, of Alexander's expedition, else he would not have failed to mention such places as Alexandria, under Kaukasos, Massaga, Nysa, Bazira, the rock Aörnos, and other localities made memorable by that expedition.

In describing the basin of the Kôphên he divides it into two distinct regions—the high region and the lower, a distinction which had been made by the contemporaries of Alexander. The high region formed the country of the P a r o p a n i s a d a i, and this Ptolemy has described in the 18th chapter of the 6th Book. He now describes the lower region which he regards as a part of India. (V. Saint-Martin, *Étude*, pp. 62-3).

The Lambatai were the inhabitants of the district now called Lamghân, a small territory lying along the northern bank of the Kâbul river bounded on the west by the Alingâr and Kunâr rivers, and on the north by the snowy mountains. Lamghân was visited in the middle of the 7th century by Hiuen Tsiang, who calls it Lan-po, and notes that its distance eastward from Kapisênê, to which before his time it had become subject, was 600 *li* (equal to 100 miles). The name of the people is met with in the *Mahâbhârata* and in the *Paurâṇik* lists under the form Lâmpâka. Cunningham would therefore correct Ptolemy's Lambatai to Lambagai by the slight change of Γ for T. A minute account of this little district is given in the *Memoirs of the Emperor Baber*, who states that it was called after Lamech, the father of Noah. The *Dictionary* of Hêmachandra, which mentions the Lampâka, gives as another name of the people that of the Muraṇḍa. Their language is Pushtu in its basis. (See Cunningham's *Geog. of Anc. India*, pp. 42-3; Saint-Martin, *Étude*, pp. 74-5; also his *L'Asie Central*, p. 48; Lassen, *Ind. Alt.*, vol. I, p. 422.

Souastênê designates the basin of the Souastos, which, as has already been noticed, is the river now called the river of Swât. The full form of the name is Śubhavastu, which by the usual mode of contraction becomes Subhâstu or Suvâstu. Souastênê is not the indigenous name of the district, but one evidently formed for it by the Greeks. It is the country now inhabited by the warlike tribes of the Yuzofzaïs which appears to have been called in ancient times with reference to the rich verdure and fertility of its valleys Udyâna, that is, 'a garden' or 'park.' It was visited by Hiuen Tsiang, who calls it the kingdom of U-chang-na.

The Daradrai:—Ptolemy has somewhat disfigured the name of these mountaineers, who are mentioned in the *Mahâbhârata* and in the *Chronicle of Kaśmîr* as the Darada. They inhabited the mountain-region which lay to the east of the Lambatai and of Souastênê, and to the north of the uppermost part of the course of the Indus along the north-west frontier of Kaśmîr. This was the region made so famous by the story of the gold-digging ants first published to the west by Hêrodotos (lib. III, c. cii), and afterwards repeated by Megasthenês, whose version of it is to be found in Strabo (lib. XV, c. i, 44) and in Arrian's *Indika* (sec. 15) and also in Pliny (lib. VI, c. xxi and lib. XI, c. xxxvi). The name of the people in Strabo is Derdai, in Pliny Dardae, and in Dionys. Periêg. (v. 1138) Dardanoi. Their country still bears their name, being called Dardistân. The Sanskrit word *darad* among other meanings has that of 'mountain.' As the regions along the banks of the Upper Indus produced gold of a good quality, which found its way to India and Persia, and other countries farther west, it has been supposed that the Indus was one of the four rivers of Paradise mentioned in the book of *Genesis*, viz., the Pishon, "which compasseth the whole land of Havilah, where there is gold; and the gold of that land is good." This opinion has been advocated by scholars of high name and authority. Havilah they take to be in a much altered form, the Sanskrit *sarôvara*, 'a lake,' with reference perhaps to the lake in Tibet called Manasarôvara. Boscawen, however, has pointed out that there was a river called the Pisanu, belonging to the region between Nineveh and Babylon, where he locates paradise.

Kaspeiria:—The name and the position concur in indicating this to be the valley of Kaśmîr, a name which, according to Burnouf, is a contraction of *Kaśyapamîra*, which is thought with good reason to be the original whence came the Kaspapyros of the old Geographer Hekataios and the Kaspatyros of Hêrodotos (lib. III, c. cii), who tells us (lib. IV, c. xliv) that it was from the city of that name and from the Paktyikan land that Skylax the Karyandian started on his voyage of discovery down the Indus in order to ascertain for Darius where that river entered the sea. It cannot be determined with certainty where that city should be located, but there can be no good reason, as Wilson has shown (in opposition to the views of Wilford, Heeren, Mannert, and Wahl) for fixing it on any other river than the Indus. "We have no traces," he says, "of any such place as Kaspatyrus west of the Indus. Alexander and his generals met with no such city, nor is there any other notice of it in this direction. On the east of the river we have some vestige of it in oriental appellations, and Kaspatyrus is connected apparently with Kaśmîr. The preferable reading of the name is Kaspa-pyrus. It was so styled by Hecataeus, and the alteration is probably an error. Now Kaśyapa-pur, the city of Kaśyapa, is, according to Sanskrit writers, the original designation of Kaśmîr; not of the province of the present day, but of the kingdom in its palmy state, when it comprehended great part of the Panjâb, and extended no doubt as far as, if not beyond, the Indus."—*Ar. Antiq.*, p. 137.

In the time of Ptolemy the kingdom of Kaśmîr was the most powerful state in all India. The dominions subject to its sceptre reached as far south as the range of the Vindhyas and embraced, together with the extensive mountain region wherein the great rivers of the Panjâb had their sources, a great part of the Panjâb itself, and the countries which lay along the courses of the Jamnâ and the Upper Ganges. So much we learn from Ptolemy's description which is quite in harmony with what is to be found recorded in the *Râjataraṅgiṇî*, regarding the period which a little preceded that in which Ptolemy wrote—that the throne of Kaśmîr was then occupied by a warlike monarch called Mêghâvahana who carried his conquests to a great distance southward (*Râjatar.* vol. III, pp. 27 sqq.) The valley proper of Kaśmîr was the region watered by the Bidaspês (Jhelam) in the upper part of its course. Ptolemy assigns to it also the sources of the Sandabal (Chenâb) and of the Rhouadîs (Râvî) and thus includes within it the provinces of the lower Himâlayan range that lay between Kaśmîr and the Satlaj.

Kylindrinê designated the region of lofty mountains wherein the Vipâśâ, the Śatadru, the Jamnâ and the Ganges had their sources. The inhabitants called K u l i n d a are mentioned in the *Mahâbhârata* in a long list there given of tribes dwelling between Mêru and Mandara and upon the Śailôdâ river, under the shadow of the Bambu forests, whose kings presented lumps of ant-gold at the solemnity of the inauguration of Yudhishṭhira as universal emperor. Cunningham would identify Kylindrinê with "the ancient kingdom of Jâlandhara which since the occupation of the plains by the Muḥammadans has been confined almost entirely to its hill territories, which were generally known by the name of Kângra, after its most celebrated fortress." Saint-Martin, however, is unable to accept this identification. A territory of the name of K u l u t a, which was formed by the upper part of the basin of the Vipâśâ, and which may be included in the Kylindrinê of Ptolemy, is mentioned in a list of the *Varâha Saṁhitâ*. Kuluta was visited by the Chinese pilgrim, Hiuen Tsiang, who transcribes the name K'iu-lu-to, a name which still exists under the slightly modified form of Koluta. (See Lassen, *Ind. Alt.* vol. I, p. 547; Wilson, *Ar. Antiq.* p. 135 n.; Saint-Martin, *Étude*, 217; Cunningham, *Geog.* pp. 136—138.

G ô r y a i a designates the territory traversed by the Gouraios or river of Ghor, which, as has already been noticed, is the affluent of the Kâbul river now called the Landaï, formed by the junction of the river of Pañjkora and the river of Swât. Alexander on his march to India passed through Gôryaia, and having crossed the River Gouraios entered the territory of the Assakênoi. The passage of the river is thus described by Arrian (*Anab.* lib. IV, c. xxv): "Alexander now advanced with a view to attack the Assakênoi, and led his army through the territory of the Gouraioi. He had great difficulty in crossing the Gouraios, the eponymous river of the country, on account of the depth and impetuosity of the stream, and also because the bottom was so strewn with pebbles that the men when wading through could hardly keep their feet." It can scarcely be doubted that the Gouraios is the Gaurî mentioned in the 6th Book of the *Mahâbhârata* along with the Suvâstu and the Kampanâ. Arrian's notion that it gave its name to the country by which it flowed has been assented to by Lassen but has been controverted by Saint-Martin, who says (p. 33), "the name of the Gouraioi did not come, as one would be inclined to believe, and as without doubt the Greeks thought, from the river of Gur which watered their territory; the numerous and once powerful tribe of Ghorî, of which a portion occupies still to this day the same district, to the west of the Landaï, can advance a better claim to the attribution of the ancient classical name." In a note to this passage he says: "Kur, with the signification of 'river,' *courant*, is a primitive term common to most of the dialects of the Indo-Germanic family. Hence the name of Kur (Greek, Κύρος, Κύῤῥος, Lat. Cyrus) common to different rivers of Asia. . . . This name (of Ghorîs or Gûrs) ought to have originally the signification of 'mountaineers.' It is at least a remarkable fact that all the mountain region adjacent to the south of the Western Hindû-kôh and its prolongation in the direction of Herât have borne or still bear the names of Gûr, Ghôr, or Ghaur, Gurkân, Gurjistân, &c. Let us add that *garayo* in Zend signifies 'mountains.'"

43. And the cities are these :—

Kaisana	120°	34° 20′
Barborana	120° 15′	33° 40′
Gôrya	122°	34° 45′
Nagara or Dionysopolis	121° 45′	33°
Drastoka	120° 30′	32° 30′

Kaisana, Barborana and Drastoka are places unknown, but as the same names occur in the list of the towns of the Paropanisadai (lib. VI, c. xviii, 4) it is not improbable, as Saint-Martin conjectures, that the repetition was not made by Ptolemy himself, but through a careless error on the part of some copyist of his works. Cunningham thinks that Drastoka may have designated a town, in one of the *darâs* or 'valleys' of the Koh-Dâman, and that Baborana may be Parwân, a place of some consequence on the left bank of the Ghorband river in the neighbourhood of Opiân or Alexandria Opiane. Kaisana he takes to be the Cartana of Pliny (lib. VI, c. xxiii) according to whom it was situated at the foot of the Caucasus and not far from Alexandria, whilst according to Ptolemy it was on the right bank of the Pânjshir river. These data, he says, point to Bêgrâm, which is situated on the right bank of the Pânjshir and Ghorband rivers immediately at the foot of the Kohistân hills, and within 6 miles of Opiân. Bêgrâm also answers the description which Pliny gives of Cartana as *Tetragonis*, or the 'square ;' for Masson, in his account of the ruins especially notices "some mounds of great magnitude, and accurately describing a square of considerable dimensions." A coin of Eukratidês has on it the legend Karisiye Nagara or city of Karisi (*Geog. of Anc. Ind.*, pp. 26—29).

Gôrya :—Saint-Martin thinks that the position of this ancient city may be indicated by the situation of Mola-gouri, a place on the right or western bank of the River Landaï, as marked in one of Court's maps in the *Jour. Beng. As. Soc.*, vol. VIII, p. 34).

Nagara or Dionysopolis :—Lassen has identified this with Nanghenhar, the Nagarahâra of Sanskrit, a place mentioned under this name in the *Paurâṇik* Geography, and also in a Buddhistic inscription thought to belong to the 9th century which was found in Behar. The city was visited by Hiuen Tsiang, who calls it Na-kie-lo-ho. It was the capital of a kingdom of the same name, which before the time of the pilgrim had become subject to Kapiśa, a state which adjoined it on the west. Its territory consisted of a narrow strip of land which stretched along the southern bank of the Kâbul river from about Jagdalak as far westward as the Khaibar Pass. The city was called also Udyânapura, that is, 'the city of gardens,' and this name the Greeks, from some resemblance in the sound translated into Dionysopolis (a purely Greek compound, signifying 'the city of Dionysos,' the god of wine), with some reference no doubt to legends which had been brought from the regions of Paropanisos by the companions of Alexander. This name in a mutilated form is found inscribed on a medal of Dionysios, one of the Greek kings, who possessed the province of what is now called Afghanistân in the 2nd century B.C. Some traces of the name of Udyânapura still exist, for, as we learn from Masson, "tradition affirms that the city on the plain of Jalâlâbâd was called A j û n a," and the Emperor Baber mentions in his *Memoirs* a place called Adinapur, which, as the same author has pointed out, is now Bala-bâgh, a village distant about 13 miles westward from Jalâlâbâd near the banks of the Surkhrud, a small tributary of the Kâbul river.

As regards the site of N a g a r a h â r a, this was first indicated by Masson, and afterwards fixed with greater precision by Mr. Simpson, who having been quartered for four months at Jalâlâbâd during the late Afghân war took the opportunity of investigating the antiquities of the neighbourhood, which are chiefly of a Buddhist character. He has given an account of his researches in a paper read before the Royal Asiatic Society, and published in the Society's *Journal* (Vol. XIII, pp. 183—207). He there states that he found at a distance of 4 or 5 miles west from Jalâlâbâd numerous remains of what must have been an ancient city, while there was no other place in all the vicinity where he could discover such marked evidences of a city having existed. The ruins in question lay along the right bank of a stream called the Surkhâb, that rushed down from the lofty heights of the Sufaid-koh, and reached to its point of junction with the Kâbul river. The correctness of the identification he could not doubt, since the word 'Nagrak,'

'Nagarat,' or 'Nagara' was still applied to the ruins by the natives on the spot, and since the site also fulfilled all the conditions which were required to make it answer to the description of the position of the old city as given by Hiuen Tsiang. (See Lassen, *Ind. Alt.*, vol. II, p. 335; Saint-Martin's *Asie Centrale*, pp. 52—56; Cunningham, *Geog. of Anc. Ind.*, pp. 44—46; Masson, *Various Journeys*, vol. III, p. 164).

44. Between the Souastos and the Indus the Gandarai and these cities:—

Proklaïs	123°	32°
Naulibi	124° 20′	33° 20′

The Gandarai:—Gandhâra is a name of high antiquity, as it occurs in one of the Vedic hymns where a wife is represented as saying with reference to her husband, "I shall always be for him a Gandhâra ewe." It is mentioned frequently in the *Mahâbhârata* and other post-Vedic works, and from these we learn that it contained the two royal cities of Takshaśilâ (Taxila) and Pushkarâvatî (Peukelaôtis) the former situated to the east and the latter to the west of the Indus. It would therefore appear that in early times the Gandhâric territory lay on both sides of that river, though in subsequent times it was confined to the western side. According to Strabo the country of the Gandarai, which he calls Gandaritis, lay between the Khoaspês and the Indus, and along the River Kôphês. The name is not mentioned by any of the historians of Alexander, but it must nevertheless have been known to the Greeks as early as the times of Hekataios, who, as we learn from Stephanos of Byzantion, calls Kaspapyros a Gandaric city. Hêrodotos mentions the Gandarioi (Book III, c. xci) who includes them in the 7th Satrapy of Darius, along with the Sattagydai, the Dadikai and the Aparytai. In the days of Aśôka and some of his immediate successors Gandhâra was one of the most flourishing seats of Buddhism. It was accordingly visited both by Fa-hian and Hiuen Tsiang, who found it to contain in a state of ruin many monuments of the past ascendancy of their faith. From data supplied by the narratives of these pilgrims Cunningham has deduced as the boundaries of Gandhâra, which they call Kien-to-lo, on the west Lamghân and Jalâlâbâd, on the north the hills of Swât and Bunir, on the east the Indus, and on the south the hills of Kâlabâgh. "Within these limits," he observes, "stood several of the most renowned places of ancient India, some celebrated in the stirring history of Alexander's exploits, and others famous in the miraculous legends of Buddha, and in the subsequent history of Buddhism under the Indo-Scythian prince Kanishka." (*Geog. of Ind.*, p. 48.) Opinions have varied much with regard to the position of the Gandarioi. Rennell placed them on the west of Baktria in the province afterwards called Margiana, while Wilson (*Ar. Antiq.*, p. 131) took them to be the people south of the Hindû-kûsh, from about the modern Kandahâr to the Indus, and extending into the Panjâb and to Kaśmîr. There is, however, no connexion between the names of Gandaria and Kandahâr.

Proklaïs is the ancient capital of Gandhâra,

situated to the west of the Indus, which was mentioned in the preceding remarks under its Sanskrit name Pushkalâvatî, which means 'abounding in the lotus.' Its name is given variously by the Greek writers as Peukelaôtis, Peukolaitis, Peukelas, and Proklaïs, the last form being common to Ptolemy with the author of the *Periplûs*. The first form is a transliteration of the Pâli Pukhalaoti; the form Peukelas which is used by Arrian is taken by Cunningham to be a close transcript of the Pâli Pukkala, and the Proklaïs of Ptolemy to be perhaps an attempt to give the Hindî name of Pokhar instead of the Sanskrit Pushkara. Arrian describes Peukelas as a very large and populous city lying near the Indus, and the capital of a prince called Astês. Ptolemy defines its position with more accuracy, as being on the eastern bank of the river of Souastênê. The *Periplûs* informs us that it traded in spikenard of various kinds, and in kostus and bdellium, which it received from different adjacent countries for transmission to the coast of India. It has been identified with Hasht-nagar (*i.e., eight cities*) which lies at a distance of about 17 miles from Parashâwar (Peshâwar). Perhaps, as Cunningham has suggested, Hasht-nagar may mean not 'eight cities' but 'the city of Astês.'

Naulibi:—"It is probable," says Cunningham, "that Naulibi is Nilâb, an important town which gave its name to the Indus; but if so it is wrongly placed by Ptolemy, as Nilâb is to the South of the Kôphês" (*Geog. of Anc. Ind.*, p. 48).

45. Between the Indus and the Bidaspês

towards the Indus the A r s a territory and these cities :—

Ithagouros ....125° 40′ 33° 20′
Taxiala ....125° 32° 15′

A r s a represents the Sanskrit U r a ś a, the name of a district which, according to Cunningham, is to be identified with the modern district of Rash in Dhantâwar to the west of Muzafarâbâd, and which included all the hilly country between the Indus and Kaśmîr as far south as the boundary of Aṭak. It was visited by Hiuen Tsiang, who calls it U-la-shi and places it between Taxila and Kaśmîr. Pliny, borrowing from Megasthenês, mentions a people belonging to these parts called the A r s a g a l i t a e. The first part of the name answers letter for letter to the name in Ptolemy, and the latter part may point to the tribe Ghilet or Ghilghit, the Gahalata of Sanskrit. (V. Saint-Martin, *Étude*, pp. 59-60). Uraśa is mentioned in the *Mahâbhârata* and once and again in the *Râjataraṅgiṇî.*

I t h a g o u r o s :—The Ithagouroi are mentioned by Ptolemy (lib. VI, c. xvi) as a people of Sêrika, neighbouring on the Issêdones and Throanoi. Saint-Martin takes them to be the Dagors or Dangors, one of the tribes of the Daradas.

T a x i a l a is generally written as Taxila by the classical authors. Its name in Sanskrit is Taksha-śilâ, a compound which means ' hewn rock' or ' hewn stone.' Wilson thinks it may have been so called from its having been built of that material instead of brick or mud, like most other cities in India, but Cunningham prefers to ascribe to the name a legendary origin. The Pâli form of the name as found in a copper-plate inscription is T a k h a s i l a, which sufficiently accounts for the Taxila of the Greeks. The city is described by Arrian (*Anab.* lib. V, c. viii) as great and wealthy, and as the most populous that lay between the Indus and the Hydaspês. Both Strabo and Hiuen Tsiang praise the fertility of its soil, and the latter specially notices the number of its springs and watercourses. Pliny calls it a famous city, and states that it was situated on a level where the hills sunk down into the plains. It was beyond doubt one of the most ancient cities in all India, and is mentioned in both of the great national Epics. At the time of the Makedonian invasion it was ruled by a prince called Taxilês, who tendered a voluntary submission of himself and his kingdom to the great conqueror. About 80 years afterwards it was taken by Aśôka, the son of Vindusâra, who subsequently succeeded his father on the throne of Magadha and established Buddhism as the state religion throughout his wide dominions. In the early part of the 2nd century B.C. it had become a province of the Græco-Baktrian monarchy. It soon changed masters however, for in 126 B.C. the Indo-Skythian Sus or Abars acquired it by conquest, and retained it in their hands till it was wrested from them by a different tribe of the same nationality, under the celebrated Kanishka. Near the middle of the first century A.D. Apollonius of Tyana and his companion Damis are said to have visited it, and described it as being about the size of Nineveh, walled like a Greek city, and as the residence of a sovereign who ruled over what of old was the kingdom of Pôros. Its streets were narrow, but well arranged, and such altogether as reminded the travellers of Athens. Outside the walls was a beautiful temple of porphyry, wherein was a shrine, round which were hung pictures on copper tablets representing the feats of Alexander and Pôros. (Priaulx's *Apollon.*, pp. 13 sqq.) The next visitors we hear of were the Chinese pilgrims Fa-hian in 400 and Hiuen Tsiang, first in 630, and afterwards in 643. To them, as to all Buddhists, the place was especially interesting, as it was the scene of one of Buddha's most meritorious acts of alms-giving, when he bestowed his very head in charity. After this we lose sight altogether of Taxila, and do not even know how or when its ruin was accomplished. Its fate is one of the most striking instances of a peculiarity observable in Indian history, that of the rapidity with which some of its greatest capitals have perished, and the completeness with which even their very names have been obliterated from living memory. That it was destroyed long before the Muḥammadan invasion may be inferred from the fact that its name has not been found to occur in any Muḥammadan author who has written upon India, even though his account of it begins from the middle of the tenth century. Even Albîrûnî, who was born in the valley of the Indus, and wrote so early as the time of Mahmûd of Ghaznî, makes no mention of the place, though his work abounds with valuable information on points of geography. The site of Taxila has been identified by

Cunningham, who has given an account of his explorations in his *Ancient Geography of India* (pp. 104—124). The ruins, he says, cover an area of six square miles, and are more extensive, more interesting, and in much better preservation than those of any other ancient place in the Panjâb. These ruins are at a place called Shâh-dhêri, which is just one mile from Kâla-ka-serai, a town lying to the eastward of the Indus, from which it is distant a three days' journey. Pliny says only a two days' journey, but he under-estimated the distance between Peukelaôtis and Taxila, whence his error.

46. Around the Bidaspês, the country of the Pandoöuoi, in which are these cities :—

Labaka	127° 30′	34° 15′
Sagala, otherwise called Euthymêdia	126° 20′	32°
Boukephala	125° 30′	30° 20′
Iômousa	124° 15′	30°

The Country of the Pandoöuoi:—The Pâṇḍya country here indicated is that which formed the original seat of the Pâṇḍavas or Lunar race, whose war with the Kauravas or Solar race is the subject of the *Mahâbhârata*. The Pâṇḍavas figure not only in the heroic legends of India, but also in its real history,—princes of their line having obtained for themselves sovereignties in various parts of the country, in Râjputâna, in the Panjâb, on the banks of the Ganges, and the very south of the Peninsula. From a passage in the *Lalitavistara* we learn that at the time of the birth of Śâkyamuni a Pâṇḍava dynasty reigned at Hastinâpura, a city on the Upper Ganges, about sixty miles to the north-east of Dehli. Megasthenês, as cited by Pliny, mentions a great Pâṇḍava kingdom in the region of the Jamnâ, of which Mathurâ was probably the capital. According to Râjput tradition the celebrated Vikramâditya, who reigned at Ujjain (the O z ê n ê of the Greeks) about half a century B. C., and whose name designates an epoch in use among the Hindûs, was a Pâṇḍava prince. From the 8th to the 12th century of our æra Pâṇḍavas ruled in Indraprastha, a city which stood on or near the site of Dehli. When all this is considered it certainly seems surprising, as Saint-Martin has observed (*Étude*, 206 n.) that the name of the Pandus is not met with up to the present time on any historic monument of the north of India except in two votive inscriptions of Buddhist *stûpas* at Bhilsa. See also *Étude*, pp. 205, 206.

L a b a k a :—"This is, perhaps," says the same author (p. 222), "the same place as a town of Lohkoṭ (Lavakôṭa in Sanskrit) which makes a great figure in the Râjput annals among the cities of the Panjâb, but its position is not known for certain. Wilford, we know not on what authority, identified it with Lâhor, and Tod admits his opinion without examining it."

S a g a l a, called also E u t h y m ê d i a :—Sagala or Sangala (as Arrian less correctly gives the name) is the Sanskrit Sâkala or Sakala, which in its Prakrit form corresponds exactly to the name in Ptolemy. This city is mentioned frequently in the *Mahâbhârata*, from which we learn that it was the capital of the Madra nation, and lay to the west of the Râvî. Arrian (*Anab.* lib. V, cc. xxi, xxii) placed it to the east of the river, and this error on his part has led to a variety of erroneous identifications. Alexander, he tells us, after crossing the Hydraôtês (Râvî) at once pressed forward to Sangala on learning that the Kathaians and other warlike tribes had occupied that stronghold for the purpose of opposing his advance to the Ganges. In reality, however, Alexander on this occasion had to deal with an enemy that threatened his rear, and not with an enemy in front. He was in consequence compelled, instead of advancing eastward, to retrace his steps and recross the Hydraôtês. The error here made by Arrian was detected by General Cunningham, who, with the help of data supplied by Hiuen Tsiang discovered the exact site which Sagala had occupied. This is as nearly as possible where *Sangla-wala-tiba* or 'Sanglala hill' now stands. This Sangala is a hill with traces of buildings and with a sheet of water on one side of it. It thus answers closely to the description of the ancient Sangala in Arrian and Curtius, both of whom represent it as built on a hill and as protected on one side from attacks by a lake or marsh of considerable depth. The hill is about 60 miles distant from Lâhor, where Alexander probably was when the news about the Kathaians reached him. This distance is such as an army by rapid marching could accomplish in 3 days, and, as we learn that Alexander reached Sangala on the evening of the third after he had left the Hydraôtês, we have here a strongly confirmative proof of the correctness of the identification. The Makedonians destroyed Sagala, but it was rebuilt by Dêmetrios, one of the Græco-Baktrian kings, who in honour of his father Euthydêmos called it Euthydêmia. From this it would appear that the reading *Euthymêdia* as given in Nobbe's and other texts, is erroneous—(see Cunningham's *Geog. of Anc. Ind.*, pp. 180—187) *cf.* Saint-Martin, pp. 103—108).

47. The regions extending thence towards the east are possessed by the Kaspeiraioi, and to them belong these cities :—

48. Salagissa	129° 30′	34° 30′
Astrassos	131° 15′	34° 15′
Labokla	128°	33° 20′
Batanagra	130°	33° 30′
Arispara	130°	32° 50′
Amakatis	128° 15′	32° 20′
Ostobalasara	129°	32°
49. Kaspeira	127°	31° 15′
Pasikana	128° 30′	31° 15′
Daidala	128°	30° 30′
Ardonê	126° 15′	30° 10′
Indabara	127° 15′	30°
Liganeira	125° 30′	29°
Khonnamagara	128°	29° 20′
50. Modoura, the *city* of the gods	125°	27° 30′
Gagasmira	126° 40′	27° 30′
Êrarasa, a Metropolis	123°	26°
Kognandaua	124°	26°

Boukephala :—Alexander, after the battle

on the western bank of the Hydaspês in which he defeated Pôros, ordered two cities to be built, one N i k a i a, so called in honour of his victory (*nikê*), and the other Boukephala, so called in honour of his favourite horse, Boukephalos, that died here either of old age and fatigue, or from wounds received in the battle. From the conflicting accounts given by the Greek writers it is difficult to determine where the latter city stood. If we follow Plutarch we must place it on the eastern bank of the Hydaspês, for he states (*Vita Alexandre*) that Boukephalos was killed in the battle, and that the city was built on the place where he fell and was buried. If again we follow Strabo (lib. XV, c. i, 29) we must place it on the west bank at the point where Alexander crossed the river which in all probability was at Dilâwar. If finally we follow Arrian we must place it on the same bank, but some miles farther down the river at Jalâlpur, where Alexander had pitched his camp, and this was probably the real site. Boukephala seems to have retained its historical importance much longer than its sister city, for besides being mentioned here by Ptolemy it is noticed also in Pliny (lib. VI, c. xx) who says that it was the chief of three cities that belonged to the Asini, and in the *Periplûs* (sec. 47) and elsewhere. N i k a i a, on the other hand, is not mentioned by any author of the Roman period except Strabo, and that only when he is referring to the times of Alexander. The name is variously written Boukephala, Boukephalos, Boukephalia, and Boukephaleia. Some authors added to it the surname of Alexandria, and in the *Peutinger*

Tables it appears as Alexandria Bucefalos. The horse Boukephalos was so named from his 'brow' being very broad, like that of an 'ox.' For a discussion on the site of Boukephala see Cunningham's *Geog. of Anc. Ind.*, pp. 159 sqq.

Iômousa is probably Jamma, a place of great antiquity, whose chiefs were reckoned at one time among the five great râjas of the north. It doubtless lay on the great highway that led from the Indus to Palibothra.

List of cities of the Kaspeiraioi:—This long list contains but very few names that can be recognized with certainty. It was perhaps carelessly transcribed by the copyists, or Ptolemy himself may have taken it from some work the text of which had been already corrupted. Be that as it may, we may safely infer from the constancy with which the figures of latitude in the list decrease, that the towns enumerated were so many successive stages on some line of road that traversed the country from the Indus to Mathurâ on the Jamnâ. Salagissa, Arispara, Pasikana, Liganeira, Khonnamagara and Kognandaua are past all recognition; no plausible conjecture has been made as to how they are to be identified.

Astrassos:—This name resembles the Atrasa of Idrîsî, who mentions it as a great city of the Kanauj Empire (*Étude*, p. 226).

Labokla:—Lassen identified this with Lâhor, the capital of the Panjâb (*Ind. Alt.*, vol. III, p. 152). Thornton and Cunningham confirm this identification. The city is said to have been founded by Lava or Lo, the son of Râma, after whom it was named Lohâwar. The *Labo* in Labo-kla must be taken to represent the name of Lava. As for the terminal *kla*, Cunningham (*Geog. of Anc. Ind.*, p. 198) would alter it to *laka* thus, making the whole name Labolaka for Lavâlaka or 'the abode of Lava.'

Batanagra:—Ptolemy places this 2 degrees to the east of Labokla, but Saint-Martin (p. 226) does not hesitate to identify it with Bhatnair (for Bhaṭṭanagara) 'the town of the Bhatis' though it lies nearly three degrees south of Lâhor. Yule accepts this identification. A different reading is Kaṭanagara.

Amakatis (v. l. Amakastis).—According to the table this place lay to the S.E. of Labokla but its place in the map is to the S.W. of it Cunningham (pp. 195—197) locates it near Shekohpur to the south of which are two ruined mounds which are apparently the remains of ancient cities. These are called Amba and Kâpi respectively, and are said to have been called after a brother and a sister, whose names are combined in the following couplet:—

Amba-Kapa pai larai
Kalpi bahin chhurâwan ai.

When strife arose 'tween Amb and Kăp
Their sister Kalpi made it up.

"The junction of the two names," Cunningham remarks, "is probably as old as the time of Ptolemy, who places a town named Amakatis or Amakapis to the west of the Râvî, and in the immediate neighbourhood of Labokla or Lâhor." The distance of the mounds referred to from Lâhor is about 25 miles.

Ostobalasara (v. l. Stobolasara) Saint-Martin has identified this with Thanesar (Sthânêśvara in Sanskrit) a very ancient city, celebrated in the heroic legends of the Pâṇḍavas. Cunningham however thinks that Thanesar is Ptolemy's Batangkaisara and suggests that we should read Satan-aisara to make the name approach nearer to the Sanskrit Sthânêśvara—the Sa-ta-ni-shi-fa-lo of Hiuen Tsiang (p. 331).

Kaspeira:—"If this name," says Saint-Martin (p. 226) "is to be applied, as seems natural, to the capital of Kaśmîr, it has been badly placed in the series, having been inserted probably by the ancient Latin copyists."

Daidala:—An Indian city of this name is mentioned by Stephanos of Byzantion, but he locates it in the west. Curtius also has a Daedala (lib. VIII, c. x), a region which according to his account was traversed by Alexander before he crossed the Khoaspês and laid siege to Mazaga. Yule in his map places it doubtfully at Dudhal on the Khaghar river to the east of Bhatneer, near the edge of the great desert.

Ardonê:—Ahroni, according to Yule, a place destroyed by Tîmûr on his march, situated between the Khaghar and Chitang rivers, both of which lose themselves in the great desert.

Indabara is undoubtedly the ancient Indraprastha, a name which in the common dialects is changed into Indabatta (Indopat), and which becomes almost Indabara in the cerebral pronunciation of the last syllable. The site of this city was in the neighbourhood of Dehli. It was the capital city of the Pâṇḍavas. The Prâkṛit form of the name is Indrabaṭṭha. (Lassen, vol. III, p. 151).

M o d o u r a, the city of the gods :—There is no difficulty in identifying this with Mathurâ (Muttra) one of the most sacred cities in all India, and renowned as the birthplace of Kṛishṇa. Its temples struck Mahmûd of Ghaznî with such admiration that he resolved to adorn his own capital in a similar style. The name is written by the Greeks *Methora* as well as *Modoura*. It is situated on the banks of the Jamnâ, higher up than Agra, from which it is 35 miles distant. It is said to have been founded by Śatrughna, the younger brother of Râma. As already mentioned it was a city of the Pâṇḍavas whose power extended far to westward.

G a g a s m i r a :—Lassen and Saint-Martin agree in recognizing this as Ajmîr. Yule, however, objects to this identification on the ground that the first syllable is left unaccounted for, and proposes Jajhar as a substitute. Gegasius, he argues, represents in Plutarch Yayâti, the great ancestor of the Lunar race, while Jajhpûr in Orissa was properly Yayâtipûra. Hence probably in Jajhar, which is near Dehli, we have the representative of Gagasmira.

E r a r a s a :—Ptolemy calls this a metropolis. It appears, says Yule, to be Girirâja, 'royal hill,' and may be Goverdhan which was so called, and was a capital in legendary times (*Ind. Antiq.*, vol. I, p. 23). Saint-Martin suggests Vârâṇasî, now Banâras, which was also a capital. He thinks that this name and the next, which ends the list, were additions of the Roman copyists.

51. Still further to the east than the Kaspeiraioi are the G y m n o s o p h i s t a i, and after these around the Ganges further north are the D a i t i k h a i with these towns :—

Konta	133° 30′	34° 40′
Margara	135°	34°
Batangkaissara and east of the river.........................	132° 40′	33° 20′
Passala	137°	34° 15′
Orza	136°	33° 20′

G y m n o s o p h i s t a i :—This Greek word means 'Naked philosophers,' and did not designate any ethnic or political section of the population, but a community of religious ascetics or hermits located along the Ganges probably, as Yule thinks in the neighbourhood of Hardwâr and also according to Benfey, of Dehli, *Indien*, p. 95. For an account of the Gymnosophists see *Ind. Antiq.*, vol. VI, pp. 242—244.

D a i t i k h a i :—This name is supposed to represent the Sanskrit *jaṭika*, which means 'wearing twisted or plaited hair.' The name does not occur in the lists in this form but Kern, as Yule states, has among tribes in the north-east "Demons with elf locks" which is represented in Wilford by *Jaṭi-dhara*.

K o n t a, says Saint-Martin (*Étude*, p. 321) is probably Kuṇḍâ on the left bank of the Jamnâ to the south-east of Saharanpûr.

M a r g a r a :—Perhaps, according to the same authority, Marhâra near the Kalindi River to the north-east of Agra.

Batangkaissara:—Yule objecting to Saint-Martin's identification of this place with Bhatkashaur in Saharanpur pargaṇa, on the ground of its being a modern combination, locates it, but doubtingly, at Kesarwa east of the Jamnâ, where the position suits fairly.

Passala:—Pliny mentions a people called Passalae, who may be recognized as the inhabitants of Pañchâla or the region that lay between the Ganges and the Jamnâ, and whose power, according to the *Mahâbhârata*, extended from the Himâlayas to the Chambal River. Passala we may assume was the capital of this important state, and may now, as Saint-Martin thinks, be represented by Bisauli. This was formerly a considerable town of Rohilkhand, 30 miles from Sambhal towards the south-east, and at a like distance from the eastern bank of the Ganges.

Orza is perhaps Sarsi situated on the Râmgaṅgâ river in the lower part of its course.

52. Below these are the Anikhai with these towns:—

Persakra	134°	32° 40′
Sannaba	135°	32° 30′
Toana to the east of the river	136° 30′	32°

53. Below these Prasiakê with these towns:—

Sambalaka	132° 15′	31° 50′
Adisdara	136°	31° 30′
Kanagora	135°	30° 40′
Kindia	137°	30° 20′
Sagala, and east of the river	139°	30° 20′

Aninakha ....137° 20′ 31° 40′
Koangka ....138° 20′ 31° 30′

Anikhai (v. ll. Nạnikhai, Manikhai):—This name cannot be traced to its source. The people it designated must have been a petty tribe, as they had only 3 towns, and their territory must have lain principally on the south bank of the Jamnâ. Their towns cannot be identified. The correct reading of their name is probably *Manikhai*, as there is a town on the Ganges in the district which they must have occupied called Manikpur. There is further a tribe belonging to the Central Himâlaya region having a name slightly similar, Manga or Mangars, and the *Âîn-i-Akbarî* mentions a tribe of Manneyeh which had once been powerful in the neighbourhood of Dehli (*Étude*, p. 322). The form Nanikha would suggest a people named in the *Mahâbhârata* and the *Purâṇas*, the Naimishas who lived in the region of the Jamnâ.

Prasiakê.—This word transliterates the Sanskrit *Prâchyaka* which means 'eastern' and denoted generally the country along the Ganges. It was the country of the Prasii, whose capital was Palibothra, now Pâtnâ, and who in the times immediately subsequent to the Makedonian invasion had spread their empire from the mouths of the Ganges to the regions beyond the Indus. The Prasiakê of Ptolemy however was a territory of very limited dimensions, and of uncertain boundaries. Though seven of its towns are enumerated Palibothra is not among them, but is mentioned afterwards as the capital of the Mandalai and placed more than 3 degrees farther south than the most southern of them all. Yule remarks upon this: "Where the tables detail cities that are in Prasiakê, cities among the Poruari, &c., we must not assume that the cities named were really in the territories named; whilst we see as a sure fact in various instances that they were not. Thus the Mandalae, displaced as we have mentioned, embrace Palibothra, which was notoriously the city of the Prasii; while Prasiakê is shoved up stream to make room for them. Lassen has so much faith in the uncorrected Ptolemy that he accepts this, and finds some reason why Prasiakê is not the land of the Prasii but something else."

Sambalaka is Sambhal, already mentioned as a town of Rohilkhand. Sambalaka or Sambhala is the name of several countries in India, but there is only this one town of the name that is met with in the Eastern parts. It is a very ancient town and on the same parallel as Dehli.

Adisdara:—This has been satisfactorily identified with Ahichhatra, a city of great antiquity, which figures in history so early as the 14th century B.C. At this time it was the capital of Northern Pañchâla. The form of the name in Ptolemy by a slight alteration becomes *Adisadra*, and this approximates closely to the original form. Another city so called belonged to Central India, and this appears in Ptolemy as Adeisathra, which he places in the country of the Bêttigoi. The meaning of the name Ahi-chhattra is 'serpent umbrella' and is explained by a local legend concerning Âdi-Râja and the serpent demon, that while the Râja was asleep a serpent formed a canopy over him with its expanded hood. The fort is sometimes called Adikoṭ, though the commoner name is Ahi-chhatar, sometimes written Ahikshêtra. The place was visited by Hiuen Tsiang. In modern times it was first visited by Captain Hodgson, who describes it as the ruins of an ancient fortress several miles in circumference, which appears to have had 34 bastions, and is known in the neighbourhood by the name of the Pâṇḍu's Fort. It was visited afterwards by Cunningham (*Anc. Geog. of Ind.*, pp. 359—363).

K a n a g o r a :—This, as Saint-Martin points out, may be a corruption for Kanagoza, a form of K a n y â k u b j a or Kanauj. This city of old renown was situated on the banks of the Kâlînadî, a branch of the Ganges, in the modern district of Farrukhâbâd. The name applies not only to the city itself but also to its dependencies and to the surrounding district. The etymology (*kanyá*, 'a girl,' and *kubja*, 'round-shouldered' or 'crooked') refers to a legend concerning the hundred daughters of Kuśanâbha, the king of the city, who were all rendered crooked by Vâyu for non-compliance with his licentious desires (see also Beal, *Buddhist Records*, vol. I, p. 209). The ruins of the ancient city are said to occupy a site larger than that of London. The name recurs in another list of towns under the form Kanogiza, and is there far displaced.

K i n d i a may be identified with Kant, an ancient city of Rohilkhand, the Shâhjahânpur of the present day. Yule hesitates whether to identify it thus or with Mirzapur on the Ganges.

S a g a l a :—"Sagala," says Saint-Martin (*Étude*,

p. 326) "would carry us to a town of Sakula or Saghêla, of which mention is made in the Buddhist Chronicles of Ceylon among the royal cities of the North of India, and which Turnour believes to be the same town as Kuśinagara, celebrated as the place where Buddha Śâkyamuni obtained *Nirvâṇa*. Such an identification would carry us to the eastern extremity of Kôśala, not far from the River Gaṇḍakî.

Koangka ought to represent the Sanskrit *kanaka*, 'gold.' Mention is made of a town called in the Buddhistic legends Kanakavatî (abounding in gold), but no indication is given as to where its locality was (*Étude*, p. 326).

54. South of this Saurabatis with these towns:—

Empêlathra	130°	30°
Nadoubandagar..................	138° 40′	29°
Tamasis	133°	29°
Kouraporeina	130°	29°

Saurabatis:—This division is placed below Prasiakê. The ordinary reading is Sandrabatis, which is a transliteration of the Sanskrit Chandravatî. The original, Saint-Martin suggests, may have been Chhattravatî, which is used as a synonym of Ahikshêtra, and applies to that part of the territory of Pañchâla, which lies to the east of the Ganges. He thinks it more than probable that Sandrabatis, placed as it is just after a group of towns, two of which belong to Ahikshêtra, does not differ from this Chhattravatî, the only country of the name known to Sanskrit Geography in the Gangetic region. None of the four towns can be identified. (See Lassen, *Ind. Alt.* vol. I, p. 602; *Étude*, p. 326). Yule, however, points out that this territory is one of those which the endeavour to make Ptolemy's names cover the whole of India has greatly dislocated, transporting it from the S. W. of Râjputâna to the vicinity of Bahâr. His map locates Sandrabitis (Chandrabati) between the River Mahî and the Ârâvalî mountains.

55. And further, all the country along the rest of the course of the Indus is called by the general name of I n d o-S k y t h i a. Of this the insular portion formed by the bifurcation of the river towards its mouth is P a t a l ê n ê, and the region above this is A b i r i a, and the region about the mouths of the Indus and Gulf of Kanthi is S y r a s t r ê n ê. The towns of Indo-Skythia are these: to the west of the river at some distance therefrom:—

56. Artoarta	121° 30′	31° 15′
Andrapana	121° 15′	30° 40′
Sabana	122° 20′	32°
Banagara	122° 15′	30° 40′
Kodrana	121° 15′	29° 20′

Ptolemy from his excursion to the Upper Ganges now reverts to the Indus and completes its geography by describing I n d o-S k y t h i a, a vast region which comprised all the countries traversed by the Indus, from where it is joined by the river of Kâbul onward to the ocean. We have already pointed out how Ptolemy's description is here vitiated by his making the combined stream of the Panjâb rivers join the Indus only one degree below its junction with the Kâbul, instead of six degrees, or half way between that point and the ocean. The egregious error he has here committed seems altogether inexcusable, for whatever may have been the sources from which he drew his information, he evidently neglected the most accurate and the most valuable of all—the records, namely, of the Makedonian invasion as transmitted in writings of unimpeachable credit. At best, however, it must be allowed the determination of sites in the Indus valley is beset with peculiar uncertainty. The towns being but very slightly built are seldom of more than ephemeral duration, and if, as often happens they are destroyed by inundations, every trace is lost of their ever having existed. The river besides frequently changes its course and leaves the towns which it abandons to sink into decay and utter oblivion.[24] Such places again as still exist after escaping these and other casualties, are now known under names either altogether different from the ancient, or so much changed as to be hardly recognizable. This instability of the nomenclature is due to the frequency with which the valley has been conquered by foreigners. The period at

[24] Aristoboulos as we learn from Strabo (lib. XV, c. i. 19) when sent into this part of India saw a tract of land deserted which contained 1,000 cities with their dependent villages, the Indus having left its proper channel, was diverted into another, on the left hand much deeper, and precipitated itself into it like a cataract so that it no longer watered the country by the usual inundation on the right hand, from which it had receded, and this was elevated above the level, not only of the new channel of the river, but above that of the (new) inundation.

which the Skythians first appeared in the valley which was destined to bear their name for several centuries has been ascertained with precision from Chinese sources. We thence gather that a wandering horde of Tibetan extraction called Yuei-chi or Ye-tha in the 2nd century B. C. left Tangut, their native country, and, advancing westward found for themselves a new home amid the pasture-lands of Zungaria. Here they had been settled for about thirty years when the invasion of a new horde compelled them to migrate to the Steppes which lay to the north of the Jaxartes. In these new seats they halted for only two years, and in the year 128 B. C. they crossed over to the southern bank of the Jaxartes where they made themselves masters of the rich provinces between that river and the Oxus, which had lately before belonged to the Grecian kings of Baktriana. This new conquest did not long satisfy their ambition, and they continued to advance southwards till they had overrun in succession Eastern Baktriana, the basin of the Kôphês, the basin of the Etymander with Arakhôsia, and finally the valley of the Indus and Syrastrênê. This great horde of the Yetha was divided into several tribes, whereof the most powerful was that called in the Chinese annals Kwei-shwang. It acquired the supremacy over the other tribes, and gave its name to the kingdom of the Yetha. They are identical with the Kushâns. The great King Kanishka, who was converted to Buddhism and protected that faith was a Kushan. He reigned in the first century of the Christian æra and ruled from Baktriana to

Kaśmîr, and from the Oxus to Surâshṭra. These Kushans of the Panjâb and the Indus are no others than the Indo-Skythians of the Greeks. In the *Râjataraṅgiṇî* they are called Sâka and Turushka (Turks). Their prosperity could not have been of very long duration, for the author of the *Periplûs*, who wrote about half a century after Kanishka's time mentions that "Minnagar, the metropolis of Skythia, was governed by Parthian princes" and this statement is confirmed by Parthian coins being found everywhere in this part of the country. Max Müller, in noticing that the presence of Turanian tribes in India as recorded by Chinese historians is fully confirmed by coins and inscriptions and the traditional history of the country such as it is, adds that nothing attests the presence of these tribes more clearly than the blank in the Brahmanical literature of India from the first century before to the 3rd after our æra. He proposes therefore to divide Sanskrit literature into two—the one (which he would call the ancient and natural) *before*, and the other (which he would call the modern and artificial) *after* the Turanian invasion. In his Indo-Skythia Ptolemy includes Patalênê, Abiria and Syrastrênê. The name does not occur in Roman authors.

Patalênê, so called from its capital Patala, was the delta at the mouth of the Indus. It was not quite so large as the Egyptian delta with which the classical writers frequently compare it. Before its conquest by the Skythians it had been subject to the Græco-Baktrian kings. Its reduction to their authority is attributed by Strabo (lib. XI, c. xii, 1) to Menander or to Dêmetrios, the son of Euthydêmos.

A b i r i a :—The country of the A b h î r a s (the Ahirs of common speech) lay to the east of the Indus, above where it bifurcates to form the delta. In Sanskrit works their name is employed to designate generally the pastoral tribes that inhabit the lower districts of the North-West as far as Sindh. That Abiria is the O p h i r of Scripture is an opinion that has been maintained by scholars of eminence.

S y r a s t r ê n ê represents the Sanskrit Surâshṭra (the modern Soraṭh) which is the name in the *Mahâbhârata* and the *Purâṇas* for the Peninsula of Gujarât. In after times it was called Valabhî. Pliny (lib. VI, c. xx) in his enumeration of the tribes of this part of India mentions the Horatae, who have, he says, a fine city, defended by marshes, wherein are kept man-eating crocodiles that prevent all entrance except by a single bridge. The name of this people is no doubt a corruption of Soraṭh. They have an inveterate propensity to sound the letter *S* as an *H*.

Ptolemy distributes into six groups the names of the 41 places which he specifies as belonging to the Indus valley and its neighbourhood. The towns of the second group indicate by their relative positions that they were successive stages on the great caravan route which ran parallel with the western bank of the river all the way from the Kôphês junction downward to the coast. The towns of the fourth group were in like manner successive stages on another caravan route, that which on the eastern side of the river traversed the country from the great confluence with the combined rivers of the Panjâb downward to the Delta. The towns of the first group (5 in number) belonged to the upper part of the valley, and were situated near the Kôphês junction. They are mentioned in a list by themselves, as they did not lie on the great line of communication above mentioned. The third group consists of the two towns which were the chief marts of commerce in the Delta. The towns of the fifth group (7 in number) lay at distances more or less considerable from the eastern side of the Delta. The towns of the sixth group were included in the territory of the Khatriaioi, which extended on both sides of the river from its confluence with the Panjâb rivers as far as the Delta. None of them can now be identified (See *Étude*, pp. 234 sqq.) and of the first group—Artoarta, Sabana, Kodrana cannot be identified.

Andrapana:—Cunningham (p. 86) thinks this is probably Draband, or Derâband, near Dera-Ismail-Khân.

Banagara (for Bana-nagara):—Banna or Banu is often cited as the name of a town and a district that lay on the line of communication between Kâbul and the Indus. It was visited both by Fa-hian and Hiuen Tsiang. The former calls the country Po-na, *i.e.*, Bana. The latter calls it Fa-la-na, whence Cunningham conjectures that the original name was Varana or Barna. It consisted of the lower half of the valley of the Kuram river, and was distant from Lamghân a

15 days' journey southward. It is one of the largest, richest and most populous districts to the west of the Indus.—(See *Geog. of Anc. Ind.*, pp. 84-86).

57. And along the river :—

Embolima	124°	31°
Pentagramma	124°	30° 20′
Asigramma	123°	29° 30′
Tiausa	121° 30′	28° 50′
Aristobathra	120°	27° 30′
Azika	119° 20′	27°

58. Pardabathra ....117° 23° 30′

Piska	116° 30′	25°
Pasipêda	114° 30′	24°
Sousikana	112°	22° 20′
Bônis	111°	21° 30′
Kôlaka	110° 30′	20° 40′

Embolima was situated on the Indus at a point about 60 miles above Aṭṭak, where the river escapes with great impetuosity from a long and narrow gorge, which the ancients mistook for its source. Here, on the western bank, rises the fort of Amb, now in ruins, crowning a position of remarkable strength, and facing the small town of Derbend, which lies on the opposite side of the river. The name of Amb suggested that it might represent the first part of the name of Emb-olima, and this supposition was raised to certitude when it was discovered that another ruin not far off, crowning a pinnacle of the same hill on which Amb is seated, preserves to this day in the tradition of the inhabitants the name of Balimah. Embolima is mentioned by Arrian (lib. IV, c. xxvii) who represents it as situated at no great distance from the rock of A o r n o s—which as Abbott has shown, was Mount Mahâban, a hill abutting on the western bank of the Indus, about eight miles west from Embolima. It is called by Curtius E c b o l i m a (*Anab.* lib. VIII, c. xii) but he gives its position wrongly—at sixteen days' march from the Indus. Ptolemy assigns to it the same latitude and longitude which he assigns to the point where the Kâbul river and Indus unite. It was erroneously supposed that Embolima was a word of Greek origin from ἐκβολή, 'the mouth of a river' conf. Cunningham, *Geog. of Anc. Ind.*, pp. 52 ff.).

P e n t a g r a m m a :—To the north of the Kôphês at a distance of about forty miles S.W. from Embolima is a place called Panjpûr, which agrees closely both in its position and the signification of its name (5 towns) with the Pentagramma of Ptolemy.

A s i g r a m m a and the five towns that come after it cannot be identified.

P a s i p ê d a :—Saint-Martin thinks this may be the Besmeïd of the Arab Geographers, which, as they tell us was a town of considerable importance, lying east of the Indus on the route from Mansûra to Multân. Its name is not to be found in any existing map; but as the Arab itineraries all concur in placing it between Rond (now Roda) and Multân, at a three days' journey from the former, and a two days' journey from the latter, we may determine its situation to have been as far down the river as Mithankôṭ, where the great confluence now takes place. If the fact that Besmeid was on the eastern side of the river staggers our faith in this identification, Saint-Martin would remind us that this part of the tables is far from presenting us with a complete or systematic treatment of the subject, and that the only way open to us of restoring some part at least of these lists is to have recourse to synonyms. He contends that when we find in the Arab itineraries (which are documents of the same nature precisely as those which Ptolemy made use of) names resembling each other placed in corresponding directions, we ought to attach more weight to such coincidences than to the contradictions real, or apparent, which present themselves in the text of our author. Analogous transpositions occur in other lists, as, for instance, in the list of places in the Narmadâ basin. Cunningham, thinking it strange that a notable place of great antiquity like Sehwân, which he identifies with Sindomana, should not be mentioned by Ptolemy under any recognizable name, hazards the conjecture that it may be either his Piska or Pasipêda. "If we take," he says, "Haidarâbâd as the most probable head of the Delta in ancient times, then Ptolemy's Sydros, which is on the eastern bank of the Indus, may perhaps be identified with the old site of Mattali, 12 miles above Haidarâbâd and his Pasipêda with Sehwan. The identification of Ptolemy's Oskana with the Oxykanus or Portikanus of Alexander and with the great mound of Mahorta of the present day is I think almost certain. If so, either Piska or Pasipêda must be Sehwân."

Sousikana:—It is generally agreed that this

is a corrupt reading for Musikana, the royal city of Musikanos, who figures so conspicuously in the records of the Makedonian Invasion, and whose kingdom was described to Alexander as being the richest and most populous in all India. Cunningham (p. 257) identifies this place with Alor, which was for many ages the capital of the powerful kingdom of Upper Sindh. Its ruins, as he informs us, are situated to the south of a gap in the low range of limestone hills which stretches southwards from Bakhar for about 20 miles until it is lost in the broad belt of sand-hills which bound the Nâra or old bed of the Indus on the west. Through this gap a branch of the Indus once flowed which protected the city on the north-west. To the north-east it was covered by a second branch of the river which flowed nearly at right angles to the other at a distance of three miles. When Alôr was deserted by the river, it was supplanted by the strong fort of Bakhar (p. 258). The same author thinks it probable that Alôr may be the Binagara of Ptolemy, as it is placed on the Indus to the eastward of Oskana, which appears to be the Oxykanus of Arrian and Curtius.

Bônis:—The table places this at the point of bifurcation of the western mouth of the river and an interior arm of it. Arab geographers mention a town called Bania in Lower Sindh, situated at the distance of a single journey below Mansurâ. This double indication would appear to suit very well with Banna, which stands at the point where the Piniarî separates from the principal arm about 25 miles above Ṭhaṭṭha. Its position is however on the eastern bank of the river. (*Etude*, pp. 238, 239.)

Kôlaka or Kôlala is probably identical with the Krôkala of Arrian's *Indika* (sec. 21), which mentions it as a small sandy island where the fleet of Nearkhos remained at anchor for one day. It lay in the bay of Karâchi, which is situated in a district called Karkalla even now.

59. And in the islands formed by the river are these towns :—

Patala	112° 30′	21°
Barbarei	113° 15′	22° 30′

60. And east of the river at some distance therefrom are these towns :—

Xodrakê	116°	24°
Sarbana	116°	22° 50′
Auxoamis	115° 30′	22° 20′
Asinda	114° 15′	22°
Orbadarou or Ordabari	115°	22°
Theophila	114° 15′	21° 10′
Astakapra	114° 40′	20° 15′

Patala as we learn from Arrian was the greatest city in the parts of the country about the mouths of the Indus. It was situated, he expressly states, at the head of the Delta where the two great arms of the Indus dispart. This indication would of itself have sufficed for its identification, had the river continued to flow in its ancient channels. It has, however, frequently changed its course, and from time to time shifted the point of bifurcation. Hence the question regarding the site of Patala has occasioned much controversy. Rennell and Vincent, followed by Burnes and Ritter, placed it at Ṭhaṭṭha; Droysen, Benfey, Saint-Martin and Cunningham, at Haidarâbâd (the Nirankoṭ of Arab writers), and McMurdo, followed by Wilson and Lassen, at a place about 90 miles to the north-east of Haidarâbâd. The last supposition is quite untenable, while the arguments in favour of Haidarâbâd, which at one time was called Pâṭalapur[25] appear to be quite conclusive. (See Saint-Martin, pp. 180 ff., Cunningham, pp. 279—287). Patala figures conspicuously in the history of the Makedonian invasion. In its spacious docks Alexander found suitable accommodation for his fleet which had descended the Indus, and here he remained with it for a considerable time. Seeing how advantageously it was situated for strategy as well as commerce, he strengthened it with a citadel, and made it a military centre for controlling the warlike tribes in its neighbourhood. Before finally leaving India he made two excursions from it to the ocean, sailing first down the western and then down the eastern arm of the river. Pâtâla in Sanskrit mythology was the name of the lowest of the seven regions in the interior of the earth, and hence may have been applied to denote generally the parts where the sun descends into the under world, the land of the west, as in contrast to Prâchayaka, the land of the east. *Pâṭala* in Sanskrit means 'the

---

[25] "The Brahmans of Sehvân have stated to us that according to local legends recorded in their Sanskrit books Kaboul is the ancient *Chichapalapoura*; Multân, Prahlâdpur; Tattah, Dêval, Haidarâbâd, Nêran, and more anciently Pâtalpuri." Dr. J. Wilson, *Journ. Bombay Asiat. Soc.*, vol. III, 1850, p. 77.

trumpet-flower,' and Cunningham thinks that the Delta may have been so called from some resemblance in its shape to that of this flower. The classic writers generally spell the name as Pattala.

Barbarei:—The position of Barbarei, like that of Patala, has been the subject of much discussion. The table of Ptolemy places it to the north of that city, but erroneously, since Barbarei was a maritime port. It is mentioned in the *Periplûs* under the name of Barbarikon, as situated on the middle mouth of the Indus. D'Anville in opposition to all the data placed it at Debal Sindhi, the great emporium of the Indus during the middle ages, or at Karâchi, while Elliot, followed by Cunningham, placed it at an ancient city, of which some ruins are still to be found, called Bambhara, and situated almost midway between Karâchi and Ṭhaṭṭha on the old western branch of the river which Alexander reconnoitred. Burnes again, followed by Ritter, placed it at Richel, and Saint-Martin a little further still to the east at Bandar Vikkar on the Hajamari mouth, which has at several periods been the main channel of the river.

Xodrakê and Sarbana or Sardana:—As the towns in this list are given in their order from north to south, and as Astakapra, the most southern, was situated on the coast of the peninsula of Gujarât, right opposite the mouth of the river Narmadâ, the position of Xodrakê and the other places in the list must be sought for in the neighbourhood of the Ṛaṇ of Kachh. Xodrakê and Sarbana have not been identified, but Yule doubtingly places the latter on the Sambhar Lake. Lassen takes Xodrake to be the capital of the Xudraka, and locates it in the corner of land between the Vitastâ and Chandrabhâgâ (*Ind. Alt.*, vol. III, p. 145).

Asinda, according to Saint-Martin, may perhaps be Sidhpur (Siddhapura), a town on the river Sarasvatî, which rising in the Âravalîs empties into the Gulf of Kachh (pp. 246-247).

Auxoamis or Axumis:—The same authority would identify this with Sûmî, a place of importance and seat of a Muḥammadan chief, lying a little to the east of the Sarasvatî and distant about twenty-five miles from the sea. Yule however suggests that Ajmir may be its modern representation.

Orbadarou or Ordabari:—Yule doubtfully identifies this with Arbuda or Mount Abû, the principal summit of the Ârâvalîs. Pliny mentions alongside of the Horatae (in Gujarât) the Odomboerae which may perhaps be a different form of the same word. The name Uḍumbara is one well-known in Sanskrit antiquity, and designated a royal race mentioned in the *Harivaṅśa*.

Theophila:—This is a Greek compound meaning 'dear to God,' and is no doubt a translation of some indigenous name. Lassen has suggested that of Sardhur, in its Sanskrit form Surâdara, which means 'adoration of the gods.' Sardhur is situated in a valley of the Rêvata mountains so celebrated in the legends of Kṛishṇa. Yule suggests Dewaliya, a place on the isthmus, which connects the peninsula with the mainland. Dr. Burgess, Thân, the chief town of a district traditionally known as Deva-Pañchâl, lying a little further west than Dewaliya. Col. Watson writes:—"The only places I can think of for Theophila are—1. Gûndi, the ancient Gundigaḍh, one and a half or two miles further up the Hathap river, of which city Hastakavapra was the port. This city was one of the halting-places of the Bhaunagar Brâhmaṇs ere they came to Gogha. It was no doubt by them considered dear to the gods. It was connected with Hastakavapra and was a city of renown and ancient. 2. Pardwa or Priya-dêva, an old village, about four or five miles west of Hathap. It is said to have been contemporary with Valabhî, and there is an ancient Jain temple there, and it is said that the Jains of Gundigaḍh had their chief temple there. 3. Dêvagana, an ancient village at the foot of the west slopes of the Khôkras about 18 miles from Hâthap to the westward."

Astakapra:—This is mentioned in the *Peri-plûs* (sec. 41), as being near a promontory on the eastern side of the peninsula which directly confronted the mouth of the Narmadâ on the opposite side of the gulf. It has been satisfactorily identified with Hastakavapra, a name which occurs in a copper-plate grant of Dhruvasêna I, of Valabhî, and which is now represented by Hathab near Bhavnagar. Bühler thinks that the Greek form is not derived immediately from the Sans-krit, but from an intermediate old Prakṛit word Hastakampra. (See *Ind. Ant.*, vol. V, pp. 204, 314.

61. Along the river are these towns:—

Panasa	122° 30′	29°
Boudaia	121° 15′	28° 15′

Naagramma	120°	27°
Kamigara	119°	26° 20′
Binagara	118°	25° 20′
Parabali	116° 30′	24° 30′
Sydros	114°	21° 20′
Epitausa	113° 45′	22° 30′
Xoana	113° 30′	21° 30′

Panasa:—The table places Panasa one degree farther south than the confluence of the Zaradros and the Indus. Ptolemy, as we have seen, egregiously misplaced this confluence, and we cannot therefore from this indication learn more than that Panasa must have been situated lower down the Indus than Pasipêda (Besmaïd) and Alexandria of the Malli which lay near the confluence. A trace of its name Saint-Martin thinks is preserved in that of Osanpur, a town on the left of the river, 21 miles below Mittankôt.

Boudaia:—According to Saint-Martin this is very probably the same place as a fort of Budhya or Bodhpur, mentioned in the Arab chronicles of the conquest of Upper Sindh and situated probably between Alôr and Mittankôṭ. Yule identifies it with Budhia, a place to the west of the Indus and south from the Bolan Pass.

Naagramma:—This Yule identifies with Naoshera, a place about 20 miles to the south of Besmaïd. Both words mean the same, 'new town.'

Kamigara:—The ruins of Arôr which are visible at a distance of four miles to the south-east of Kori, are still known in the neighbourhood under the name of Kaman. If to this word we add the common Indian affix *nagar*—'city,' we have a near approach to the Kamigara of Ptolemy.

Binagara:—This some take to be a less correct form than Minnagar given in the *Periplûs*, where it is mentioned as the metropolis of Skythia, but under the government of Parthian princes, who were constantly at feud with each other for the supremacy. Its position is very uncertain. Cunningham would identify it with Alôr. Yule, following McMurdo, places it much further south near Brâhmanâbâd, which is some distance north from Haidarâbâd. The *Periplûs* states that it lay in the interior above Barbarikon (sec. 38).

Xoana:—Yule suggests that this may be Sewana, a place in the country of the Bhauliṅgas, between the desert and the Arâvalîs.

62. The parts east of Indo-Skythia along the coast belong to the country of Larike, and here in the interior to the west of the river Namados is a mart of commerce, the city of Barygaza .... 113° 15′ 17° 20′

63. To the east of the river:—

Agrinagara	118° 15′	22° 30′
Siripalla	118° 30′	21° 30′
Bammogoura	116°	20° 45′
Sazantion	115° 30′	20° 30′
Zêrogerei	116° 20′	19° 50′
Ozênê, the capital of Tiastanes	117°	20°
Minagara	115° 10′	19° 30′
Tiatoura	115° 50′	18° 50′
Nasika	114°	17°

L a r i k ê :—Lârdêśa was an early name for the territory of Gujarât and the Northern Koṅkaṇ. The name long survived, for the sea to the west of that coast was in the early Muḥammadan time called the sea of Lâr, and the language spoken on its shores was called by Mas'ûdi, Lâri (Yule's *Marco Polo*, vol. II, p. 353, n.). Ptolemy's Larikê was a political rather than a geographical division and as such comprehended in addition to the part of the sea-board to which the name was strictly applicable, an extensive inland territory, rich in agricultural and commercial products, and possessing large and flourishing towns, acquired no doubt by military conquest.

B a r y g a z a, now Bharôch, which is still a large city, situated about 30 miles from the sea on the north side of the river Narmadâ, and on an elevated mound supposed to be artificial, raised about 80 feet above the level of the sea. The place is repeatedly mentioned in the *Periplûs*. At the time when that work was written, it was the greatest seat of commerce in Western India, and the capital of a powerful and flourishing state. The etymology of the name is thus explained by Dr. John Wilson (*Indian Castes*, vol. II, p. 113): "The Bhârgavas derive their designation from Bhârgava, the adjective form of Bhṛigu, the name of one of the ancient Ṛishis. Their chief habitat is the district of Bharôch, which must have got its name from a colony of the school of Bhṛigu having been early established in this Kshêtra, probably granted to them by some conqueror of the district. In the name Barugaza given to it by Ptolemy, we have a

Greek corruption of Bhṛigukshêtra (the territory of Bhṛigu) or Bhṛigukachha, 'the tongue-land' of Bhṛigu." The illiterate Gujarâtis pronounce Bhṛigukshêtra as Bargacha, and hence the Greek form of the name.

Agrinagara:—This means 'the town of the Agri.' Yule places it at Âgar, about 30 miles to the N. E. of Ujjain.

Siripalla:—A place of this name (spelt Sêripala) has already been mentioned as situated where the Namados (Narmadâ) changes the direction of its course. Lassen therefore locates it in the neighbourhood of Haump, where the river turns to southward.

Bammogoura:—In Yule's map this is identified with Pavangarh, a hill to the north of the Narmadâ.

Sazantion:—This may perhaps be identical with Sajintra, a small place some distance north from the upper extremity of the Bay of Khambât.

Zêrogerei:—This is referred by Yule to Dhâr, a place S. W. of Ozênê, about one degree.

Ozênê:—This is a transliteration of Ujjayinî, the Sanskrit name of the old and famous city of Avanti, still called Ujjain. It was the capital of the celebrated Vikramâditya, who having expelled the Skythians and thereafter established his power over the greater part of India, restored the Hindû monarchy to its ancient splendour. It was one of the seven sacred cities of the Hindûs, and the first meridian of their astronomers. We learn from the *Mahâvaṅśa* that Aśôka, the grandson of Chandragupta (Sandrakottos) was sent by his father the king of

Pâṭaliputra (Patna) to be the viceroy of Ujjain, and also that about two centuries later (B.C. 95) a certain Buddhist high priest took with him 40,000 disciples from the Dakkhinagiri temple at Ujjain to Ceylon to assist there in laying the foundation stone of the great Dâgaba at Anurâdhapura. Half a century later than this is the date of the expulsion of the Skythians by Vikramâditya, which forms the æra in Indian Chronology called *Saṁvat* (57 B.C.) The next notice of Ujjain is to be found in the *Periplûs* where we read (Sec. 48) " Eastward from Barygaza is a city called Ozênê, formerly the capital where the king resided. From this place is brought down to Barygaza every commodity for local consumption or export to other parts of India, onyx-stones, porcelain, fine muslins, mallow-tinted cottons and the ordinary kinds in great quantities. It imports from the upper country through Proklaïs for transport to the coast, spikenard, kostos and bdellium." From this we see that about a century and a half after Vikramâditya's æra Ujjain was still a flourishing city, though it had lost something of its former importanee and dignity from being no longer the residence of the sovereign. The ancient city no longer exists, but its ruins can be traced at the distance of a mile from its modern successor. Ptolemy tells us that in his time Ozênê was the capital of Tiastanês. This name transliterates Chashṭâna, one which is found on coins and the cave temple inscriptions of Western India. This prince appears probably to have been the founder of the Kshatrapa dynasty of Western India (see *Ind. Alt.*, vol. III, p. 171).

M i n a g a r a is mentioned in the *Periplûs*, where its name is more correctly given as M i n-n a g a r, *i.e.*, 'the city of the M i n' or Skythians. This Minagara appears to have been the residence of the sovereign of Barygaza. Ptolemy places it about 2 degrees to the S. W. of Ozênê. Yule remarks that it is probably the Manekir of Mas'û-di, who describes it as a city lying far inland and among mountains. Benfey doubts whether there were in reality two cities of this name, and thinks that the double mention of Minnagar in the *Periplûs* is quite compatible with the supposition that there was but one city so called. (*Indien*, p. 91).

T i a t o u r a :—This would transliterate with Chittur, which, however, lies too far north for the position assigned to Tiatoura. Yule suggests, but doubtingly, its identity with Chandur. This however lies much too far south.

N a s i k a has preserved its name unaltered to the present day, distant 116 miles N. E. from Bombay. Its latitude is 20° N., but in Ptolemy only 17°. It was one of the most sacred seats of Brâhmaṇism. It has also important Buddhistic remains, being noted for a group of rock-temples. The word *nâsikâ* means in Sanskrit 'nose.'

64. The parts farther inland are possessed by the P o u l i n d a i A g r i o p h a g o i, and beyond them are the K h a t r i a i o i, to whom belong these cities, lying some east and some west of the Indus :—

Nigranigramma	124°	28° 15′
Antakhara	122°	27° 20′
Soudasanna	123°	26° 50′

Syrnisika	121°	26° 30′
Patistama	121°	25°
Tisapatinga	123°	24° 20′

The 'Poulindai' Agriophagoi are described as occupying the parts northward of those just mentioned. Pulinda is a name applied in Hindû works to a variety of aboriginal races. Agriophagoi is a Greek epithet, and indicates that the Pulinda was a tribe that subsisted on raw flesh and roots or wild fruits. In Yule's map they are located to the N. E of the Raṇ of Kachh, lying between the Khatriaioi in the north and Larikê in the south. Another tribe of this name lived about the central parts of the Vindhyas.

Khatriaioi:—According to Greek writers the people that held the territory comprised between the Hydraôtês (Râvî) and the Hyphasis (Biyas) were the Kathaioi, whose capital was Sangala. The *Mahâbhârata*, and the Pâli Buddhist works speak of Sangala as the capital of the Madras, a powerful people often called also the Bâhîkas. Lassen, in order to explain the substitution of name, supposes that the mixture of the Madras with the inferior castes had led them to assume the name of Khattrias (Kshatriya, the warrior caste), in token of their degradation, but this is by no means probable. The name is still found spread over an immense area in the N. W. of Iṇdia, from the Hindû-kôh as far as Bengal, and from Nêpâl to Gujarât, under forms slightly variant, Kathîs, Kattis, Kathîas, Kattris, Khatris, Khetars, Kattaour, Kattair, Kattaks, and others.

One of these tribes, the Kâṭhis, issuing from the lower parts of the Panjâb, established themselves in Surâshṭra, and gave the name of Kâṭhiâvad to the great peninsula of Gujarât.' (*Étude*, p. 104).

The six towns mentioned in section 64 can none of them be identified.

65. But again, the country between Mount Sardônyx and Mount Bêttigô belongs to the Tabasoi, a great race, while the country beyond them as far as the Vindhya range, along the eastern bank of the Namados, belongs to the Prapiôtai, who include the Rhamnai, and whose towns are these :—

Kognabanda	...120° 15′	23°
Ozoabis	...120° 30′	23° 40′
Ostha	...122° 30′	23° 30′
Kôsa, where are diamonds	...121° 20′	22° 30′

Tabasoi is not an ethnic name, but designates a community of religious ascetics, and represents the Sanskrit *Tâpasâs*, from *tapas* 'heat' or 'religious austerity.' The haunts of these devotees may be assigned to the valley of the Tâptî or Tâpî (the Nanagouna of Ptolemy) to the south of the more western portion of the Vindhyas that produced the sardonyx.

Prapiôtai:—Lassen locates this people, including the subject race called the Rhamnai, in the upper half of the Narmadâ valley. From the circumstance that diamonds were found near Kôsa, one of their towns, he infers that their territory extended as far as the Upper Varadâ, where diamond mines were known to have existed. Kôsa was probably situated in the neighbourhood of Baital, north of the sources of the Tâptî and the Varadâ.

Rhamnai:—The name of this people is one of the oldest in Indian ethnography. Their early seat was in the land of the Ôreîtai and Arabitai beyond the Indus, where they had a capital called Rhambakia. As they were connected by race with the Brahui, whose speech must be considered as belonging to the Dekhan group of languages, we have here, says Lassen (*Ind. Alt.* vol. III, p. 174), a fresh proof confirming the view that before the arrival of the Aryans all India, together with Gedrôsia, was inhabited by the tribes of the same widely diffused aboriginal race, and that the Rhamnai, who had at one time been settled in Gedrôsia, had wandered thence as far as the Vindhya mountains. Yule conjectures that the Rhamnai may perhaps be associated with Râmagiri, now Râmtek, a famous holy place near Nâgpûr. The towns of the Prapiôtai, four in number, cannot with certainty be identified.

66. About the Nanagouna are the Phyllitai and the Bêttigoi, including the Kandaloi along the country of the Phyllitai and the river, and the Ambastai along the country of the Bêttigoi and the mountain range, and the following towns :—

67. Agara	129° 20′	25°
Adeisathra	128° 30′	24° 30′
Soara	124° 20′	24°
Nygdosora	125°	23°
Anara	122° 30′	22° 20′

The Phyllitai occupied the banks of the Tâptî lower down than the Rhamnai, and extended northward to the Sâtpura range. Lassen considers their name as a transliteration of Bhilla, with an appended Greek termination. The Bhills are a well-known wild tribe spread to this day not only on the Upper Narmadâ and the parts of the Vindhya chain adjoining, but wider still towards the south and west. In Ptolemy's time their seats appear to have been further to the east than at present. Yule thinks it not impossible that the Phyllitai and the Drilophyllitai may represent the Puḷinda, a name which, as has already been stated, is given in Hindû works to a variety of aboriginal races. According to Caldwell (*Drav. Gram.*, p. 464) the name *Bhilla* (*vil, bil*) means 'a bow.'

Bêttigoi is the correct reading, and if the name denotes, as it is natural to suppose, the people living near Mount Bêttigô, then Ptolemy has altogether displaced them, for their real seats were in the country between the Koimbatur Gap and the southern extremity of the Peninsula.

Kandaloi:—Lassen suspects that the reading here should be Gondaloi, as the Gonds (who are nearly identical with the Khands) are an ancient race that belonged to the parts here indicated. Yule, however, points out that Kuntaladêśa and the Kanṭalas appear frequently in lists and in inscriptions. The country was that, he adds, of which Kalyâṇ was in after days the capital (Elliot, *Jour. R. As. S.* vol. IV, p. 3).

Ambastai:—These represent the Ambashṭha

of Sanskrit, a people mentioned in the Epics, where it is said that they fought with the club for a weapon. In the *Laws of Manu* the name is applied to one of the mixed castes which practised the healing art. A people called Ambautai are mentioned by our author as settled in the east of the country of the Paropanisadai. Lassen thinks these may have been connected in some way with the Ambastai. Their locality is quite uncertain. In Yule's map they are placed doubtfully to the south of the sources of the Mahânadî of Orissa.

Of the four towns, Agara, Soara, Nygdosora and Anara, in section 67, nothing is known.

Adeisathra:—It would appear that there were two places in Ancient India which bore the name of Ahichhattra, the one called by Ptolemy Adisdara (for Adisadra), and the other as here, Adeisathra. Adisdara, as has been already shown, was a city of Rohilkhand. Adeisathra, on the other hand, lay near to the centre of India. Yule quotes authorities which seem to place it, he says, near the Vindhyas or the Narmadâ. He refers also to an inscription which mentions it as on the Sindhu River, which he takes to be either the Kâli-sindh of Mâlwâ, or the Little Kâli-sindh further west, which seems to be the Sindhu of the *Mêghadûta*. Ptolemy, singularly enough, disjoins Adeisathra from the territory of the Adeisathroi, where we would naturally expect him to place it. Probably, as Yule remarks, he took the name of the people from some Pauranik ethnic list and the name of the city from a traveller's route, and thus failed to make them fall into proper relation to each other.

68. Between Mount Bôttigô and Adeisathros are the Sôrai nomads, with these towns:—

Sangamarta	133°	21°
Sôra, the capital of Arkatos	130°	21°

69. Again to the east of the Vindhya range is the territory of the (Biolingai or) Bôlingai, with these towns:—

Stagabaza or Bastagaza	133°	28° 30′
Bardaôtis	137° 30′	28° 30′

Sôra designates the northern portion of the Tamiḻ country. The name in Sanskrit is Chôla, in Telugu Choḷa, but in Tamiḻ Sôṛa or Chôṛa. Sôṛa is called the capital of Arkatos. This must be an error, for there can be little doubt that Arkatos was not the name of a prince, but of a city, the Ârkâḍ of the present day. This is so suitably situated, Caldwell remarks, as to suggest at once this identification, apart even from the close agreement as far as the sound is concerned. The name is properly Âr-kâḍ, and means 'the six forests.' The Hindûs of the place regard it as an ancient city, although it is not mentioned by name in the *Purâṇas* (*Drav. Gram.*, Introd. pp. 95, 96). There is a tradition that the inhabitants of that part of the country between Madras and the Ghâṭs including Ârkâḍ as its centre were Kuṛumbars, or wandering shepherds, for several centuries after the Christian æra. Cunningham takes Arkatos to be the name of a prince, and inclines to identify Sôra with Zora or Jora (the Jorampur of the maps) an old town lying immediately under the walls of

Karnul. The Sôrai he takes to be the Suari (*Geog.* p. 547).

Biolingai or Bôlingai:—Ptolemy has transplanted this people from their proper seats, which lay where the Arâvalî range slopes westward towards the Indus, and placed them to the east of the Vindhyas. He has left us however the means of correcting his error, for he makes them next neighbours to the Pôrvaroi, whose position can be fixed with some certainty. Pliny (lib. VI, c. xx) mentions the Bolingae and locates them properly. According to Pâṇini, Bhauliṅgi was the seat of one of the branches of the great tribe of the Śalvas or Śâlvas.

Stagabaza:—Yule conjectures this may be Bhôjapûr, which he says was a site of extreme antiquity, on the upper stream of the Bêtwâ, where are remains of vast hydraulic works ascribed to a king Bhôja (*J. A. S. Beng.* vol. XVI, p. 740). To account for the first part of the name *staga* he suggests the query: Taṭaka-Bhôja, the 'tank' or 'lake' of Bhoja?

Bardaôtis:—This may be taken to represent the Sanskrit Bhadrâvatî, a name, says Yule, famed in the Epic legends, and claimed by many cities. Cunningham, he adds, is disposed to identify it with the remarkable remains (pre-Ptolemaic) discovered at Bharâod, west of Rêwâ.

70. Beyond these is the country of the Pôrouaroi with these towns:—

Bridama	134° 30′	27° 30′
Tholoubana	136° 20′	27°
Malaita	136° 30′	25° 50′

71. Beyond these as far as the Ouxentos range are the Adeisathroi with these towns:—

Maleiba	140°	27° 20′
Aspathis........................	138° 30′	25° 20′
Panassa	137° 40′	24° 30′
Sagêda, the Metropolis	133°	23° 30′
Balantipyrgon	136° 30′	23° 30′

Pôrouaroi (Pôrvaroi):—This is the famous race of the Pauravas, which after the time of Alexander was all predominant in Râjasthâna under the name of the Pramâras. The race figures conspicuously both in the legendary and real history of the North of India. It is mentioned in the hymns of the *Veda*, and frequently in the *Mahâbhârata*, where the first kings of the Lunar race are represented as being Pauravas that reigned over the realms included between the Upper Ganges and the Yamunâ. The later legends are silent concerning them, but they appear again in real history and with fresh distinction, for the gallant Pôros, who so intrepidly contended against Alexander on the banks of the Hydaspês, was the chief of a branch of the Paurava whose dominions lay to the west of that river, and that other Pôros who went on an embassy to Augustus and boasted himself to be the lord paramount of 600 vassal kings was also of the same exalted lineage. Even at the present day some of the noblest houses reigning in different parts of Râjasthân claim to be descended from the Pauravas, while the songs of the national bards still extol the vanished grandeur and the power and glory of this ancient race. Saint-Martin locates the Pôrouaroi of the text in the west of Upper India, in the very heart of the Râjpût country, though the table would lead us to place them much farther to the east. In the position indicated the name even of the Pôrouaroi is found almost without alteration in the Purvar of the inscriptions, in the Pôravars of the Jain clans, as much as in the designation spread everywhere of Povars and of Pouârs, forms variously altered, but still closely approaching the classic Paurava (*Étude*, pp. 357 sqq.)

The names of the three towns assigned to the Pôrvaroi,—B r i d a m a, T h o l o u b a n a and M a l a i t a designate obscure localities, and their position can but be conjectured. Saint-Martin suggests that the first may be Dildana, the second Doblana, and the third Plaita, all being places in Râjputâna. Yule, however, for Bridama proposes Bardâwaḍ, a place in a straight line from Indôr to Nimach, and for M a l a i t a,—Maltaun; this place is in the British territory of Sagâr and Narmadâ, on the south declivity of the Naral Pass.

A d e i s a t h r o i:—It has already been pointed out that as Ptolemy has assigned the sources of the Khabêris (the Kâvêrî) to his Mount Adeisathros, we must identify that range with the section of the Western Ghâṭs which extends immediately northward from the Koimbatur Gap. He places Adeisathros however in the central parts of India, and here accordingly we must look for the cities of the eponymous people. Five are mentioned, but S a g ê d a only, which was the metropolis, can be identified with some certainty. The name represents the Sâkêta of Sanskrit. Sâkêta was another name for A y ô d h y â on the Sarayû, a city of vast extent and famous as the capital of the kings of the Solar race and as the residence for some years of Śâkyamuni, the founder of Buddhism. The Sagêda of our text was however a different city, identified by Dr. F. Hall with Têwar, near Jabalpûr, the capital of the Chêdi, a people of Bandêlakhand renowned in Epic poetry. Cunningham thinks it highly probable that the old form of the name of this people was Chaṅgêdi and may be preserved in the Sagêda of Ptolemy and in the Chi-ki-tho of Hiuen Tsiang in Central India, near the Narmadâ. He says:— "The identification which I have proposed of Ptolemy's Sagêda Metropolis with Chêdi appears to me to be almost certain. In the first place, Sagêda is the capital of the Adeisathroi which I take to be a Greek rendering of Hayakshêtra or the country of the Hayas or Haihayas. It adjoins the country of the Bêttigoi, whom I would identify with the people of Vakâṭaka, whose capital was Bhândak. One of the towns in their country, situated near the upper course of the Sôn, is named Balantipyrgon, or Balampyrgon. This I take to be the famous Fort of Bândogaṛh, which we know formed part of the Chêdi dominions. To the north-east was Panassa, which most probably preserves the name of some town on the Parṇâsâ or Banâs River, a tributary which joins the Sôn to the north-east of Bândogaṛh. To the north of the Adeisathroi, Ptolemy places the Pôrouaroi or Parihârs, in their towns named Tholoubana, Bridama, and Malaita. The first I would identify with Boriban (Bahuriband) by reading Oöloubana or Voloubana. The second must be Bilhâri; and the last may be Lameta, which gives its name to the Ghâṭ on the Narmadâ, opposite Têwar, and may thus stand for Tripura itself. All these identifications hold so well together, and mutually support each other, that I have little doubt of their correctness." *Archæolog. Surv. of Ind.* vol. IX, pp. 55—57.

Panassa:—This in Yule's map is doubtfully placed at Panna, a decaying town in Bandelakhand with diamond mines in the neighbourhood. In the same map Baland is suggested as the representative of Balantipyrgon.

72. Farther east than the Adeisathroi towards the Ganges are the Mandalai with this city:—

Asthagoura ....142° 25°

73. And on the river itself these towns:—

Sambalaka ....141° 29° 30′
Sigalla ....142° 28°
Palimbothra, the Royal residence ....143° 27°
Tamalitês ....144° 30′ 26° 30′
Oreophanta ....146° 30′ 24° 30′

74. In like manner the parts under Mount Bêttigô are occupied by the Brakhmanai Magoi as far as the Batai with this city:—

Brakhmê ....128° 19°

75. The parts under the range of Adeisathros as far as the Arouraioi are occupied by the Badiamaioi with this city:—

Tathilba ....134° 18° 50′

76. The parts under the Ouxentos range are occupied by the Drilophyllitai, with these cities:—

Sibrion	139°	22° 20′
Opotoura	137° 30′	21° 40′
Ozoana	138° 15′	20° 30′

Mandalai:—The territory of the Mandalai lay in that upland region where the Sôn and the Narmadâ have their sources. Here a town situated on the latter river still bears the name Mandalâ. It is about 50 miles distant from Jabalpûr to the south-east, and is of some historic note. Ptolemy has, however, assigned to the Mandalai dominions far beyond their proper limits, for to judge from the towns which he gives them they must have occupied all the right bank of the Ganges from its confluence with the Jamnâ downwards to the Bay of Bengal. But that this is improbable may be inferred from the fact that Palimbothra (Pâṭnâ) which the table makes to be one of their cities, did not belong to them, but was the capital of Prasiakê, which, as has already been remarked, is pushed far too high up the river. Tamalitês, moreover, which has been satisfactorily identified with Tamluk, a river port about 35 miles S. W. from Calcutta possessed, according to Wilford, a large territory of its own. The table also places it only half a degree more to the southward than Palimbothra, while in reality it is more than 3 or 4 deg. Cunningham inclines to identify with the Mandalai the Mundas of Chutia Nâgpur, whose language and country, he says, are called

Mundala, and also with the Malli of Pliny (lib. VI. c. xxi.)—*Anc. Geog. of Ind.*, pp. 508, 509.

Sambalaka:—A city of the same name attributed to Prasiakê (sec. 53) has been already identified with Sambhal in Rohilkhand. The Sambalaka of the Mandalai may perhaps be Sambhalpur on the Upper Mahânadî, the capital of a district which produces the finest diamonds in the world.

Sigalla:—This name has a suspicious likeness to Sagala, the name of the city to the west of Lâhor, which was besieged and taken by Alexander, and which Ptolemy has erroneously placed in Prasiakê (sec. 53).

Palimbothra:—The more usual form of the name is Palibothra, a transcription of Pâliputra, the spoken form of Pâṭaliputra, the ancient capital of Magadha, and a name still frequently applied to the city of Pâṭnâ which is its modern representative. In the times of Chandragupta (the Sandrokottos of the Greeks) and the kings of his dynasty, Palibothra was the capital of a great empire which extended from the mouths of the Ganges to the regions beyond the Indus. Remains of the wooden wall by which the city, as we learn from Strabo, was defended, were discovered a few years ago in Pâṭnâ (by workmen engaged in digging a tank) at a depth of from 12 to 15 feet below the surface of the ground. Palimbothra, as we have noticed, did not belong to the Mandalai but to the Prasioi.

Tamalitês represents the Sanskrit Tâmraliptî, the modern Tamluk, a town lying in a low and damp situation on a broad reach or bay of the Rûpnârâyaṇ River, 12 miles above its junction with the Hughlî mouth of the Ganges. The Pâli form of the name was Tâmalitti, and this accounts for the form in Greek. Pliny mentions a people called Taluctae belonging to this part of India, and the similarity of the name leaves little doubt of their identity with the people whose capital was Tamluk. This place, in ancient times, was the great emporium of the trade between the Ganges and Ceylon. We have already pointed out how wide Ptolemy was of the mark in fixing its situation relatively to Palimbothra.

Brakhmanai Magoi:—Mr. J. Campbell has suggested to me that by Brakhmanai Magoi may be meant 'sons of the Brâhmaṇs,' that is, Canarese Brâhmaṇs, whose forefathers married women of the country, the word *magoi* representing the Canarese *maga*, 'a son.' The term, he says, is still in common use, added to the name of castes, as Haiga-Makalu (*makalu*—plural of *maga*) *i.e.* Haiga Brâhmaṇs. Lassen supposed that Ptolemy, by adding *Magoi* to the name of these Brâhmaṇs, meant to imply either that they were a colony of Persian priests settled in India, or that they were Brâhmaṇs who had adopted the tenets of the Magi, and expresses his surprise that Ptolemy should have been led into making such an unwarrantable supposition. The country occupied by these Brâhmaṇs was about the upper Kâvêrî, and extended from Mount Bêttigô eastward as far as the Batai.

Brakhmê:—"Can this," asks Caldwell, "be Brahmadêśam, an ancient town on the Tâmraparṇî, not far from the foot of the Podigei Mount (Mt. Bêttigô) which I have found referred to in several ancient inscriptions?"

Badiamaioi:—There is in the district of Belgaum a town and hill-fort on the route from Kalâdgi to Balâri, not far from the Mâlprabhâ, a tributary of the Kṛishṇâ, called Badâmi, and here we may locate the Bâdiamaioi. Tathilba, their capital, cannot be recognized.

Drilophyllitai:—These are placed by Ptolemy at the foot of the Ouxentos, and probably had their seats to the south-west of that range. Their name indicates them to have been a branch of the Phyllitai, the Bhills, or perhaps Pulindas. Lassen would explain the first part of their name from the Sanskrit *dṛiḍha* (strong) by the change of the *ḍh* into the liquid. Ozoana, one of their three towns is, perhaps, Seoni, a place about 60 miles N. E. from Nâgpur.

77. Further east than these towards the Ganges are the Kokkonagai with this city:—

Dôsara	142° 30′	22° 30′

78. And on the river farther west:—

Kartinaga	146°	23°
Kartasina	146°	21° 40′

79. Under the Maisôloi the Salakênoi towards the Oroudian (or Arouraian) Mountains with these cities:—

Bênagouron	140°	20° 15′
Kastra	138°	19° 30′
Magaris	137° 30′	18° 20′

80. Towards the Ganges River the Sabarai, in whose country the diamond is found in great abundance, their towns are :—

Tasopion	140° 30′	22°
Karikardama....................	141°	20° 15′

81. All the country about the mouths of the Ganges is occupied by the Gangaridai with this city :—

Gangê, the Royal residence...	146°	19° 15′

Kokkonagai :—Lassen locates this tribe in Chutia Nâgpur, identifying Dôsara with Doesâ in the hill country, between the upper courses of the Vaitaraṇî and Suvarṇarêkha. He explains their name to mean the people of the mountains where the *kôka* grows,—*kôka* being the name of a kind of palm-tree. Yule suggests that the name may represent the Sanskrit Kâkamukha, which means 'crow-faced,' and was the name of a mythical race. He places them on the Upper Mahânadî and farther west than Lassen. The table gives them two towns near the Ganges.

Kartinaga and Kartasina :—The former, Yule thinks, may be Karṇagaṛh near Bhâgalpur, perhaps an ancient site, regarding which he refers to the *Jour. R. As. Soc.* vol. XVIII, p. 395; Kartasina he takes to be Karṇasôṇagaṛh, another ancient site near Berhampur (*J. R. A. S.* N. S., vol. VI, p. 248 and *J. As. S. Beng.*, vol. XXII, p. 281).

Salakênoi :—This people may be located to the west of the Gôdâvarî, inland on the north-western borders of Maisôlia. Their name, Lassen thought (*Ind. Alt.*, vol. III, p. 176) might be connected with the Sanskrit word *Sâla*, the Sâl tree. Yule suggests that it may represent the Sanskrit Saurikîrṇa. None of their towns can be recognized.

Sabarai:—The Sabarai of Ptolemy Cunningham takes to be the Suari of Pliny, and he would identify both with the aboriginal Śavaras or Suars, a wild race who live in the woods and jungles without any fixed habitations, and whose country extended as far southward as the Pennâr River. These Śavaras or Suars are only a single branch of a widely spread race found in large numbers to the S. W. of Gwalior and Narwar and S. Râjputâna, where they are known as Surrius. Yule places them farther north in Dôsarênê, towards the territory of Sambhalpur, which, as we have already remarked, produced the finest diamonds in the world. Their towns have not been identified.

Gangaridai:—This great people occupied all the country about the mouths of the Ganges. Their capital was Gangê, described in the *Periplûs* as an important seat of commerce on the Ganges. They are mentioned by Virgil (*Georg.* III, l. 27), by Valerius Flaccus (*Argon.* lib. VI, l. 66), and by Curtius (lib. IX, c. ii) who places them along with the Pharrasii (Prasii) on the eastern bank of the Ganges. They are called by Pliny (lib. VI, c. lxv) the Gangaridae Calingae, and placed by him at the furthest extremity of the Ganges region, as is indicated by the expression *gens novissima*, which he applies to them. They must have been a powerful people, to judge from the military force which Pliny reports them to have maintained, and their territory could scarcely have been restricted to the marshy jungles at the mouth of the river now known as the Sundarbans, but must have comprised a considerable portion of the province of Bengal. This is the view taken by Saint-Martin. Bengal, he says, represents, at least in a general way, the country of the Ganga-ridae, and the city which Pliny speaks of as their capital, Parthalis can only be Vardhana, a place which flourished in ancient times, and is now known as Bardhwân. The name of the Gangari-dai has nothing in Sanskrit to correspond with it, nor can it be a word, as Lassen supposed, of purely Greek formation, for the people were mentioned under this name to Alexander by one of the prin-ces in the North-west of India. The synonymous term which Sanskrit fails to supply is found among the aboriginal tribes belonging to the region occupied by the Gangaridai, the name being pre-served almost identically in that of the Goṅghrîs of S. Bahâr, with whom were connected the Gaṅgayîs of North-western, and the Gaṅgrâr of Eastern Bengal, these designations being but variations of the name which was originally common to them all.

Gangê:—Various sites have been proposed for Gangê. Heeren placed it near Duliapur, a village about 40 miles S. E. of Calcutta on a branch of the Isamatî River; Wilford at the confluence of the Ganges and Brahmaputra, where, he says, there was a town called in Sanskrit Hastimalla, and in the spoken dialect Hâthimalla, from elephants being picquetted there; Murray at

Chittagong; Taylor on the site of the ancient Hindu Capital of Baṅga (Bengal) which lies in the neighbourhood of Soṇargâoṅ (Suvarṇagrâma), a place 12 miles to the S. E. of Ḍhakka; Cunningham at Jêsor; and others further west, near Calcutta, or about 30 miles higher up the Hughlî, somewhere near Chinsurâ. Another Gangê is mentioned by Artemidoros above or to the N. W. of Palibothra, and this Wilford identifies with Prayâg, *i.e.*, Allahâbâd, but Groskurd with Anupshahr.

Ptolemy now leaves the Gangetic regions and describes the inland parts of the territories along the Western Coast of the Peninsula.

82. In the parts of Ariakê which still remain to be described are the following inland cities and villages: to the west of the Bênda these cities :—

Malippala	119° 30′	20° 15′
Sarisabis	119° 30′	20°
Tagara	118°	19° 20′
Baithana (the royal seat of [Siro] Ptolemaios or Polemaios)...	117°	18° 30′
Deopali or Deopala	115° 40′	17° 50′
Gamaliba	115° 15′	17° 20′
Omênogara	114°	16° 20′

83. Between the Bênda and Pseudostomos :

Nagarouris (or Nagarouraris)	120°	20° 15′
Tabasô	121° 30′	20° 40′
Indê	123°	20° 45′
Tiripangalida....................	121° 15′	19° 40′

Hippokoura, the royal seat of Baleokouros	119° 45′	19° 10′
Soubouttou	120° 15′	19° 10′
Sirimalaga	119° 20′	18° 30′
Kalligeris	118°	18°
Modogoulla	119°	18°
Petirgala	117° 45′	17° 15′
Banaouasei	116°	16° 45′

Seven cities are enumerated in Ariakê, as lying to the west of the Bênda, and regarding four of these, Malippala, Sarisabis, Gamaliba and Omênogara, nothing is known. The *Periplûs* (sec. 51) notices Tagara and Baithana in a passage which may be quoted: "In Dakhinabades itself there are two very important seats of commerce, Paithana towards the south of Barygaza, from which it is distant a twenty days' journey, and eastward from this about a ten days' journey is another very large city, Tagara. From these marts goods are transported on waggons to Barygaza through difficult regions that have no road worth calling such. From Paithana great quantities of onyx-stones and from Tagara large supplies of common cotton-cloth, muslins of all kinds, mallow-tinted cottons and various other articles of local production imported into it from the maritime districts."

Baithana is the Paithana of the above extract, and the Paiṭhân of the present day, a town of Haidarâbâd, or the territory of the Nizam, on the left bank of the river Gôdâvarî, in latitude 19° 29′ or about a degree further north than it is placed by Ptolemy. Paithana is the Prâkrit form of the Sanskrit P r a t i s h ṭ h â n a, the name of the capital of Śâlivâhana. Ptolemy calls it the capital of Siroptolemaios or Siropolemaios, a name which represents the Sanskrit Śrî-Pulômâvit, the Puḷumâyi of the Nasik Cave and Amarâvati Stûpa Inscriptions, a king of the great Andhra dynasty.

T a g a r a :—The name is found in inscriptions under the form Tagarapura (*J. R. A. S.* vol. IV, p. 34). Ptolemy places it to the north-east of Baithana and the *Periplûs*, as we see from the extract, to the east of it at the distance of a ten days' journey. Wilford, Vincent, Mannert, Ritter and others take it to be Dêvagaḍh, now Daulatâbâd, which was the seat of a sovereign even in 1293, and is situated not far from Élura, so famous for its excavated temples. But if Baithana be Paiṭhan, Tagara cannot be Dêvagaḍh, unless the distance is wrongly given. There is, moreover, nothing to show that Dêvagaḍh was connected with the Tagarapura of the inscriptions. Paṇḍit Bhagvânlâl identified Tagara with Junnar, a place of considerable importance, situated to the north of Pûṇâ. He pointed out that the Sanskrit name of Tagara was Trigiri a compound meaning 'three hills,' and that as Junnar stood on a high site between three hills this identification was probably correct. Junnar however lies to the westward of Paiṭhan. Yule places Tagara at Kulburga, which lies to the south-east of Paiṭhan, at a distance of about 150 miles, which would fairly represent a ten days' journey, the distance given in the *Periplûs*. Grant Duff would identify it with a place near Bhîr on the Gôdâvarî, and Fleet with Kolhâpur. The Silahâra princes or chiefs who formed three distinct branches of a dynasty that ruled over two parts of the Koṅkaṇ and the country about Kolhâpur style themselves, 'The Lords of the excellent city of Tagara.' If, says Prof. Bhaṇḍârkar, the name of Tagara has undergone corruption, it would take the form, according to the laws of Prâkrit speech, of Târur or Têrur, and he therefore asks 'can it be the modern Dârur or Dhârur in the Nizam's dominions, 25 miles east of Grant Duff's Bhîr, and 70 miles S.E. of Paiṭhan?' (see Muller's *Geog. Græc. Minor.* vol. I, p. 294, n.; Elphinstone's *History of India*, p. 223; Burgess, *Arch. Surv. W. Ind.* vol. III, p. 54; and *Bombay Gazetteer*, vol. XIII, pt. ii, p. 423, n.). Mr. Campbell is of opinion that the maritime districts from which local products were brought to Tagara and thence exported to Barygaza, lay on the coast of Bengal, and not on the Koṅkaṇ coast, from which there was easy transit by sea to the great northern emporium in the Gulf of Khambât, while the transit by land through Tagara could not be accomplished without encountering the most formidable obstacles.

Deopali:—This name means 'the city of God,' and Deopali may therefore perhaps be Dêvagaḍh, the two names having the same meaning.

Tabasô:—This would seem to be a city of the Tabasoi, already mentioned as a large community, of Brâhmaṇ ascetics.

Hippokoura:—A town of this name has already been mentioned as a seaport to the south of Simylla. This Hippokoura lay inland, and was the capital of the southern parts of Ariakês, as Paithana was the capital of the northern. Its position is uncertain. Yule places it doubtfully at Kalyâṇ, a place about half a degree to the west of Bidar, and at some distance south from the river Mañjirâ. Ptolemy calls it the capital of Baleokouros. Bhândârkar conjectures this to have been the Viḷivâyakura, a name found upon two other Andhra coins discovered at Kôlhâpur. There is no other clue to its identification, but see Lassen, *Ind. Alt.* vol. III, pp. 179, 185.

Sirimalaga may perhaps be Mâlkhêḍ, a town in Haidarâbâd, situated on a tributary of the Bhîmâ, in lat. 17° 8′ and long. 77° 12′. The first part of the word *Siri* probably represents the Sanskrit honorific prefix *śrî*.

Kalligeris:—Perhaps Kaṇhagiri, a place about ½ a degree to the south of Mûdgal.

Modogoulla:—There can be little doubt that this is Mûdgal, a town in the Haidarâbâd districts,—lat. 16° 2′, long. 76° 26′,—N. W. from Balâri. Petirgala cannot be identified.

Banaouasei:—This place is mentioned in the *Mahâvaṅso*, in the Pâli form Wanawâsi, by which a city or district is designated. Banaouasei must beyond doubt have been the capital of this country, and is identical with the modern Banavâsî, situated on the upper Varadâ, a tributary of the Tuṅgabhadrá. Saint-Martin thinks that it was the city visited by Hiuen Tsiang, and called by him Kon-kin-na-pu-lo, *i.e.*, Koṅkaṇapura; Cunningham is of opinion that both the bearing and the distance point to Ânagundi, but Dr. Burgess suggests Kôkanûr for Kôn-kin-na-pu-lo.

84. The inland cities of the Pirates are these:—

Olokhoira	114°	15°
Mousopallê, the metropolis	115° 30′	15° 45′

85. Inland cities of Limyrikê, to the west of the Pseudostomos are these:—

Naroulla	117° 45′	15° 50′
Kouba	117°	15°
Paloura	117° ′51	14° 40′

86. Between the Pseudostomos and the Baris, these cities:—

Pasagê	124° 50′	19° 50′
Mastanour	121° 30′	18° 40′
Kourellour	119°	17° 30′
Pounnata, where is beryl	121° 20′	17° 30′
Aloê	120° 20′	17°
Karoura, the royal seat of Kêrobothros	119°	16° 20′
Arembour	121°	16° 20′
Bideris	119°	15° 50′
Pantipolis	118°	15° 20′
Adarima	119° 30′	15° 40′
Koreour	120°	15°

87. Inland town of the Aïoi:—

Morounda	121° 20′	14° 20′

The dominion of the sea appears to have satisfied the ambition of the pirates, as they possessed on shore only a narrow strip of territory enclosed between the line of coast and the western declivities of the Ghâṭs. Their capital, Mousopallê, Yule places at Miraj, a town near the Kṛishṇâ, but doubtfully. Their other town, Olokhoira, is probably Khêḍâ, a town in the district of Ratnagiri in lat. 17° 44′ long. 73° 30′. As Khêḍâ is the name of several other places in this part of the country, *Olo*, whatever it may mean, may have been in old times prefixed to this particular Khêḍâ for the sake of distinction.

Kouba:—This is generally taken to be Goa or Govâ, the capital of the Portuguese possessions in India, and there can be little doubt of the correctness of the identification. The two towns Naroulla and Paloura, which Ptolemy places with Kouba to the west of the Pseudostomos, cannot be identified. To judge from his figures of longitude, Paloura lay 15′ farther east than Kouba, but as he makes the coast run eastward instead of southward, it must be considered to have lain south of Kouba. The name is Tamiḻ, and means, according to Caldwell (*Introd.* p. 104) 'Milk town.' It is remarkable, he observes, how many names of places in Southern India mentioned by Ptolemy end in *οὐρ* or *οὐρα* = 'a town.' There are 23 such places in all.

Pasagê:—According to Yule's map this represents Palsagi, the old name of a place now called Halsi, south-east of Goa, from which it is distant somewhat under a degree.

Mastanour and Kourellour cannot be identified.

Pounnata has not yet been identified, though Ptolemy gives a sort of clue in stating that it produced the beryl. Yule places it in his map near Seringapatam. (See *Ind. Ant.* vol. XII, p. 13).

Aloê:—This may be Yellapur, a small town in North Canara, in lat. 14° 56′ long. 74° 43′.

K a r o u r a :—" Karoura," says Caldwell, " is mentioned in Tamiḻ traditions as the ancient capital of the Chêra, Kêra, or Kêrala kings, and is generally identified with Karûr, an important town in the Koimbatur district, originally included in the Chêra kingdom. It is situated on the left bank of the river Amarâvatî, a tributary of the Kâvêrî, near a large fort now in ruins. Ptolemy notes that Karoura was the capital of Kêrobothros, *i.e.*, Kêralaputra(Cherapati?) Karûra means 'the black town,' and I consider it identical with Kâragam, and Kaḍâram, names of places which I have frequently found in the Tamiḻ country, and which are evidently the poetical equivalents of Karûr. The meaning of each of the names is the same. Ptolemy's word Karoura represents the Tamiḻ name of the place with perfect accuracy " (*Introd.* pp. 96, 97).

A r e m b o u r :—Lassen compares this name with Oorumparum, but the situation of the place so called (lat. 11° 12′ long. 76 ° 16′) does not suit well the position of Arembour as given by Ptolemy.

B i d e r i s :—Perhaps Erod or Yirodu in the district of Koimbatur (lat. 11° 20′ long. 77° 46′) near the Kâvêrî.

P a n t i p o l i s, according to Yule, represents the obsolete name Pāntiyapura, which he places at Hangal, in the Dhârwâḍ district.

M o r o u n d a :—This is the only inland city of the Aïoi named by Ptolemy. It has not been identified.

The concluding tables enumerate the inland towns belonging to the districts lying along the Eastern Coast of the Peninsula.

88. Inland cities of the Kareoi:—

Mendôla	123°	17° 40′
Sêlour	121° 45′	16° 30′
Tittoua	122°	15° 20′
Mantittour	123°	15° 10′

89. Inland cities of the Pandionoi:—

Tainour	124° 45′	18° 40′
Peringkarei	123° 20′	18°
Korindiour	125°	17° 40′
Tangala or Taga	123° 30′	16° 50′
Modoura, the royal city of Pandion	125°	16° 20′
Akour	124° 45′	15° 20′

90. Inland cities of the Batoi:—

Kalindoia	127° 40′	17° 30
Bata........................	126° 30′	17°
Talara	128°	16° 45′

Inland cities of the Kareoi:—none of the four named in the table can be identified.

Peringkarei:—This town has preserved its name almost without change, being now known as Peruṅgari, on the river Vaigaï, about 40 miles lower down its course than Madurâ. With regard to this name, Caldwell remarks that if it had been written Peruṇgkarei it would have been perfectly accurate Tamiḻ, letter for letter. The meaning is 'great shore,' and *perum* 'great' becomes *peruṅg* before *k*, by rule. Ptolemy places a town called Tainour at the distance of less than a degree to the north-east of Peringkarei. The direction would suit Tanjor, but the distance is more than a degree. Ptolemy has however placed his Peringkarei quite in a wrong position with regard to Madurâ.

T a n g a l a or T a g a :—There can be little doubt that this is now represented by Diṇḍugal, an important and flourishing town lying at a distance of 32 miles north by west from Madurâ.

M o d o u r a :—This is now called Madurâ or Madurai—on the banks of the River Vaigai. It was the second capital of the Southern Pâṇḍyas; we have already noticed it in the description of the territory of this people.

B a t a :—This may perhaps be Paṭṭukôṭṭa, a small town not very far inland from the northern end of the Argolic Gulf (Palk's Passage). The other two towns of the Batoi cannot be recognized. As Pudukôṭṭa is the capital of the Tondiman Râja, Lassen has suggested its identity with Bata. It is upwards of 20 miles farther inland than Paṭṭukôṭṭa.

91. Inland cities of the Paralia of the S ô r ê t a i :—

Kaliour	129°	17° 20′
Tennagora	132°	17°
Eikour	129°	16° 40′
Orthoura, the royal city of Sôrnagos	130°	16° 20′
Berê	130° 20′	16° 15′
Abour	129°	16°
Karmara	130° 20′	15° 40′
Magour	130°	15° 15′

92. The inland cities of the Arvarnoi are these:—

Kerauge	133°	16° 15′
Phrourion	132°	15°
Karigê	132° 40′	15°
Poleour........................	131° 30′	14° 40′
Pikendaka	131° 30′	14°
Iatour	132° 30′	14°
Skopoloura	134° 15′	14° 35′
Ikarta	133° 30′	13° 40′
Malanga, the royal city of Basaronagos..................	133°	13°
Kandipatna	133° 30°	12° 20′

93. The inland cities of the Maisôloi:—

Kalliga........................	138°	17°
Bardamana	136° 15′	15° 15′
Koroungkala	135°	15°
Pharytra or Pharetra.........	134° 20′	13° 20′
Pityndra, the metropolis ...	135° 20′	12° 30′

Orthoura:—Of the eight inland cities named as belonging to the maritime territory of the Sôrêtai, only two—Abour and the capital, have been identified. Abour is Âmbûrdurg in N. Arkaṭ, lat. 12° 47′, long. 78° 42′. Regarding Orthoura Cunningham says: "Chôḷa is noticed by Ptolemy, whose *Orthura regia Sornati* must be Uriûr, the capital of Soranâtha, or the king of the Soringae, that is the Sôras, Chôras or Chôlas. Uṛaiyûr is a few miles south-south-east of Tiruchbinâpalli. The Soringae are most probably the Syrieni of Pliny, with their 300 cities, as they occupied the coast between the Pandae and the Derangae or Dra. vidians."—*Anc. Geog. of Ind.*, p. 551.

Phrourion:—This is a Greek word signifying 'a garrisoned fort,' and may perhaps be meant as a translation of an indigenous name having that signification, as Durga, 'a hill-fort,' a common affix to names of places in the Peninsula.

Karigô:—This should no doubt be read Karipô under which form it can be at once identified with Kaḍapâ, a place lying 5 miles from the right bank of the Northern Pennâr on a small tributary of that river.

Pikendaka:—*Konda* is a frequent termination in the names of towns in this part of India. The letters of Pikendaka may have been transposed in copying, and its proper form may have been Pennakonda, the name of a town in the district of Balâri (lat. 14°5′ long. 77° 39′).

Iatour:—From Yule's map it would appear there is a place lying a degree westward from Kaḍapâ which still bears this name, Yêtûr.

Malanga:—In our notice of Melangê it was pointed out that Cunningham had fixed the locality of Malanga near Élur, a place some distance inland about half way between the Kṛishṇâ and the Gôdâvarî towards their embouchures, and in the neighbourhood of which are the remains of an old capital named Veṅgî. With regard to the king's name Bassaronaga, he thinks that this may be identified with the Pali Majêrika-nâga of the *Mahâwaṅso* and thus Ptolemy's Malaṅga would become the capital of the Nâgas of Majerika, *Anc. Geo. of Ind.*, (pp. 539, 540). In Yule's map Malanga is placed conjecturally about two degrees farther south at Velur, near the mouth of the Pennâr.

Of the five cities attributed to the Maisôloi, only Koroungkala can be recognized. It appears to be the place now known as Worankal, the mediæval capital of Telingana. It has but few tokens remaining to attest its former grandeur.

Pityndra, the capital of Maisôlia, was probably Dhanakaṭaka now Dharaṇikôṭa, about 20 miles above Bêjwâḍâ on the Kṛishṇâ.

94. Islands lying near the part of India which projects *into the ocean* in the Gulf of Kanthi :—

Barakê	111°	18°

95. And along the line of coast as far as the Kolkhic Gulf :—

Milizêgyris (or Milizigêris)..	110°	12° 30′
Heptanêsia	113°	13°
Trikadiba	113° 30′	11°
Peperinê	115°	12° 40′
Trinêsia	116° 20′	12°
Leukê	118°	12°
Nanigêris	122°	12°

96. And in the Argaric Gulf :—

Kôry	126° 30′—13°

Barakê :—This is the name given in the *Periplûs* to the Gulf of Kachh, called by our author the Gulf of Kanthi, a name which to this day is applied to the south coast of Kachh The *Periplûs* does not mention Barakê as an island, but says that the Gulf had 7 islands. Regarding

Barakê, Dr. Burgess says: "Yule places Barakê at Jaggat or Dwârakâ; Lassen also identifies it with Dwârakâ, which he places on the coast between Purbandar and Miyânî, near Śrînagar. Mula-Dwârakâ, the original site, was further east than this, but is variously placed near Mâdhupur, thirty-six miles north-west from Sômanâth-Paṭṭan, or three miles south-west from Kôḍinâr, and nineteen miles east of Sômanâth. This last spot is called Mula-Dwârakâ to this day." (*Târîkh-i-Sôraṭh*, Introd. p. 7).

Milizêgyris occurs in the *Periplûs* as Melizeigara, which may be identified with Jayagaḍ or Sîdi-Jayagaḍ, which would appear to be the Sigerus of Pliny (lib. vi, c. 26).

Heptanêsia (*or group of* 7 *islands*) probably corresponded to the Sesikrienai of the *Periplûs*, which may be the Burnt Islands of the present day, among which the Viṅgôrlâ rocks are conspicuous.

Trikadiba or 'the island Trika,'—*diba* being the Sanskrit word *dvîpa*, 'an island.'

Peperinê:—This, to judge from the name, should be an island somewhere off the coast of Cottonara, the great pepper district, as stated by Pliny (lib. VI, c. xxvi).

Trinêsia (*or group of* 3 *islands*):—Ptolemy places it off the coast of Limyrikê between Tyndis and Mouziris, but nearer the former.

Leukê:—This is a Greek word meaning 'white.' The island is placed in the *Periplûs* off the coast where Limyrikê begins and in Ptolemy near where it ends.

Nanigêris:—To judge from Ptolemy's

figures he has taken this to be an island lying between Cape Kumârî (Comorin) and Taprobanê (Ceylon).

Kôry:—It has already been noticed that Kôry was both the name of the Island of Râmêśvaram and of the promontory in which it terminated.

Cap. 2.

Position of India beyond the Ganges.

1. India beyond the Ganges is bounded on the west by the river Ganges; on the north by the parts of Skythia and Sêrikê already described, on the east by the Sinai along the Meridian, which extends from the furthest limits of Sêrikê to the Great Gulf, and also by this gulf itself, on the south by the Indian Ocean and part of the Green Sea which stretches from the island of Menouthias in a line parallel to the equator, as far as the regions which lie opposite to the Great Gulf.

India beyond the Ganges comprised with Ptolemy not only the great plain between that river and the Himâlayas, but also all south-eastern Asia, as far as the country of the Sinai (China). Concerning these vast regions Ptolemy is our only ancient authority. Strabo's knowledge of the east was limited in this direction by the Ganges, and the author of the *Periplûs*, who was a later and intermediate writer, though he was aware that inhabited countries stretched far beyond that limit even onwards to the eastern end of the world, appears to have learned little more about them than the mere fact of their existence. Ptolemy, on the other hand, supplies us with much information regarding them. He traces the line of coast as far as the Gulf of Siam (his Great Gulf) enumerating the tribes, the trading marts, the river mouths and the islands that would be passed on the way. He has also a copious nomenclature for the interior, which embraces its inhabitants, its towns, its rivers, and its mountain ranges. His conceptions were no doubt extremely confused and erroneous, and his data, in many instances, as inconsistent with each other as with the reality. Still, his description contains important elements of truth, and must have been based upon authentic information. At the same time an attentive study of his nomenclature and the accompanying indications has led to the satisfactory identification of a few of his towns, and a more considerable number of the rivers and mountains and tribes which he has specified.

His most notable error consisted in the supposition that the eastern parts of Asia were connected by continuous land with the east coast of Africa, so that, like Hipparkhos, he conceived the Indian Ocean to resemble the Mediterranean in being surrounded on all sides by land. He makes accordingly the coast of the Sinai, beyond the Gulf of Siam, turn toward the south instead of curving up towards the north. Again he represents the Malay Peninsula (his Golden Khersonese) which does not project so far as to reach the equator, extend to 4 degrees southward from it, and he mentions neither the Straits of Malacca nor the great island of Sumatra, unless indeed his Iabadios be this island, and not Java, as is generally supposed. By the Green Sea (Πρασώδης θάλασσα) which formed a part of the southern boundary is meant the southern part of the Indian Ocean which stretched eastward from Cape Prasum (Cape Delgado) the most southern point on the east coast of Africa known to Ptolemy. The island of Menouthias was either Zanzibar or one of the islands adjacent to it. It is mentioned by the author of the *Periplûs*.

In his description of India beyond the Ganges Ptolemy adheres to the method which he had followed in his account of India within the Ganges. He therefore begins with the coast, which he describes from the Eastern Mouth of the Ganges to the Great Promontory where India becomes conterminous with the country of the Sinai. The mountains follow, then the rivers, then the towns in the interior, and last of all the islands.

2. The seacoast of this division is thus described. In the Gangetic Gulf beyond the Mouth of the Ganges called Antibolei:—

The coast of the Airrhadoi:—.

Pentapolis	150°	18°
Mouth of River Katabêda...	151° 20′	17°
Barakoura, a mart	152° 30′	16°
Mouth of the River Tokosanna	153°	14° 30′

Wilford, probably misled by a corrupt reading, took the name of the Airrhadoi to be another form of Antibole. He says (*Asiat. Research.*, Vol. XIV, p. 444) "Ptolemy says that the easternmost branch of the Ganges was called Antibolê or Airrhadon. This last is from the Sanskrit Hradâna; and is the name of the Brahmaputra. Antibole was the name of a town situated at the confluence of several large rivers to the S. E. of Ḍhakka and now called Feringibazar." By the Airrhadoi, however, are undoubtedly meant the Kirâta. With regard to the position here assigned to them Lassen thus writes (*Ind. Alt.*, vol. III, pp. 235-237):—" By the name K i r r a d i a Ptolemy designates the land on the coast of further India from the city of Pentapolis, perhaps the present Mirkanserai in the north, as far as the mouth of the Tokosanna or Arakan river. The name of this land indicates that it was inhabited by the Kirâta, a people which we find in the great Epic settled in the neighbourhood of the Lauhitya, or Brahmaputra, consequently somewhat further to the north than where Ptolemy locates them. Hence arises the question whether the Kirâta who, as we know, belong to the Bhoṭa, and are still found in Nêpâl had spread themselves to such a distance in earlier times, or whether their name has been erroneously applied to a different people. The last assumption is favoured by the account in the *Periplûs*, according to which ships sailing northward from Dôsarênê, or the country on both sides of the Vaitaraṇi, arrived at the land of the wild flat-nosed Kirradai, who like the other savage tribes were men-eaters. Since the author of that work did not proceed beyond Cape Comorin, and applied the name of Kirâta to a people which lived on the coast to the S. W. of the Ganges, it is certain that he had erroneously used this name to denote the wild and fabulous races. Ptolemy must have followed him or other writers of the kind, and to the name Kirâta has given a signification which did not originate with himself. Although the Kirâta, long before the time in which he lived, had wandered from their northern Fatherland to the Himâlaya and thence spread themselves to the regions on the Brahmaputra, still it is not to be believed that they should have possessed themselves of territory so far south as Chaturgrâma (Chittagong) and a part of Arakan. We can therefore scarcely be mistaken if we consider the inhabitants of this territory at that time as a people belonging to further India, and in fact as tribal relatives of the Tamerai, who possessed the mountain region that lay back in the interior, as I shall hereafter show. I here remark that between the name of the city Pentapolis, *i.e. five cities*, and the name of the most northern part of Kirradia, Chaturgrâma, *i.e. four cities*, there is a connexion that can scarcely be mistaken, since Chaturgrâma could not originally have denoted a country, but only a place which later on became the capital, though it was originally only the capital of four village communities over which a common headship was possessed, while Pentapolis was the seat of a headship over five towns or rather villages, as it can scarcely be believed that the rude tribes of Kirradia were civilized enough to possess towns. A confirmation of this view is offered by the circumstance that the Bunzu, who must have been descendants of a branch of the Tamerai, live in villages under headships. We must further state that according to the treatises used by Ptolemy the best *Malabathrum* was got from Kirradia. I see no reason to doubt the correctness of this statement, although the trees from which this precious oil and spice were prepared and which are different kinds of the laurel, do not appear at the present day to be found in this country, since, according to the testimony of the most recent writers the botanical productions of Arakan at least have not as yet been sufficiently investigated. It can, however, be asserted that in Silhet, which is not very remote from Chaturgrâma, *Malabathrum* is produced at this very day." Saint-Martin expresses similar views. He writes (*Étude*, pp. 343, 344). "The Kirrhadia of Ptolemy, a country mentioned also in the *Periplûs* as lying west from the mouths of the Ganges and the Skyritai of Megasthenes are cantons of Kirâta, one of the branches of the aboriginal race the widest spread in Gangetic India, and the most anciently known. In different passages of the *Purâṇas* and of the epics their name is applied in a general manner to the barbarous tribes of the eastern frontiers of Âryavarta, and it has preserved itself in several quarters, notably in the eastern districts of Nêpâl. There is a still surviving tradition in Tripurî (Tipperah), precisely where Ptolemy places his Kirrhadia, that the first name of the country was Kirât (*J. A. S. Beng.*, Vol. XIX., Long, *Chronicles of Tripurâ*, p. 536.) The Tamerai were a tribe of the same family."

Mouth of the River Katabêda:—This may be the river of Chittagong called the Karmaphulî. The northern point of land at its mouth is, according to Wilford (*Asiat. Research.* vol. XIV, p. 445) called Paṭṭana, and hence he thinks that Chaṭgrâm or Chaturgrâm (Chittagong) is the Pentapolis of Ptolemy for Paṭṭanphulli, which means 'flourishing seat.' The same author has proposed a different identification for the Katabêda River. "In the district of Sandowê," he says, "is a river and a town called in modern maps Sedoa for Saindwa (for Sandwîpa)" and in Ptolemy S a d u s and S a d a. Between this river and Arakan there is another large one concealed behind the island of Cheduba, and the name of which is Kâtâbaidâ or Kâtâbaiza. This is the river Katâbêda of Ptolemy, which, it is true, he has placed erroneously to the north of Arakan, but as it retains its name to this day among the natives, and as it is an uncommon one in that country, we can hardly be mistaken. As that part of the country is very little frequented by seafaring people the Kâttâbaidâ is not noticed in any map or sea chart whatever. It was first brought to light by the late Mr. Burrows, an able astronomer, who visited that part of the coast by order of Government. In the language of that country *kâtû* is a fort and Byeitzâ or Baidzâ is the name of a tribe in that country." (*Asiat. Res.*, vol. XIV, pp. 452, 453).

B a r a k o u r a :—This mart is placed in Yule's map at Râmâi, called otherwise Râmu, a town lying 68 miles S.S.E. of Chittagong.

Mouth of the T o k o s a n n a :—This river Wilford and Lassen (*Ind. Alt.*, vol. III, p. 237) identified with the Arakan river. Yule prefers the Nâf, which is generally called the Teke-nâf, from the name of a tribe inhabiting its banks.

3. That of the Silver country (Argyra).

Sambra, a city..................	153° 30′	13° 45′
Sada, a city	154° 20°	11° 20′
Mouth of the River Sados...	153° 30′	12° 30′
Bêrabonna, a mart	155° 30′	10° 20′
The mouth of the River Têmala	157° 30′	10°
Têmala, a city	157° 30′	9°
The Cape beyond it	157° 20′	8°

4. That of the Bêsyngeitai Cannibals on the Sarabakic Gulf where are—

Sabara, a city	159° 30′	8° 30′
Mouth of the River Bêsynga	162° 20′	8° 25′
Bêsynga, a mart.......	162°	9°
Bêrabai, a city	162° 20′	6°
The Cape beyond it	159°	4° 40′

Arakan is no doubt the Silver Country, but the reason why it should have been so designated is not apparent, since silver has never so far as is known, been one of its products. It appears to have included part of the province of Pegu, which lies immediately to the south of it.

Sada:—This town is mentioned in that part of Ptolemy's introductory book (ch. xiii, § 7) of which a translation has been given, as the first port on the eastern side of the Gangetic Gulf at which ships from Paloura on the opposite coast touched before proceeding to the more distant ports of the Golden Khersonese and the Great Gulf. It cannot be with certainty identified. "It may perhaps have been Ezata, which appears in Pegu legend as the name of a port between Pegu and Bengal."—Yule, quoting *J. A. S. Beng.*, vol. XXVIII, p. 476.

B ê r a b o n n a :—The same authority suggests that this may be Sandowê, which Wilford proposed to identify with Sada.

T ê m a l a is the name of a town, a river, and a cape. In the introductory book (c. xiii, § 8) it is called Tamala, and said to lie to the south-east of Sada, at a distance of 3500 stadia. Yule would identify it, though doubtfully, with Gwa. Lassen again places it at Cape Negrais, which is without doubt the promontory which Ptolemy says comes after Têmala.

The S a r a b a k i c Gulf is now called the Gulf of Martaban :—The name (Bêsyngytai) of the cannibals is partly preserved in that of Bassein, which designates both a town and the river which is the western arm of the Irâwadî. Ptolemy calls this river the B ê s y n g a. The emporium of the same name Lassen takes to be Rangûn, but the similarity of name points to its identification with Bassein, an important place as a military position, from its commanding the river.

B ê r a b a i :—Beyond this Ptolemy has a promontory of the same name, which may be Barago Point. The names at least are somewhat similar and the position answers fairly to the requirements. Lassen took Bêrabai, the town, to be Martaban.

5. That of the G o l d e n K h e r s o n e s e (Χρυσῆς Χερσονήσου)

Takôla, a mart	160°	4° 15′
The Cape beyond it	158° 40′	2° 40′

Mouth of the River Khrysoanas	159°	1°
Sabana, a mart	160°	3°S.L.
Mouth of the River Palandos	161°	2°S.L.
Cape Maleou Kôlon	163°	2°S.L.
Mouth of the River Attaba	164°	1°S.L.
Kôli, a town	164° 20′	on the equator
Perimoula	163° 15′	2° 20′
Perimoulik Gulf	168° 30′	4° 15′

The Golden Khersonese denotes generally the Malay Peninsula, but more specially the Delta of the Irâwaḍî, which forms the province of Pegu, the Suvarnabhumi (Pali form,—*Sovannabhumi*) of ancient times. The Golden Region which lies beyond this, in the interior, is Burmâ, the oldest province of which, above Ava, is still, as Yule informs us, formally styled in State documents Sonaparânta, *i. e.* 'Golden Frontier.'[26]

Takôla:—Rangûn, as Yule points out, or a port in that vicinity, best suits Ptolemy's position with respect to rivers, &c.,[27] while at the same

[26] Thornton notices in his *Gazetteer of India* (s. v. *Burmah*) that when Colonel Burney was the resident in Ava, official communications were addressed to him under the authority of the "Founder of the great golden city of precious stones; the possessor of mines of gold, silver, rubies, amber and noble serpentine."

[27] Dr. Forchammer in his paper on *the First Buddhist Mission to Suvannabhûmî*, pp. 7, 16, identifies Takôla with the Burman Kola or Kula-taik and the Talaing Taîkkulâ, the ruins of which are still extant between the present Ayetthima and Kinyua, now 12 miles from the sea-shore, though it was an important seaport till the 16th century.—J. B.

time Thakalai is the legendary name of the founder of Rangûn Pagoda. There was, however, he says, down to late mediæval times, a place of note in this quarter called Takkhala, Takola, or Tagala, the exact site of which he cannot trace, though it was apparently on the Martaban side of the Sitang estuary.

Mouth of the Khrysoana River:—This must be the Eastern or Rangûn mouth of the Irâwaḍî, for, as Yule states on the authority of Dr. F. Mason, Hmâbi immediately north of Rangûn was anciently called Suvarṇanadî, *i. e.* 'Golden River,' and this is the meaning of Khrysoana.

Sabana:—This may be a somewhat distorted form of Suvarṇa, 'golden-coloured,' and the mart so called may have been situated near the mouth of the Saluen River. Yule therefore identifies it with Satung or Thatung. Lassen assigns it quite a different position, placing it in one of the small islands lying off the southern extremity of the Peninsula.

Cape Maleou Kôlon:—Regarding this Yule says, "Probably the Cape at Amherst. Mr. Crawford has noticed the singular circumstance that this name is pure Javanese, signifying "Western Malays." Whether the name Malay can be so old is a question; but I observe that in Bastian's *Siamese Extracts*, the foundation of Takkhala is ascribed to the Malays." Lassen places it much further south and on the eastern coast of the Peninsula, identifying it with Cape Romania (*Ind. Alt.*, vol. III, p. 232).

Kôli:—In the *Proceedings of the Royal Geographical Society*, vol. IV, p. 639 ff, Colonel

Yule has thrown much light on Ptolemy's description of the coast from this place to Kattigara by comparing the glimpse which it gives us of the navigation to China in the 1st or 2nd century of our era with the accounts of the same navigation as made by the Arabs seven or eight centuries later. While allowing that it would be rash to dogmatize on the details of the trans-gangetic geography, he at the same time points out that the safest guide to the true interpretation of Ptolemy's data here lies in the probability that *the nautical tradition was never lost.* He calls attention also to the fact that the names on the route to the Sinae are many of them Indian, specifying as instances Sabana, Pagrasa, R. Sôbanos, Tîpônobastê, Zaba, Tagora, Balonga, Sinda, Aganagara, Brama, Ambastas, Rabana, River Kottiaris, Kokkonagara, &c. At Kôli the Greek and Arab routes first coincide, for, to quote his words, "I take this Kôli to be the Kalah of the Arabs, which was a month's sail from Kaulam (Quilon) in Malabar, and was a place dependent on the Mahârâja of Zâbaj (Java or the Great Islands) and near which were the mountains producing tin. Ko-lo is also mentioned in the Chinese history of the T'ang dynasty in terms indicating its position somewhere in the region of Malaka. Kalah lay on the sea of Shalâhiṭ (which we call Straits of Malaka), but was not very far from the entrance to the sea of Kadranj, a sea which embraced the Gulf of Siam, therefore I presume that Kalah was pretty far down the Malay Peninsula. It may, however, have been Kadah, or Quedda as we write it, for it was 10 days' voyage from Kalah to Tiyûmah (Batûmah, Koyûmah). Now the Sea of Kadranj was entered, the Perimulic Gulf of Ptolemy."

Perimulic Gulf:—Pliny mentions an Indian promontory called Perimula where there were very productive pearl fisheries (lib. VI, c. 54), and where also was a very busy mart of commerce distant from Patala, 620 Roman miles (lib. VI, c. 20). Lassen, in utter disregard of Pliny's figures indicating its position to be somewhere near Bombay, placed it on the coast of the Island of Manâr. In a note to my translation of the *Indika* of Megasthenes I suggested that Perimula may have been in the Island of Salsette. Mr. Campbell's subsequent identification of it however with Simylla (Tiamula) where there was both a cape and a great mart of trade I think preferable, and indeed quite satisfactory. But, it may be asked, how came it to pass that a place on the west coast of India should have the same name as another on the far distant Malay coast. It has been supposed by way of explanation that in very remote times a stream of emigration from the south-eastern shores of Asia flowed onward to India and other western countries, and that the names of places familiar to the emigrants in the homes they had left were given to their new settlements. There is evidence to show that such an emigration actually took place. Yule places the Malay Perimula at Pahang. The Perimulic Gulf is the Gulf of Siam, called by the Arabs, as already stated, the Sea of Kadranj. Lassen takes it to be only an indentation of the

Peninsular coast by the waters of this Gulf, which in common with most other writers he identifies with Ptolemy's Great Gulf.

6. That of the Lêstai (Robber's country).

Samaradê........................	163°	4° 50′
Pagrasa........................	165°	4° 50′
Mouth of the River Sôbanos	165° 40′	4° 45′
(Fontes Fluvii)[23]	162° 30′	13°
Pithônobastê, a mart	166° 20′	4° 45′
Akadra..	167°	4° 45′
Zabai, the city..................	168° 40′	40° 45′

7. That of the Great Gulf.

The Great Cape where the Gulf begins	169° 30′	4° 15′
Thagora	168°	6°
Balonga, a Metropolis	167° 30′	7°
Throana	167°	8° 30′
Mouth of the River Doanas.	167°	10°
(Sources of a river)[23]	163°	27°
Kortatha, a metropolis	167°	12° 30′
Sinda, a town	167° 15′	16° 40′
Pagrasa	167° 30′	14° 30′
Mouth of the River Dôrias.	168°	15° 30′
(Sources of a river)[23].........	163°	27°
or (Tab. Geog.)	162°	20° 28′
Aganagara	169°	16° 20′
Mouth of the River Sêros ...	171° 30′	17° 20′
(Sources of a river)[23]...	170° (½ add. Tab.)	32°
(Another source)[23] ...	173° (½ add. Tab.)	30°
(The confluence)[23]	171°	27°

---

[23] Additions of the Latin Translator.

The end of the Great Gulf
towards the Sinai .... 173° 17° 20′

S a m a r a d ê :—This coincides with Samarat, the Buddhistic classical name of the place commonly called Ligor (*i. e. Nagara*, 'the city'), situated on the eastern coast of the Malay Peninsula and subject to Siam.

Mouth of the River S ô b a n o s :—Sôbanos is the Sanskrit Suvarṇa, in its Pali form Sobaṇṇa, which means 'golden.' One of the old cities of Siam, in the Meinam basin was called Sobanapuri, *i. e.* 'Gold-town.'

P i t h ô n a b a s t ê, Yule thinks, may correspond to the Bungpasoi of our maps at the mouth of the large navigable river Bangpa-Kong. It is at the head of the Gulf of Siam eastward of Bankok.

A k a d r a :—Yule would identify this with the Kadranj of the Arabs, which he places at Chantibon on the eastern coast of the gulf.

Z a b a i :—This city, according to Ptolemy, lay to the west of the Doanas, or Mekong river, and Yule therefore identifies it with the seaport called Ṣanf or Chanf by the Arab navigators. Ṣanf or Chanf under the limitations of the Arabic alphabet represents C h a m p â, by which the southern extremity of Cochin-China is designated. But Champâ lies to the south of the Mekong river, and this circumstance would seem to vitiate the identification. Yule shows, however, that in former times Champâ was a powerful state, possessed of a territory that extended far beyond its present limits. In the travels of Hiuen Tsiang (about A. D. 629) it is called Mahâchampâ. The locality of the ancient port of Zabai or Champâ is probably therefore to be sought on the west coast of Kambôja, near the Kampot, or the Kang-kao of our maps. (See *Ind. Ant.*, vol. VI, pp. 228-230).

By the Great Gulf is meant the Gulf of Siam, together with the sea that stretches beyond it towards China. The great promontory where this sea begins is that now called Cape Kamboja.

Sinda was situated on the coast near Pulo Condor, a group of islands called by the Arabs Sandar-Fulât and by Marco Polo Sondur and Condur. Yule suggests that these may be the Satyrs' Islands of Ptolemy, or that they may be his Sinda.

8. The mountains in this division are thus named :—

Bêpyrrhos, whose extremities lie in	148°	34°
and	154°	26°
and Maiandros, whose extremities lie in	152°	24°
and	160°	16°
and Damassa (or Dobassa), whose extremities lie in	162°	23°
and	166°	33°
and the western part of Sêmanthinos, whose extremities lie in	170°	33°
and	180°	26°

Bêpyrrhos :—The authorities are pretty well agreed as to the identification of this range. "Bêpyrrhos," says Lassen (*Ind. Alt.*, vol I., pp. 549-50) "answers certainly to the Himâlaya from the sources of the Sarayû to those of the Tista." "Ptolemy," says Saint-Martin (*Etude*, p. 337) "applies to a portion of the Himâlayan chain the name of Bêpyrrhos, but with a direction to the south-east which does not exist in the axis of this grand system of mountains. In general, his notions about the Eastern Himâlayas are vague and confused. It is the rivers which he indicates as flowing from each group, and not the position which he assigns to the group itself that can serve us for the purpose of identification. He makes two descend from Bêpyrrhos and run to join the Ganges. These rivers are not named, but one is certainly the Kauśikî and the other ought to be either the Gandakî or the Tîsta." Yule remarks, "Ptolemy shows no conception of the great Brahmaputra valley. His Bêpyrrhos shuts in Bengal down to Maeandrus. The latter is the spinal range of Arakan (Yuma), Bêpyrrhos, so far as it corresponds to facts, must include the Sikkim Himâlaya and the Gâro Hills. The name is perhaps Vipula—'vast,' the name of one of the mythical cosmic ranges but also a specific title of the Himâlaya."

Mount Maiandros:—From this range descend all the rivers beyond the Ganges as far as the Bêsynga or Bassein river, the western branch of the Irâwaḍî. It must therefore be the Yuma chain which forms the eastern boundary of Arakan, of which the three principal rivers are the Mayu, the Kula-dan and the Lê-myo. According to Lassen Maiandros is the graecized form of Mandara, a sacred mountain in Indian mythology.

Dobassa or Damassa range:—This range contributes one of the streams which form the great river Doanas, Bêpyrrhos which is further to the west, contributing the other confluent. A single glance at the map, Saint-Martin remarks (*Étude*, p. 338), clearly shows that the reference here is to the Brahmaputra river, whose indigenous name, the Dihong, accounts readily for the word Doanas. It would be idle, he adds, to explain where errors so abound, what made Ptolemy commit the particular error of making his Doanas run into the Great Gulf instead of joining the eastern estuary of the Ganges. The Dobassa Mountains, I therefore conclude, can only be the eastern extremity of the Himâlaya, which goes to force itself like an immense promontory into the grand elbow which the Dihong or Brahmaputra forms, when it bends to the south-east to enter Asâm. If the word Dobassa is of Sanskrit origin, like other geographical appellations applied to these eastern regions, it ought to signify the 'mountains that are obscure,'—Tâmasa Parvata. Yule (quoting *J. A. S. Beng.* vol. XXXVII, pt. ii, p. 192) points out that the Dimasas are mentioned in a modern paper on Asâm, as a race driven down into that valley by the immigration of the Bhôtiyas. This also points to the Bhôtân Himâlayas as being the Damassa range, and shows that of the two readings, Dobassa and Damassa, the latter is preferable.

Mount Sêmanthinos is placed 10 degrees further to the east than Maiandros, and was regarded as the limit of the world in that direction. Regarding these two Sanskrit designations, Saint-Martin, after remarking that they are more mythic than real, proceeds to observe: "These Oriental countries formed one of the horizons of the Hindu world, one of the extreme regions, where positive notions transform themselves gradually into the creations of mere fancy. This disposition was common to all the peoples of old. It is found among the nations of the east no less than in the country of Homer. Udayagiri,—the mountain of the east where the sun rises, was also placed by the Brahmanik poets very far beyond the mouths of the Ganges. The Sêmanthinôs is a mountain of the same family. It is the extreme limit of the world, it is its very girdle (*Samanta* in Sanskrit). In fine, Purânik legends without number are connected with Mandara, a great mountain of the East. The fabulous character of some of these designations possesses this interest with respect to our subject, that they indicate even better than notions of a more positive kind the primary source of the information which Ptolemy employed. The Maiandros, however, it must be observed, has a definite locality assigned it, and designates in Ptolemy the chain of heights which cover Arakan on the east."

9. From Bêpyrrhos two rivers discharge into the Ganges, of which the more northern has its sources in .... 148° 33°
and its point of junction with
the Ganges in .... 140° 15′ 30° 20′
The sources of the other
river are in .... 142° 27°
and its point of junction with
the Ganges in .... 144° 26°

10. From Maiandros descend the rivers beyond the Ganges as far as the Bêsynga River,

but the river Sêros flows from the range of Sêmanthinos from two sources, of which the most western lies in .... 170° 30′ 32°
and the most eastern in...... 173° 30′ 30°
and their confluence is in ... 171° 27°

11. From the Damassa range flow the Daonas and Dôrias (the Doanas runs as far as to Bêpyrrhos)
and the Dôrias rises in ...... 164° 30′ 28°

Of the two streams which unite to form the Doanas that from the Damassa range rises in .... 162° 27° 30′
that from Bêpyrrhos rises in 153° 27° 30′
The two streams unite in ... 160° 20′ 19°

The river Sôbanas which flows from Maiandros rises in .... 163° 30′ 13°

12. The rivers which having previously united flow through the Golden Khersonese from the mountain ridges, without name, which overhang the Khersonese—the one flowing into the Khersonese first detaches from it the Attabas in about .... 161° 2° 20′
and then the Khrysoanas in about 161° 1° 20′
and the other river is the Palandas.

Nearly all the rivers in the foregoing table have already been noticed, and we need here do little more than remind the reader how they have been identified. The two which flow from Bêpyrrhos into the Ganges are the Kauśikî and the Tîsta. The Bêsynga is the Bassein River or Western branch of the Irâwaḍî. The Sêros enters the sea further eastward than any of the other rivers, probably in Champâ, the Zaba of Ptolemy, while Lassen identifies it with the Mekong. The D a o n a s is no doubt the Brahmaputra, though Ptolemy, taking the estuary of the Mekong or Kamboja river to be its mouth, represents it as falling into the Great Gulf. It was very probably also, to judge from the close resemblance of the names when the first two letters are transposed, the Oidanes of Artemidôros, who, according to Strabo (lib. XV, c. i, 72), describes it as a river that bred crocodiles and dolphins, and that flowed into the Ganges. Curtius (lib. VIII, c. 9) mentions a river called the Dyardanes that bred the same creatures, and that was not so often heard of as the Ganges, because of its flowing through the remotest parts of India. This must have been the same river as the Oidanes or Doanas, and therefore the Brahmaputra. The D ô r i a s is a river that entered the Chinese Sea between the Mekong Estuary and the Sêros. The S o b a n a s is perhaps the river Meinam on which Bangkok, the Siamese capital, stands. The A t t a b a s is very probably the Tavoy river which, though its course is comparatively very short, is more than a mile wide at its mouth, and would therefore be reckoned a stream of importance. The similarity of the names favours this identification. The K h r y s o a n a is the eastern or Rangûn arm of the Irâwaḍî. The P a l a n d a s is probably the Salyuen River.

Ptolemy now proceeds to describe the interior of Transgangetic India, and begins with the tribes or nations that were located along the banks of the Ganges on its eastern side.

13. The regions of this Division lying along the course of the Ganges on its eastern side and furthest to the north are inhabited by the G a n g a n o i, through whose dominions flows the river Sarabos, and who have the following towns :—

Sapolos..........................	139° 20′	35°
Storna	138° 40′	34° 40′
Heorta	138° 30′	34°
Rhappha	137° 40′	33° 40′

For G a n g a n o i should undoubtedly be read T a n g a n o i, as Taṅgaṇa was the name given in the heroic ages to one of the great races who occupied the regions along the eastern banks of the upper Ganges. Their territory probably stretched from the Râmgaṅgâ river to the upper Sarayû, which is the Sarabos of Ptolemy. Their situation cannot be more precisely defined, as none of their towns named in the table can with certainty be recognized. " Concerning the people themselves," says Saint-Martin (*Étude*, pp. 327, 328) " we are better informed. They are represented in the *Mahâbhârata* as placed between the Kirâta and the Kulinda in the highlands which protected the plains of Kôsala on the north. They were one of the barbarous tribes, which the Brahmanic Âryans, in pushing their conquests to the east of the Ganges and Jamnâ, drove back into the Himâlayas or towards the Vindhyas. It is principally in the Vindhya regions that the descendants of the Taṅgaṇa of classic times are now to be found. One of the Râjput tribes, well-known in the present day under the name of Taṅk or Toṅk is settled in Rohilkhand, the very district where the *Mahâbhârata* locates the Taṅgaṇa and Ptolemy his Tanganoi. These Taṅk Râjputs extend westward to a part of the Doâb, and even as far as Gujarât, but it is in the race of the Daṅgayas, spread over the entire length of the Vindhya Mountains and the adjacent territory from the southern borders of the ancient Magadha to the heart of Mâlwa to the north of the lower Narmadâ, it is in this numerous race, subdivided into clans without number, and which is called according to the districts inhabited Dhaṅgîs, Dhâṅgars, Doṅga, &c. that we must search for the point of departure of the family and its primordial type. This type, which the mixture of Âryan blood has modified and ennobled in the tribes called Râjput, preserves its aboriginal type in the mass of mountain tribes, and this type is purely Mongolian, a living commentary on the appellation of Mlechha, or Barbarian, which the ancient Brahmanic books apply to the Taṅgaṇa." (Conf. *Bṛih. Saṁh.* IX, 17; X, 12; XIV, 12, 29; XVI, 6; XVII, 25; XXXI, 15 *Râmâyaṇa* IV, 44, 20).

The towns, we have said, cannot be identified with certainty, but we may quote Wilford's views as to what places now represent them. He says (*Asiat. Research.* vol. XIV, p. 457): "The Bân or Śaraban river was formerly the bed of the Ganges and the present bed to the eastward was also once the Bân or Śaraban river. This Ptolemy mistook for the Râmagaṅgâ, called also the Bân, Śaraban and Śarâvatî river, for the four towns which he places on its banks, are either on the old or the new bed of the Ganges. S t o r n a and S a p o l o s are Hastnaura, or Hastina-nagara on the old bed, and Sabal, now in ruins, on the eastern bank of the new bed, and is commonly called Sabalgaṛh. Hastinâpur is 24 miles S. W. of Dârânagar, and 11 to the west of the present Ganges; and it is called Hastnawer in the *Ayin Akbari*. Heorta is Awartta or Hardwâr. It is called Arate in the *Peutinger* tables, and by the Anonymous of Ravenna."

14. To the south of these are the Maroundai who reach the Gangarîdai, and have the following towns on the east of the Ganges:—

Boraita	142° 20′	29°
Kôrygaza	143° 30′	27° 15′
Kondôta	145°	26°
Kelydna	146°	25° 30′
Aganagora	146° 30′	22° 30′
Talárga	146° 40′	21° 40′

The Maroundai occupied an extensive territory, which comprised Tirhut and the country southward on the east of the Ganges, as far as the head of its delta, where they bordered with the Gangarîdai. Their name is preserved to this day in that of the Mûndas, a race which originally belonged to the Hill-men of the North, and is now under various tribal designations diffused through Western Bengal and Central India, "the nucleus of the nation being the Ho or Hor tribe of Singhbhûm.[29] They are probably the Monedes of

[29] *J. A. S. B.*, vol. XXXV, p. 168. The Mûnda tribes as enumerated by Dalton, *id.* p. 158, are the Kuars of Ilichpur, the Korewas of Sirguja and Jaspur, the Kherias of Chutia Nâgpur, the Hor of Singhbhum, the Bhumij of

whom Pliny speaks, in conjunction with the Suari. That they were connected originally with the Muraṇḍa, a people of Lampâka (Lamghân) at the foot of the Hindu-Kôh mentioned in the inscription on the Allâhâbâd pillar, along with the Śaka, as one of the nations that brought tributary gifts to the sovereign of India, is sufficiently probable[30]; but the theory that these Muraṇḍa on being expelled from the valleys of the Kôphês by the invasion of the Yetha, had crossed the Indus and advanced southwards into India till they established themselves on the Ganges, in the kingdom mentioned by Ptolemy, is, as Saint-Martin has clearly proved (*Étude*, pp. 329,330) utterly untenable, since the sovereign to whom the Muraṇḍa of the north sent their gifts was Samudragupta, who reigned subsequently to the time of Ptolemy, and they could not therefore have left their ancestral seats before he wrote. Saint-Martin further observes that not only in the case before us but in a host of analogous instances, it is certain that tribes of like name with tribes in India are met with throughout the whole extent of the region north of the Indus, from the eastern extremity of the Himâlaya as far as the Indus and the Hindu-Kôh, but this he points out is attributable to causes more general than the partial migration of certain tribes. The *Vayu Purâṇa* mentions the Muraṇḍa among the Mlechha tribes which gave kings to

---

Mânbhûm Dhalbhûm, and the Sântals of Mânbhûm Singhbhum, Katak, Hâzâribâgh and the Bhâgalrpur hills. The western branches are the Bhills of Mâlwa and Kânhdêś and the Kôlis of Gujarât.

[30] *Mahâbh.* vii, 4847; Reinaud, *Mém. sur l'Inde*, p. 353 Lassen, *Ind. Alt.*, vol. II, p. 877.—Ed.

India during the period of subversion which followed the extinction of the two great Aryan dynasties. See Cunningham, *Anc. Geog. of Ind.*, pp. 505-509, also Lassen, *Ind. Alt.*, vol. III, pp. 136f. 155—157, and vol. II, p. 877n.

Regarding the towns of the Maroundai, we may quote the following general observations of Saint-Martin (*Etude*, pp. 331, 332). "The list of towns attributed to the Maroundai would, it might be expected, enable us to determine precisely what extent of country acknowledged in Ptolemy's time the authority of the Muraṇḍa dynasty, but the corruption of many of the names in the Greek text, the inexactitude or insufficiency of the indications and, in fine, the disappearance or change of name of old localities, render recognition often doubtful, and at times impossible." He then goes on to say: "The figures indicating the position of these towns form a series almost without any deviation of importance, and betoken therefore that we have an itinerary route which cuts obliquely all the lower half of the Gangetic region. From Boraita to Kelydna this line follows with sufficient regularity an inclination to S. E. to the extent of about 6 degrees of a great circle. On leaving Kelydna it turns sharply to the south and continues in this direction to Talarga, the last place on the list, over a distance a little under four degrees. This sudden change of direction is striking, and when we consider that the Ganges near Râjmahal alters its course just as sharply, we have here a coincidence which suggests the enquiry whether near the point where the Ganges so suddenly bends, there is a place having a name something like

Kelydna, which it may be safely assumed is a bad transcription into Greek of the Sanskrit Kâlinadî ('black river') of which the vulgar form is Kâlindî. Well then, Kâlindî is found to be a name applied to an arm of the Ganges which communicates with the Mahânandâ, and which surrounds on the north the large island formed by the Mahânandâ and Ganges, where once stood the famous city of Gâuḍa or Gaur, now in ruins. Gauḍa was not in existence in Ptolemy's time, but there may have been there a station with which if not with the river itself the indication of the table would agree. At all events, considering the double accordance of the name and the position, it seems to me there is little room to doubt that we have there the locality of Kelydna. The existing town of Mâldâ, built quite near the site of Gauṛ, stands at the very confluence of the Kâlindî and Mahânandâ. This place appears to have preserved the name of the ancient Malada of the Purânik lists, very probably the Molindai of Megasthenês. This point being settled, we are able to refer thereto the towns in the list, both those which precede and those which follow after. We shall commence with the last, the determination of which rests on data that are less vague. These are Aganagara and Talarga. The table, as we have seen, places them on a line which descends towards the sea exactly to the south of Kelydna. If, as seems quite likely, these indications have been furnished to Ptolemy by the designating of a route of commerce towards the interior, it is natural to think that this route parted from the great emporium of the Ganges (the Gangê Regia of Ptolemy, the

Ganges emporium of the Periplûs) which should be found, as we have already said, near where Hûghli now stands. From Kelydna to this point the route descends in fact exactly to the south, following the branch of the Ganges which forms the western side of the delta. The position of Aghadîp Agadvîpa) on the eastern bank of the river a little below Katwâ, can represent quite suitably Aganagora (Aganagara); while Talarga may be taken to be a place some leagues distant from Calcutta, in the neighbourhood of Hûghli. . . . . The towns which precede Kelydna are far from having the same degree of probability. We have nothing more here to serve for our guidance than the distances taken from the geographical notations, and we know how uncertain this indication is when it has no check to control it. The first position above Kelydna is Kondota or Tondota; the distance represented by an arc of two degrees of a great circle would conduct us to the lower Bagamatî (Bhagavatî). Korygaza or Sorygaza (distant ½ degree) would come to be placed perhaps on the Gandakî, perhaps between the Gandakî and the lower Sarayû; last of all Boraita, at two degrees from Korygaza, would conduct us to the very heart of ancient Kôsala, towards the position of the existing town of Bardâ. We need scarcely add, in spite of the connexion of the last two names, that we attach but a faint value to determinations which rest on data so vague." Boraita may be, however, Bharêch in Auḍh, as Yule has suggested, and with regard to Korygaza, it may be observed that the last part of the name may represent the Sanskrit *kachha*, which means *a marsh* or *place near a marsh*, and hence Korygaza may be Gorakhpur, the situation of which is notably marshy.

15. Between the Imaös and Bêpyrrhos ranges the T a k o r a i o i are farthest north, and below them are the K o r a n g k a l o i, then the P a s s a l a i, after whom to the north of Maiandros are the T i l a d a i, such being the name applied to the B ô s e i d a i, for they are short of stature and broad and shaggy and broad-faced, but of a fair complexion.

T a k o r a i o i :—This tribe occupied the valleys at the foot of the mountains above Eastern Kôsalâ and adjoined the Tanganoi. The Tanganas are mentioned among the tribes of the north in the lists of the *Brihat Samhitâ* (IX, 17; X, 12; XIV, 29). They have left numerous descendants in different parts of Gangetic India. A particular clan in Rohilkhand not far from the seats of the Takoraioi preserves still the name under the form Dakhaura (Elliot's *Supplementary Glossary of Indian terms*, p. 360), and other branches are met with near the Jamnâ and in Râjputâna. Towards the east again the Dekra form a considerable part of the population of Western Asâm (*J. A. S. Beng.*, vol. XVIII, p. 712).

K o r a n g k a l o i ;—These are probably of the same stock, if not actually the same people, as Korankâra of the *Purâṇas* (*Asiat. Research.*, vol. VIII), and the Kyankdanis of Shêkavati. Their position is near the sources of the Gaṇḍak.

P a s s a l a i :—The Passalai here mentioned are not to be confounded with the Passalai of the Doâb.

In the name is easily to be recognized the Vaiśâli of Hiuen Tsiang, which was a small kingdom stretching northward from the Ganges along the banks of the river Gaṇḍak. The capital had the same name as the kingdom, and was situated in the immediate neighbourhood of Hâjipur, a station near the junction of the Gaṇḍak and Ganges, where a great fair is annually held, distant from Pâtna about 20 miles. "Here we find the village of Besârh, with an old ruined fort, which is still called Raja Bisal-ka-garh, or the fort of Raja Visala, who was the reputed founder of the ancient Vaiśâli." (Cunningham, *Anc. Geog. of Ind.*, p. 443).

Tiladai :—We here leave the regions adjoining the Ganges, and enter the valleys of the Brahmaputra. The Tiladai are called also Bêsadai or Basadai. Ptolemy places them above the Maiandros, and from this as well as his other indications, we must take them to be the hill-people in the vicinity of Silhet, where, as Yule remarks, the plains break into an infinity of hillocks, which are specially known as *tîla*. It is possible, he thinks, that the Tiladai occupied these *tîlas*, and also that the Tiladri hills (mentioned in the *Kshetra Samâsa*) were the same Tîlas. The same people is mentioned in the *Periplûs*, but under the corrupt form of Sêsatai. The picture drawn of them by the author of that work corresponds so closely with Ptolemy's, that both authors may be supposed to have drawn their information from the same source. We may quote (in the original) what each says of them :—

Periplûs : ἔθνος τι, τῷ μεν σώματι κολοβοὶ καὶ

σφόδρα πλατυπρόσωποι, ἐννοίαις δὲ λῷστοι αὐτοὺς [δὲ] λέγεσθαί [φασι] Σησάτας, παρομοίους ἀνημέροις.

Ptolemy : εἰσὶ γὰρ κολοβοὶ, καὶ πλατεῖς, καὶ δασεῖς, και πλατυπρόσωποι, λευκοὶ μέντοι τὰς χρόας.

Description of the regions which extend from the Brahmaputra to the Great Gulf.

16. Beyond Kirrhadia, in which they say the best *Malabathrum* is produced, the Zamîrai, a race of cannibals, are located near Mount Maiandros.

17. Beyond the Silver Country, in which there are said to be very many silver mines, (μέταλλα ἀσήμου), is situated in juxtaposition to the Bêsyngeitai, the Gold Country (Χρυσῆ χώρα), in which are very many gold mines, and whose inhabitants resemble the Zamîrai, in being fair-complexioned, shaggy, of squat figure, and flat-nosed.

Kirrhadia :—This has been already noticed. With reference to its product *Malabathrum*, which is not betel, but consists of the leaves of one or more kinds of the cinnamon or cassia-tree. I may quote the following passage from the *J. A. S. Beng.*, vol. XVI, pp. 38-9 :—" *Cinnamomum albiflorum* is designated *taj*, *tejpat* in Hindustani, the former name being generally applied to the leaf and the latter to the bark of the tree ; *taj*, *tejpata*, or *tejapatra*, by all which names this leaf is known, is used as a condiment in all parts of India. It is indigenous in Silhet, Asâm, Rungpur (the Kirrhadia of Ptolemy), and in the valleys of the mountain-range as far as Masuri. The dry branches and leaves are brought annually in large quantities from the former place, and sold at a fair, which is held at Vikramapura. *Taj*, however, is a name that is also given in the eastern part of Bengal to the bark of a variety of *Cinnamomum Zeylanicum* or *Cassia lignea*, which abounds in the valleys of Kachâr, Jyntiya and Asâm." The word *Malabathrum* is a compound of *tamala* (the Sanskrit name of *Cinnamomum albiflorum*) and *pâtra*, 'a leaf.' Another derivation has been suggested *mâlâ*, 'a garland,' and *pâtra* 'a leaf.' (Lassen, *Ind. Alt.*, vol. I, p. 283 seq., and conf. Dymock's *Veget. Mat. Med*, p. 553).

The following interesting passage describes the mode in which the Bêsadai trade in this article with the Chinese. I translate from the *Periplûs*, cap. 65:—"On the confines of Thina is held an annual fair attended by a race of men called the Sêsatai, who are of a squat figure, broad-faced, and in appearance like wild beasts, though all the same they are quite mild and gentle in their disposition. They resort to this fair with their wives and children, taking great loads of produce packed in mats like the young leaves of the vine. The fair is held where their country borders on that of the Thinai. Here, spreading out the mats they use them for lying on, and devote several days to festivity. This being over, they withdraw into their own country and the Thinai, when they see they have gone, come forward and collecting the mats, which had been purposely left behind, extract first from the Calami (called Petroi), of which they were woven, the sinews and fibres, and then taking the leaves fold them double and roll them up into balls through which they pass the fibres of the

Calami. The balls are of three kinds, and are designated according to the size of the leaf from which they are made, *hadro, meso* and *mikrosphairon*. Hence there are three kinds of *Malabathrum*, and these are then carried into India by the manufacturers.

Z a m î r a i :—A various reading is Zamerai. It has been already stated that this was a tribe of the same family as the Kirâta, beside whom they are named in the great geographical catalogue of the *Mahâbhârata*. Ramifications of the Zamîrai still exist under the names of Zamarias, Tomara, &c., in the midst of the savage districts which extend to the S. and S.E. of Magadha, and to the west of the Sôn.

The silver country, it has already been noticed, is Arakan, and the gold country and copper country, Yule remarks, correspond curiously even in approximate position with the Sonaparânta (golden frontier land), and Zampadîpa of Burmese state-documents. The Malay peninsula, taken generally, has still many mines both of the precious and the useful metals.

18. And, again, between the ranges of B ê p y r r h o s and D a m a s s a, the country furthest north is inhabited by the A n i n a k h a i (or Aminakhai), south of these the I n d a-p r a t h a i, after these the I b ê r i n g a i, then the D a b a s a i (or Damassai ?), and up to Maiandros the N a n g a l o g a i, which means "the World of the Naked" (γυμνῶν κόσμος).

19. Between the D a m a s s a range and the frontiers of the S i n a i are located furthest

north the Kakobai; and below them the Basanârai.

20. Next comes the country of Khalkîtis, in which are very many copper mines. South of this, extending to the Great Gulf the Koudoutai, and the Barrhai, and, after them the Indoi, then the Doânai, along the river of the same name.

21. To these succeeds a mountainous country adjoining the country of Robbers (Λῃστῶν) wherein are found elephants and tigers. The inhabitants of the Robber country are reported to be savages (θηριώδεις), dwelling in caves, and that have skins like the hide of the hippopotamus, which darts cannot pierce through.

Aninakhai:—The position Ptolemy assigns to them is the mountain region to the north of the Brahmaputra, corresponding to a portion of Lower Asâm.

Indaprathai:—This is a purely Hindu name. In Sanskrit documents and in inscriptions mention is made of several towns in the provinces of the Ganges, which had taken the name of the old and famous Indraprastha (the modern Dehli), and we may conclude that the Indaprâthai of the East were a Brahmanic settlement. In subsequent times Sanskrit designations spread further down into the Dekhan with the cultus, either of the Brahmans or the Buddhists. Instances in point are Modura and Kosamba, which have been already noticed. The

Indaprâthai appear to have established themselves in the districts S. of the Brahmaputra, and of the Aninakhai.

Ibêringai and Dabasai or Damassai:—The Damassai (now the Dimasas as already noticed), occupied the region extending from their homonymous mountains to the Brahmaputra, but further to the east than the Aninakhai and Ibêringai.

Nangalogai:—Many tribes still existing on the hills, east and north-east of Silhet, are called Nâgas. This name, which is given correctly in Ptolemy as Nanga, is the Indian word for *naked*, and according to Yule it is written *Nanga* in the Musalman History of Asâm. The absolute nakedness of both sexes, he says, continues in these parts to the present day. The latter half of the name *lóg* (Sanskrit *lôk*), is the Indian term for *people*, *mankind*, or *the world*, as Ptolemy has it.

With regard to the other tribes enumerated, Saint Martin remarks (*Étude*, pp. 345-6):—"The Ibêringai are still a tribe of the north just as the Dabassaê, perhaps on the mountains of the same name. There is still a tribe of Dhobas in Dinajpur, one of the districts of the north-east of Bengal, on the confines of the ancient Kâmarûpa. To the east of the Dobassa mountains, towards the frontiers of the Sinai, the tribe of the Kakobai is found to a surety in that of the Khokus, who occupy the same districts. The Basannarae, in a locality more southern, are very probably the Bhanzas, a tribe of the mountains to the south of Tippera, east of the mouth of the Brahmaputra. In the Koudoutai and the Barrhai, it is easy to recognize, though Ptolemy carries them too far into the south, the Kolitas and the Bhars or Bhors, two of the most notable parts of the population of Western Asâm, and of the districts of Bengal that belong to Kâmarûpa. The Doânai or Daonai are perpetuated in the Zaôn of Eastern Asâm; and the name of the Lôstae, the last of the list, corresponds to all appearance to that of the Lepchhas, a well-known mountain race on the confines of Sikkîm to the west of the Tistâ." For notices of the tribes which he has thus identified with those of Ptolemy, he refers to the *Journal of the Asiatic Society of Bengal*, vols. VI, IX, XIV, and XVIII. His identification of the Lêstai with the Lepchhas is in every way unfortunate. That the name Λησταὶ is not a transcript of any indigenous name, but the Greek name for *robbers* or *pirates*, is apparent from the fact alone that the η has the iôta subscribed. The Lepchhas, moreover, live among mountains, far in the interior, while Ptolemy locates his Lôstai along the shores of the Gulf of Siam.

Ptolemy gives next a list of 33 towns in the interior by way of supplement to those already mentioned as situated along the course of the Ganges, followed by a list of the towns in the Golden Khersonese :—

22. The inland towns and villages of this division (Transgangetic India), in addition to those mentioned along the Ganges are called :—

Sêlampoura	148° 30′	33° 20′
Kanogiza	143°	32°

Kassida	146°	31° 10′
Eldana	152°	31°
Asanabara	155°	31° 30′
Arkhinara	163°	31°
Ourathênai	170°	31° 20′
Souanagoura	145° 30′	29° 30′
Sagôda or Sadôga	155° 20′	29° 20′
Anina	162°	29°
Salatha	165° 40′	28° 20′
23. Rhadamarkotta, in which is much *nard*...	172°	28′
Athênagouron	146° 20′	27°
Maniaina (or Maniataia)	147° 15′	24° 40′
Tôsalei, a metropolis ...	150°	23° 20′
Alosanga	152°	24° 15′
Adeisaga	159° 30′	23°
Kimara	170°	23° 15′
Parisara	179°	21° 30′
Tougma, a metropolis...	152° 30′	22° 15′
Arisabion	158° 30′	22° 30′
Posinara	162° 15′	22° 50′
Pandasa	165°	21° 20′
Sipibêris (or Sittêbêris).	170°	23° 15′
Triglypton, called also Trilingon, capital of the kingdom	154°	18°

In this part the cocks are said to be bearded, and the crows and parrots white.

24. Lariagara	162° 30′	18° 15′
Rhingibêri	166°	18°
Agimoitha	170° 40′	18° 40′
Tomara	172°	18°

Dasana or Doana	165°	15° 20′
Mareoura, a metropolis, called also Malthoura	158°	12° 30′
Lasippa (or Lasyppa)...	161°	12° 30′
Barenkora (or Barenathra	164° 30′	12° 50′

25. In the Golden Khersonese—

Balongka	162°	4° 40′
Kokkonagara	160°	2°
Tharrha	162°	1° 20′ S.
Palanda	161°	1° 20′ S.

Regarding the foregoing long list of inland towns, the following general observations by Saint-Martin are instructive: "With Ptolemy, unfortunately," he says (*Étude*, pp. 348-9) "the correspondence of names of towns in many instances, is less easy to discover than in the case of the names of peoples or tribes. This is shown once again in the long-enough list which he adds to the names of places already mentioned under the names of the people to which they respectively belonged. To judge from the repetitions in it and the want of connexion, this list appears to have been supplied to him by a document different from the documents he had previously used, and it is precisely because he has not known how to combine its contents with the previous details that he has thus given it separately and as an appendix, although thereby obliged to go again over the same ground he had already traversed. For a country where Ptolemy had not the knowledge of it as a whole to guide him, it would be unjust to reproach him with this want of connexion in his materials, and the confusion therefrom resulting; but this absence, almost absolute, of connexion does only render the task of the critic all the more laborious and unwelcome and there results from it strange mistakes for those who without sufficiently taking into account the composition of this part of the Tables, have believed they could find in the relative positions which the places have there taken a sufficient means of identification. It would only throw one into the risk of error to seek for correspondences to these obscure names, (of which there is nothing to guarantee the correctness, and where there is not a single name that is assigned to a definite territory,) in the resemblances, more or less close, which could be furnished by a topographical dictionary of India."

Sêlampoura:—This suggests Sêlempur, a place situated at some distance north of the Dêva or lower Sarayû. The identity of the names is our only warrant for taking them as applying to one and the same town; but as the two places which follow belong to the same part of the country, the identification is in some measure supported. Sêlempur is situated on a tributary of the Sarayû, the little Gandak.

Kanogiza:—This is beyond doubt the famous city of Kanyakubja or Kanauj, which has already been noticed under the list of towns attributed to Prasiakê, where the name is given as Kanagora. Ptolemy, while giving here the name more correctly has put the city hopelessly out of its position with reference to the Ganges, from which he has removed it several degrees, though it stood upon its banks. Among Indian cities it ranks next in point of antiquity to Ayôdhyâ in Audh, and it was for many centuries the Capital of North-Western India. It was then a stately city, full of incredible wealth, and its king, who was sometimes styled the Emperor of India, kept a very splendid court. Its remains are 65 miles W.N.W. from Lakhnau. The place was visited by Hiuen Tsiang in 634 A.D. Pliny (*H. N.* lib. VI, c. 21) has Calinipaxa. Conf. Lassen, *Ind. Alt.* vol. I, p. 158; *Mahâbh.* III, 8313; *Râmâyaṇa*, I, 34, 37.

Kassida:—Here we have another case of a recurrence of the same name in an altered form. In Sanskrit and in inscriptions Kâśî is the ordinary name of Bânâras. How Ptolemy came to lengthen the name by affixing *da* to it has not been explained. Ptolemy has mutilated Vâranâsî into Erarasa, which he calls a metropolis, and assigns to the Kaspeiraioi. Such is the view taken by Saint-Martin, but Yule, as we have seen, identifies Erarasa with Govardhan (Girirâja). He also points out, on the authority of Dr. F. Hall that Vâranâsî was never used as a name for Bânâras.

Souanagoura:—M. Saint-Martin (*Étude*, p. 351) thinks this is a transcript of the vulgar form of Suvarṇanagara, and in this name recognizes that of one of the ancient capitals of Eastern Bengal, Suvarṇagrâma (now Sôṇargâon, about 12 miles from Ḍhakka), near the right bank of the Lower Brahmaputra.

Sagôda:—There can be no doubt of the identity of this place with Ayôdhyâ, the capital of Kôśala, under the name of Sâkêta or Sagêda. Sâkyamuni spent the last days of his life in this city, and during his sojourn the ancient name of Ayôdhyâ gave place to that of Sâkêta, the only one current. Hindu lexicographers give Sâkêta and Kôsala (or Kôśala) as synonyms of Ayôdhyâ. The place is now called Audh, and is on the right bank of the Sarayû or Ghâghrâ, near Faizâbâd, a modern town, built from its ruins. At some distance north from Audh is the site of Śrâvastî, one of the most celebrated cities in the annals of Buddhism. For the identity of Sâkêta with Ayôdhyâ and also Viśakha see Cunningham, *Geog. of Anc Ind.*, pp. 401 sqq.

Rhadamarkotta (v. l. Rhandamarkotta). Saint-Martin has identified this with Rangâmatî, an ancient capital situated on the western bank of the lower Brahmaputra, and now called Udêpur (Udayapura,—*city of sunrise*). Yule, who agrees with this identification, gives as the Sanskrit form of the name of the place, Rangamṛitika. The passage about *Nard* which follows the mention of Rhadamarkotta in the majority of editions is, according to Saint-Martin (*Étude*, p. 352 and note), manifestly corrupt. Some editors, correct πολλή, *much*, into πόλεις, *cities*, and thus Nardos becomes the name of a town, and Rhadamarkotta the name of a district, to which Nardos and the towns that come after it in the Table belong. On this point we may quote a passage from Wilford, whose views regarding Rhadamarkotta were different. He says (*Asiat. Research.* vol. XIV, p. 441), Ptolemy has delineated tolerably well the two branches of the river of Ârâ and the relative situation of two towns upon them, which still retain their ancient name, only they are transposed. These two towns are Urathêna, and Nardos or Nardon; Urathena is Rhâdana, the ancient name of Amarapur, and Nardon is Narteuh on the Kayn-dween. . . ." He says that "Narteuh was situated in the country of Rhandamarkoṭa, literally, the Fort of Randamar, after which the whole country was designated."

Tôsalei, called a Metropolis, has become of great importance since recent archæological discoveries have led to the finding of the name in the Aśôka Inscriptions on the Dhauli rock. The inscription begins thus: " By the orders of Dêvanampiya (beloved of the gods) it is enjoined to the public officers charged with the administration of the city of Tôsali," &c. Vestiges of a larger city have been discovered not far from the site of this monument, and there can be no doubt that the Tôsali of the inscription was the capital in Aśôka's time of the province of Orissa, and continued to be so till at least the time of Ptolemy. The city was situated on the margin of a pool called Kôsalâ-Gangâ, which was an object of great religious veneration throughout all the country. It is pretty certain that relative to this circumstance is the name of Tosala-Kôsâlakas, which is found in the *Brahmânda Purâṇa*, which Wilford had already connected with the Tôsalê of Ptolemy. He had however been misled by the 2nd part of the word to locate the city in N. Kôsalâ, that is Audh. An obvious objection to the locating of Tôsalê in Orissa is that Ptolemy assigns its position to the eastern side of the Ganges, and Lassen and Burnouf have thus been led to conclude that there must have been two cities of the name. Lassen accordingly finds for Ptolemy's Tôsalê a place somewhere in the Province of Ḍhâkkâ. But there is no necessity for this. If we take into account that the name of Tôsalê is among those that are marked as having been added to our actual Greek texts by the old Latin translators (on what authority we know not) we shall be the less surprised to find it out of its real place. (Saint-Martin, *Étude*, pp. 353-4, citing *J. A. S. Beng.*, vol. VII, pp. 435 and 442; Lassen, *Ind. Alt.*, vol. II, p. 256, and vol. III, p. 158; and *Asiat. Research*. vol. VIII, p. 344).

Alosanga:—The geographical position of Alosanga places it a quarter degree to the north of the upper extremity of Mount Maiandros. "By a strange fatality," says Wilford (*Asiat. Res. ut s.*, p. 390) "the northern extremity of Mount Maindros in Ptolemy's maps is brought close to the town of Alosanga, now Ellasing on the Lojung river, to the north-west of Ḍhakka This mistake is entirely owing to his tables of longitude and latitude."

Tougma:—In Yule's map this is identified, but doubtfully, with Tagaung, a place in Khrysê (Burma) east from the Irâwaḍî and near the tropics.

Triglypton or Trilingon:—Opinions vary much as to where this capital was situated. Wilford says (*Asiat. Research*. vol. XIV, p. 450-2): "Ptolemy places on the Tokosanna, the Metropolis of the country, and calls it Trilingon, a true Sanskrit appellation. Another name for it, says our author, was Triglypton, which is an attempt to render into Greek the meaning of Trilinga or

Trai-liṅga, the three 'lingas' of Mahâdêva; and this in Arakan is part of an extensive district in the *Purâṇas*, called Tri-pura, or the three towns and townships first inhabited by three Daityas. These three districts were Kamilâ, Chattala and Burmânaka, or Raśâng, to be pronounced Ra-shânh, or nearly so; it is now Arâkan. Kamilla alone retains the name of Tripura, the two other districts having been wrested from the head Râja. Ptolemy says that in the country of the Trilinga, there were white ravens, white parrots, and bearded cocks. The white parrot is the *kâkâtwâ*; white ravens are to be seen occasionally in India . . . Some say that this white colour might have been artificial . . . . The bearded cocks have, as it were, a collar of reversed feathers round the neck and throat, and there only, which gives it the appearance of a beard. These are found only in the houses of native princes, from whom I procured three or four; and am told that they came originally from the hills in the N. W. of India." Lassen has adopted a somewhat similar view. He says (*Ind. Alt.*, vol. III, p. 238-9): "Triglyphon was probably the capital of the Silver country, Arâkan of the present day. It lies, according to Ptolemy's determination, one degree further east and 3½ degrees further north than the mouths of the Arâkan river. The mouths are placed in the right direction, only the numbers are too great. It may be added that the foundation of this city, which was originally called Vaiśâlî, belongs to earlier times than those of Ptolemy, and no other capital is known to us in this country. The Greek name which means 'thrice cloven,' *i.e.*, 'three-forked' or 'a trident' suits likewise with Arakan, because it lies at the projections of the delta, and the Arâkan river, in the lower part of its course, splits into several arms, three of which are of superior importance. Ptolemy's remark that the cocks there are bearded and the ravens and parrots white, favours this view, for according to Blyth (*J. A. S. Beng.*, vol. XV, p. 26) there is found in Arâkan a species of the Bucconidae, which on account of their beards are called by the English 'barbets,' and on the same authority we learn that what is said of the ravens and parrots is likewise correct." Cunningham again, says (*Anc. Geog. of Ind.*, pp. 518-9): "In the inscriptions of the Kalachuri, or Haihaya dynasty of Chêdi, the Râjas assume the titles of "Lords of Kâliñjarapura, and of Trikaliṅga." Trikaliṅga, or the three Kaliṅgas, must be the three kingdoms Dhanakaṭaka, or Amarâvatî, on the Kṛishṇâ, Andhra or Waraṅgol, and Kaliṅga, or Râjamahêndri. "The name of Trikaliṅga is probably old, as Pliny mentions the Macco-Calingæ and the Gangarides-Calingae as separate peoples from the Calingae, while the *Mahâbhârata* names the Kaliṅgas three separate times, and each time in conjunction with different peoples. As Trikaliṅga thus corresponds with the great province of Tèlingana, it seems probable that the name of Têlingana may be only a slightly contracted form of Trikalingâna, or the three Kaliṅgas. I am aware that the name is usually derived from Tri-liṅga,' or the three *phalli* of Mahâdêva. But the mention of Macco-Calingae and Gangarides-

Calingae by Pliny would seem to show that the three Kaliṅgas were known as early as the time of Megasthenês, from whom Pliny has chiefly copied his Indian Geography. The name must therefore be older than the Phallic worship of Mahâdêva in Southern India." Caldwell observes (*Dravid. Gram.*, Introd., p. 32) that though Triliṅgon is said to be on the Ganges, it may have been considerably to the south of it, and on the Gôdâvarî, which was always regarded by the Hindus as a branch of the Ganges, and is mythologically identical with it. The Andhras and Kaliṅgas, the two ancient divisions of the Telugu people are represented by the Greeks as Gangetic nations. It may be taken as certain that Triglyphon, Triliṅga or Modogalinga was identical with Telingâna or Trilingam, which signifies the country of the *three liṅgas*. The Telugu name and language are fixed by Pliny and Ptolemy as near the mouths of the Ganges or between the Ganges and the Gôdâvarî. Modo or Modoga is equivalent to *mûdu* of modern Telugu. It "means *three*." Yule again places Trilingon on the left bank of the Brahmaputrâ, identifying it with Tripura (Tippera), a town in the district of the same name, 48 miles E.S.E. of Ḍhakka.

Rhingibêri:—Saint-Martin and Yule, as we have seen, place Rangâmatî on the Brahmaputrâ at Udipur. Wilford, however, had placed it near Chitagaoṅ, and identified it with Ptolemy's Rhinggibêri. "Ptolemy," he says (*Asiat. Res.*, vol. XIV, p 439); "has placed the source of the Dorias" (which in Wilford's opinion is the Dumurâ or Dumriyâ, called in the lower part of its course the

Karmaphuli) "in some country to the south of Salhata or Silhet, and he mentions two towns on its banks: Pandassa in the upper part of its course, but unknown; in the lower part Rhingibêri, now Rangâmatî near Châtgâv (Chitagaoṅ), and Reang is the name of the country on its banks. On the lesser Dumurâ, the river Chingri of the *Bengal Atlas*, and near its source, is a town called there Reang. Rangâmati and Rangâ-bâṭi, to be pronounced Rangabari, imply nearly the same thing."

Tomara was no doubt a place belonging to the Zamîrai or Tamarai, who were located inland from Kirrhadia, and inhabited the Garô Hills.

Mareoura or Malthoura:—In Yule's map this metropolis is located, but doubtfully, to the west of Tougma (Tagauṅ) near the western bank of the Khyendwen, the largest confluent of the Irâwaḍî.

Bareukora (or Bareuathra) is in Yule's map identified with Ramû, a place in the district of Chitagaoṅ, from which it is 68 miles distant to the S.S.E. Wilford identified it with Phalgun, another name for which, according to the *Kshetra Samaśa* was Pharuïgâra, and this he took to be Ptolemy's Bareukora. Phalgun he explains to be the Palong of the maps.

Kokkonagara:—Yule suggests for this Pegu. "It appears," he says, "from Târanâtha's history of Buddhism (ch. xxxix.) that the Indo-Chinese countries were in old times known collectively as Koki. In a Ceylonese account of an expedition against Râmaniyâ, supposed to be Pegu, the army captures the city of Ukkaka, and in it the Lord of Râmaniyâ. Kokkonagara again, is perhaps the Kâkula of Ibn Batuta, which was certainly a city on the Gulf of Siam, and probably an ancient foundation from Kaliṅga, called after Śrî-kâkola there."

T h a r r a :—The same authority identifies this with Tharâwati at the head of the delta of the Irâwaḍî. It is one of the divisions of the Province of Pegu.

Ptolemy's description of Transgangetic India now closes with the Islands.

26. The islands of the division of India we have been describing are said to be these:

Bazakata .... 149° 30′ 9° 30′
[Khalinê........................ 146° 9° 20′]

In this island some say there is found in abundance the murex shell-fish (κόχλος) and that the inhabitants go naked, and are called A g i n n a t a i.

27. There are three islands called S i n d a i, inhabited by Cannibals, of which the centre lies in.................... 152° 8° 40′ S.
Agathou daimonos ... 145° 15′ on the equator.

28. A group of five islands, the B a r o u s a i, whose inhabitants are said to be cannibals, and the centre of which lies in 152° 20′ 5° 20′ S. A group of three islands, the S a b a d e i b a i, inhabited by cannibals, of which the centre lies in .... 160° 8° 30′ S.

B a z a k a t a may perhaps be the island of Cheduba, as Wilford has suggested. Lassen takes it to be an island at the mouth of the Bassein river, near Cape Negrais, called Diamond Island. Its inhabitants are called by Ptolemy the Aginnatai, and represented as going naked. Lassen, for Aginnatai would therefore read Apinnatai, "because *apinaddha* in Sanskrit means unclothed;" but *apinaddha* means 'tied on,' clothed. Yule thinks it may perhaps be the greater of the two Andâmân islands. He says (*Proc. Roy. Geog. Soc.* vol. IV, 1882, p. 654); "Proceeding further the (Greek) navigator reaches the city of Kôli or Kôlis, leaving behind him the island of Bazakota, 'Good Fortune' (Ἀγαθοῦ Δαίμονος) and the group of the Barusæ. Here, at Kôli, which I take to be a part of the Malay peninsula, the course of the first century Greek, and of the ninth century Arab, come together." Bazakota and the Island of Good Fortune may be taken as the Great and the Little Andâmân respectively. The Arab relation mentions in an unconnected notice an island called Malhân between Serendib and Kalah, *i.e.*, between Ceylon and the Malay Peninsula, which was inhabited by black and naked cannibals. "This may be another indication of the Andâmân group, and the name may have been taken from Ptolemy's Maniolae, which in his map occupy the position in question." And again: "Still further out of the way (than the Andâmâns) and difficult of access was a region of mountains containing mines of silver. The landmarks (of the Arab navigator) to reach these was a mountain called Alkhushnâmi ('the Auspicious'). "This land of silver mines is both by position and by this description identified with the Argyrê of Ptolemy. As no silver is known to exist in that region (Arakan) it seems probable that the Arab indications to that effect were adopted from the Ptolemaic charts. And this leads me to suggest that the Jibal Khushnâmi also was but a translation of the Αγαθοῦ δαίμονος νῆσος, or isle of Good Fortune, in those maps, whilst I have thought also that the name Andâmân might have been adopted from a transcript of the same name in Greek as Αγ. δαίμον."

Khalinê in Yule's map is read as Saline, and identified with the Island of Salang, close to the coast in the latitude of the Nikobar Islands.

The Sindai Islands are placed by Ptolemy about as far south as his island of Iabadios (Java) but many degrees west of them. Lassen says (*Ind. Alt.*, vol. III, pp. 250-1) that the northmost of the three islands must be Pulo-Rapat, on the coast of Sumatra, the middle one the more southern, Pulo Pangor, and the island of Agatho-Daimon, one of the Salat Mankala group. The name of Sindai might imply, he thinks, that Indian traders had formed a settlement there. He seems to have regarded the Island of Agatho-Daimon as belonging to the Sindai group, but this does not appear to me to be sanctioned by the text. Yule says: "Possibly Sundar-Fulât, in which the latter word seems to be an Arabized plural of the Malay *Pulo* 'island' is also to be traced in Sindae Insulae, but I have not adopted this in the map."

The Barousai Islands:—"The (Arab) navigators," says Yule in his notes already referred to, "crossing the sea of Horkand with the west monsoon, made land at the islands of Lanja-Lañka, or Lika-Bâlûs, where the naked inhabitants came off in their canoes bringing ambergris and cocoa-nuts for barter, a description which with the position identifies these islands with the Nikobars, Nekaveram of Marco Polo, Lâka-Vâram of Rashîdu'd-dîn, and, I can hardly hesitate to say, with the Barusae Islands of Ptolemy."

Sabadeibai Islands:—The latter part of this name represents the Sanskrit *dwîpa*, 'an island.' The three islands of this name are probably those lying east from the more southern parts of Sumatra.

29. The island of Iabadios (or Sabadios) which means the island of Barley. It is said to be of extraordinary fertility, and to produce very much gold, and to have its capital called Argyrê (Silver-town) in the extreme west of it.
It lies in .... 167° 8° 30′ S.
and the eastern limit lies in ...169° 8° 10′ S.

30. The Islands of the Satyrs, three in number, of which the centre is in 171° 2° 30′ S. The inhabitants are said to have tails like those with which Satyrs are depicted.

31. There are said to be also ten other islands forming a continuous group called Maniolai, from which ships fastened with iron nails are said to be unable to move away, (perhaps on account of the magnetic iron in the islands) and hence they are built with wooden bolts. The inhabitants are called Maniolai, and are reputed to be cannibals.

The island of Iabadios:—*Yava*, the first part

of this name, is the Sanskrit word for 'barley,' and the second part like *deiba, diba, diva,* and *div* or *diu,* represents *dvîpa,* 'an island.' We have here therefore the Island of Java, which answers in most respects to Ptolemy's description of it. The following note regarding it I take from Bunbury's *History of Ancient Geography* (pp. 643-4): "The name of Java has certainly some resemblance with Iabadius, supposing that to be the correct form of the name, and, what is of more consequence, Ptolemy adds that it signifies 'the island of barley,' which is really the meaning of the name of Java. The position in latitude assigned by him to the island in question (8½ degrees of south latitude) also agrees very well with that of Java: but his geographical notions of these countries are in general so vague and erroneous that little or no value can be attached to this coincidence. On the other hand, the abundance of gold would suit well with Sumatra, which has always been noted on that account, while there is little or no gold found in Java. The metropolis at its western extremity would thus correspond with Achîn, a place that must always have been one of the principal cities of the island. In either case he had a very imperfect idea of its size, assigning it a length of only about 100 Geog. miles, while Java is 9° or 540 G. miles in length, and Sumatra more than 900 G. miles. It seems not improbable that in this case, as in several others, he mixed up particulars which really referred to the two different islands, and applied them to one only: but it is strange that if he had any information concerning such islands as Sumatra and Java, he should have no notion that they were of very large size, at the same time that he had such greatly exaggerated ideas of the dimensions of Ceylon." Mannert took Iabadios to be the small island of Banka on the S.E. of Sumatra. For the application of the name of Java to the Island of Sumatra, see Yule's *Marco Polo*, vol. II, p. 266, note 1.

Regarding the Islands of the Satyrs, Lassen says (*Ind. Alt.*, vol. III, p. 252): The three islands, called after the Satyrs, mark the extreme limits of the knowledge attained by Ptolemy of the Indian Archipelago. The inhabitants were called Satyrs because, according to the fabulous accounts of mariners, they had tails like the demi-gods of that name in Greek mythology. Two of these must be Madura and Bali, the largest islands on the north and east coasts of Java, and of which the first figures prominently in the oldest legends of Java; the second, on the contrary, not till later times. The third island is probably Lombok, lying near Bali in the east. A writer in Smith's *Dictionary of Classical Geography* thinks these islands were perhaps the Anamba group, and the Satyrs who inhabited them apes resembling men. Yule says in the notes:—" Sandar-Fulât we cannot hesitate to identify with Pulo Condor, Marco Polo's Sondur and Condur. These may also be the Satyrs' islands of Ptolemy, but they may be his Sindai, for he has a Sinda city on the coast close to this position, though his Sindai islands are dropped far way. But it would not be difficult to show that Ptolemy's islands have been located almost at random, or as from a pepper-castor."

Ptolemy locates the Maniolai Islands, of which he reckons ten, about 10 degrees eastward from Ceylon. There is no such group however to be found in that position, or near it, and we may safely conclude that the Maniolai isles are as mythical as the magnetic rocks they were said to contain. In an account of India, written at the close of the 4th or beginning of the 5th century, at the request either of Palladius or of Lausius, to whom Palladius inscribed his *Historia Lausiaca*, mention is made of these rocks: "At Muziris," says Priaulx, in his notice of this account[31] "our traveller stayed some time, and occupied himself in studying the soil and climate of the place and the customs and manners of its inhabitants. He also made enquiries about Ceylon, and the best mode of getting there, but did not care to undertake the voyage when he heard of the dangers of the Sinhalese channel, of the thousand isles, the Maniolai which impede its navigation, and the loadstone rocks which bring disaster and wreck on all iron-bound ships." And Masû'di, who had traversed this sea, says that ships sailing on it were not fastened with iron nails, its waters so wasted them. (*The Indian Travels of Apollonius of Tyana, &c.*, p. 197). After Ptolemy's time a different position was now and again assigned to these rocks, the direction in which they were moved being more and more to westward. Priaulx (p. 247), uses this

---

[31] Wilford (*As. Res.* vol. XIV, pp. 429-30), gives the fable regarding these rocks from the *Chaturvarga Chintâmani*, and identifies them with those near Pârindra or the lion's place in the lion's mouth or Straits of Singapur.

as an argument in support of his contention that the Roman traffic in the eastern seas gradually declined after 273 A.D., and finally disappeared. How, otherwise, he asks, can we account for the fact that the loadstone rocks, those myths of Roman geography, which, in Ptolemy's time, the flourishing days of Roman commerce, lay some degrees eastward of Ceylon, appear A.D. 400 barring its western approach, and A.D. 560 have advanced up to the very mouth of the Arabian Gulf. But on the Terrestrial Globe of Martin Behem, Nuremberg A.D. 1492, they are called M a n i l l a s, and are placed immediately to the north of Java Major. Aristotle speaks of a magnetic mountain on the coast of India, and Pliny repeats the story. Klaproth states that the ancient Chinese authors also speak of magnetic mountains in the southern sea on the coasts of Tonquin and Cochin-China, and allege regarding them that if foreign ships which are bound with plates of iron approach them, such ships are there detained, and can in no case pass these places. (Tennant's *Ceylon*, vol. I, p. 444 n.) The origin of the fable, which represents the magnetic rocks as fatal to vessels fastened with iron nails, is to be traced to the peculiar mode in which the Ceylonese and Malays have at all times constructed their boats and canoes, these being put together without the use of iron nails; the planks instead being secured by wooden bolts, and stitched together with cords spun from the fibre of the cocoanut. "The Third Calender,' in the *Arabian Nights Entertainment*, gives a lively account of his shipwreck upon the Loadstone Mountain, which he tells us was entirely covered towards the sea with the nails that belonged to the immense number of ships which it had destroyed.

Cap. 3.

Position of the Sinai.

[11*th Map of Asia.*]

1. The Sinai are bounded on the north by the part of Sêrikê already indicated, on the east and south by the unknown land, on the west by India beyond the Ganges, along the line defined as far as the Great Gulf and by the Great Gulf itself, and the parts immediately adjacent thereto, and by the Wild Beast Gulf, and by that frontier of the Sinai around which are placed the Ikhthyophagoi Aithiopes, according to the following outline :—

2. After the boundary of the Gulf on the side of India the mouth of

the river Aspithra	170°	16°
Sources of the river on the eastern side of the Sêmanthinos range	180°	26°
Bramma, a town...............	177°	12° 30′
The mouth of the river Ambastes	176°	10°
The sources of the river......	179° 30′	15°
Rhabana, a town...............	177°	8° 30′
Mouth of the river Sainos ...	176° 20′	6° 30′
The Southern Cape	175° 15′	4°
The head of Wild Beast Gulf	176°	2°

The Cape of Satyrs	175°	on the line
Gulf of the Sinai[32]	178°	2° 20′

3. Around the Gulf of the Sinai dwell the fish-eating Aithiopians.

Mouth of the river Kottiaris	177° 20′	7°	S.
Sources of the river	180° 40′	2°	S.
Where it falls into the river Sainos..............	180°	on the line.	
Kattigara, the port of the Sinai......................	177°	8° 30′	S.

4. The most northern parts are possessed by the Sêmanthinoi, who are situated above the range that bears their name. Below them, and below the range are the Akadrai, after whom are the Aspithrai, then along the Great Gulf the Ambastai, and around the gulfs immediately adjoining the Ikhthyophagoi Sinai.

5. The interior towns of the Sinai are named thus :—

Akadra	178° 20′	21° 15′	
Aspithra	175°	16°	
Kokkonagara	179° 50′	2°	S.
Sarata	180° 30′	4°	S.

6. And the Metropolis

Sinai or Thînai	180° 40′	3°	S.

which they say has neither brazen walls nor anything else worthy of note. It is encompassed on the side of Kattigara towards the west by

[32] Latin Translator.

the unknown land, which encircles the Green Sea as far as Cape Prason, from which begins, as has been said, the Gulf of the Batrakheian Sea, connecting the land with Cape Rhapton, and the southern parts of Azania.

It has been pointed out how egregiously Ptolemy misconceived the configuration of the coast of Asia beyond the Great Gulf, making it run southward and then turn westward, and proceed in that direction till it reached the coast of Africa below the latitude of Zanzibar. The position, therefore of the places he names, cannot be determined with any certainty. By the Wild Beast Gulf may perhaps be meant the Gulf of Tonquin, and by the Gulf of the Sinai that part of the Chinese Sea which is beyond Hai-nan Island. The river Kottiaris may perhaps be the river of Canton. Thinai, or Sinai, may have been Nankin, or better perhaps Si-guan-fu, in the province of Shen-si, called by Marco Polo, by whom it was visited, Ken-jan-fu. "It was probably," says Yule (Marco Polo, vol. II, p. 21) "the most celebrated city in Chinese history and the capital of several of the most potent dynasties. In the days of its greatest fame it was called Chaggan." It appears to have been an ancient tradition that the city was surrounded by brazen walls, but this Ptolemy regarded as a mere fable. The author of the *Periplûs* (c. 64), has the following notice of the place :—" There lies somewhere in the interior of Thina, a very great city, from which silk, either raw or spun or woven into cloth is carried overland to Barygaza through Baktria or by the Ganges to Limyrikê . . . . Its situation is under the Lesser Bear." Ptolemy has placed it 3 degrees south of the equator!!

Cap. 4.

Position of the Island of Taprobane.

[*Map of Asia* 12.]

1. Opposite Cape Kôry, which is in India, is the projecting point of the Island of Taprobanê, which was called formerly Simoundou, and now Salikê. The inhabitants are commonly called Salai. Their heads are quite encircled with long luxuriant locks, like those of women. The country produces rice, honey, ginger, beryl, hyacinth[33] and has mines of every sort—of gold and of silver and other metals. It breeds at the same time elephants and tigers.

2. The point already referred to as lying opposite to Kôry is called North Cape (Boreion Akron) and lies ...... 126° 12° 30′

3. The descriptive outline of the rest of the island is as follows:—

After the North Cape which is situated in	126°	12° 30′
comes Cape Galiba............	124°	11° 30′
Margana, a town	123° 30′	10° 20′

[33] In one of the temples, says Kosmos, is the great hyacinth, as large as a pine-cone, the colour of fire and flashing from a distance, especially when catching the beams of the sun, a matchless sight.

Iôgana, a town	123° 20′	8° 50′
Anarismoundon, a cape......	122°	7° 45′
Mouth of the River Soana...	122° 20′	6° 15′
Sources of the river	124° 30′	3°
Sindokanda, a town	122°	5°
Haven of Priapis	122°	3° 40′
4. Anoubingara	121°	2° 40′
Headland of Zeus	120° 30′	1°
Prasôdês Bay	121°	2°
Noubartha, a town	121°40′	on the Line.
Mouth of the river Azanos...	123° 20′	1° S.
The sources of the river......	126°	1° N.
Odôka, a town.................	123°	2° S.
Orneôn, (Birds' Point) a headland	125°	2° 30′ S.
5. Dagana, a town sacred to the Moon	126°	2° S.
Korkobara, a town	127° 20′	2° 20′ S.
Cape of Dionysos	130°	1° 30′ S.
Kêtaion Cape	132° 30′	2° 20′ S.
Mouth of the river Barakês	131° 30′	1° N.
Sources of the river	128°	2° N.
Bôkana, a town	131°	1° 20′ N.
The haven of Mardos or Mardoulamnê	131°	2° 20′ N.
6. Abaratha, a town ...	131°	3° 15′ N.
Haven of the Sun (Heliou limên).....................	130°	4°
Great Coast (Aigialos Megas)	130°	4° 20′

Prokouri, a town	131°	5° 20′
The haven of Rizala	130° 20′	6° 30′
Oxeia, a headland	130°	7° 30′
Mouth of the river Gangês	129°	7° 20′
The sources of the river...	127°	7° 15′
Spatana Haven	129°	8°
7. Nagadiba or Nagadina, a town	129°	8° 30′
Pati Bay	128° 30′	9° 30′
Anoubingara, a town......	128° 20′	9° 40′
Modouttou, a mart.........	128°	11° 20′
Mouth of the river Phasis	127°	11° 20′
The sources of the river...	126°	8°
Talakôry (or Aakotê,) a mart	126° 20′	11° 20′

After which the North Cape.

8. The notable mountains of the island are those called G a l i b a, from which flow the Phasis and the Ganges, and that called M a l a i a, from which flow the Soanas and the Azanos and the Barakês, and at the base of this range, towards the sea, are the feeding grounds of the elephants.

9. The most northern parts of the Island are possessed by the G a l i b o i and the M o u d o u t t o i, and below these the A n o u r o g r a m m o i and the N a g a d i b o i, and below the Anourogrammoi the S o a n o i, and below the Nagadiboi the S e n n o i, and below these the S a n d o k a n d a i, towards the west, and below these towards the feeding grounds of the elephants

the Boumasanoi, and the Tarakhoi, who are towards the east, below whom are the Bôkanoi and Diordouloi, and furthest south the Rhogandanoi, and the Nageiroi.

10. The inland towns in the island are these:—

Anourogrammon, the royal residence	124° 10′	8° 40′
Maagrammon, the metropolis	127°	7° 20′
Adeisamon	129°	5°
Podoukê	124°	3° 40′
Oulispada	126° 20′	40′
Nakadouba	128° 30′	on the Line.

11. In front of Taprobanê lies a group of islands which they say number 1378. Those whose names are mentioned are the following:—

Ouangalia (or Ouangana)	120° 15′	11° 20′
Kanathra	121° 40′	11° 15′
Aigidiôn	118°	8° 30′
Ornêon	119°	8° 30′
Monakhê................	116°	4° 15′
Amminê	117°	4° 30′
12. Karkos..............	118°	40′ S.
Philêkos	116° 30′	2° 40′ S.
Eirênê	120°	2° 30′ S.
Kalandadroua	121°	5° 30′ S.
Abrana	125°	4° 20′ S.
Bassa	126°	6° 30′ S.
Balaka................	129°	5° 30′ S.

Alaba	131°	4° S.
Goumara	133°	1° 40′ S.
13. Zaba..................	135°	on the Line.
Zibala	135°	4° 15′ N.
Nagadiba.....................	135°	8° 30′
Sonsouara	135°	11° 15′

14. Let such then be the mode of describing in detail the complete circuit of all the provinces and satrapies of the known world, and since we indicated in the outset of this compendium how the known portion of the earth should be delineated both on the sphere and in a projection on a plane surface exactly in the same manner and proportion as what is traced on the solid sphere, and since it is convenient to accompany such descriptions of the world with a summary sketch, exhibiting the whole in one comprehensive view, let me now therefore give such a sketch with due observance of the proper proportion.

This island of Taprobanê has changed its name with notable frequency. In the *Râmâyaṇa* and other Sanskrit works it is called Laṅkâ, but this was an appellation unknown to the Greeks. They called it at first Antichthonos, being under the belief that it was a region belonging to the opposite portion of the world (Pliny, lib. VI, c. xxii). In the time of Alexander, when its situation was better understood, it was called Taprobanê. Megasthenês mentions it under this name, and remarks that it was divided (*into two*) by a river, that its inhabitants were called Palaeogoni and that it produced more gold and pearls of large size than India. From our author we learn that the old name of the island was Simoundou, and that Taprobanê, its next name, was obsolete in his time, being replaced by Salikê. The author of the *Periplûs* states, on the other hand, that Taprobanê was the old name of the island, and that in his time it was called Palai Simoundou. The section of his work however in which this statement occurs (§ 61) is allowed to be hopelessly corrupt. According to Pliny, Palaesimundus was the name of the capital town, and also of the river on whose banks it stood. How long the island continued to be called Salikê does not appear, but it was subsequently known under such names as Serendivus, Sirlediba, Serendib, Zeilan, and Sailan, from which the transition is easy to the name which it now bears, Ceylon.

With regard to the origin or derivation of the majority of these names the most competent scholars have been divided in their opinions. According to Lassen the term Palaiogonoi was selected by Megasthenês to designate the inhabitants of the island, as it conveyed the idea entertained of them by the Indians that they were Râkshasas, or giants, 'the sons of the progenitors of the world.' To this it may be objected that Megasthenês did not intend by the term to describe the inhabitants, but merely to give the name by which they were known, which was different from that of the island. Schwanbeck again suggested that the term might be a transliteration of Pâli-janâs, a Sanskrit compound, which he took to mean "men of the sacred doctrine" (*Ind. Ant.*, vol. VI, p. 129, n.) But, as Priaulx has pointed out (*Apollon.*

of Tyana, p. 110), this is an appellation which could scarcely have been given to others than learned votaries of Buddhism, and which could scarcely be applicable to a people who were not even Buddhist till the reign of Aśôka, who was subsequent to Chandragupta, at whose court Megasthenês acquired his knowledge of India. Besides, it has been pointed out by Goldstücker (*l.c.* n. 59) that *Pali* has not the meaning here attributed to it. He adds that the nearest approach he could find to Palaiogonoi is—*pâra* 'on the other side of the river' and *janâs* 'a people'; Pârajanâs, therefore, 'a people on the other side of the river.' Tennent, in conclusion, takes the word to be a Hellenized form of *Pali-putra*, 'the sons of the Pâli,' the first Prasian colonists of the island. A satisfactory explanation of Palai-Simoundou has not yet been hit on. That given by Lassen, Pâli-Simanta, or Head of the Sacred Law, has been discredited. We come now to Taprobanê. This is generally regarded as a transliteration of Tâmraparṇî, the name which Vijaya, who, according to tradition, led the first Indian colony into Ceylon, gave to the place where he first landed, and which name was afterwards extended to the whole island. It is also the name of a river in Tinneveli, and it has, in consequence, been supposed that the colonists, already referred to, had been, for some time, settled on its banks before they removed to Ceylon. The word means 'Copper-coloured leaf.' Its Pâli form is Tambapaṇṇi (see *Ind. Ant.*, Vol. XIII, pp. 33f.) and is found, as has been before noticed, in the inscription of Aśôka on the Girnâr rock. Another name, applied to it by Brahmanical writers, is Dwîpa-Râvaṇa, *i.e.*, 'the island of Râvaṇa, whence perhaps Taprobanê.' Salikê, Serendivas, and other subsequent names, are all considered to be connected etymologically with Sîṁhala (colloquially Sîlam), the Pâli form of Sîhala, a derivative from *siṁha*, 'a lion,' *i.e.* 'a hero'—the hero Vijaya. According to a different view these names are to be referred to the Javanese *sela*, 'a precious stone,' but this explanation is rejected by Yule (*Marco Polo*, vol. II, p. 296, n. 6). For Salikê, Tennent suggests an Egyptian origin, Siela-keh, *i.e.*, 'the land of Siela.'

Little more was known in the west respecting the island beyond what Megasthenês had communicated until the reign of the Emperor Claudius, when an embassy was sent to Rome by the Sinhalese monarch, who had received such astonishing accounts of the power and justice of the Roman people that he became desirous of entering into alliance with them. He had derived his knowledge of them from a castaway upon his island, the freedman of a Roman called Annius Plocamus. The embassy consisted of 4 members, of whom the chief was called Rachia, an appellation from which we may infer that he held the rank of a Râja. They gave an interesting, if not a very accurate, account of their country, which has been preserved by Pliny (*Nat. Hist.* lib. VI). Their friendly visit, operating conjointly with the discovery of the quick passage to and from the East by means of the monsoon, gave a great impetus to commercial enterprise, and the rich marts, to which access had thus been opened, soon began to be frequented by the galleys of the West. Ptolemy, living in Alexandria, the great entrepôt in those days of the Eastern traffic, very probably acquired from traders arriving from Ceylon, his knowledge concerning it, which is both wonderfully copious, and at the same time, fairly accurate, if we except his views of its magnitude, which like all his predecessors he vastly over-estimated. On the other hand, he has the merit of having determined properly its general form and outline, as well as its actual position with reference to the adjoining continent, points on which the most vague and erroneous notions had prevailed up to his time, the author of the *Periplûs* for instance describing the island as extending so far westward that it almost adjoined Azania in Africa. The actual position of Ceylon is between 5° 55′ and 9° 51′ N. lat., and 79° 42′ and 81° 55′ E. long. Its extreme length from north to south is 271½ miles, its greatest width 137½ miles, and its area about one-sixth smaller than that of Ireland. Ptolemy however made it extend through no less than 15 degrees of latitude and 12 of longitude. He thus brought it down more than two degrees south of the equator, while he carried its northern extremity up to 12½° N. lat., nearly 3 degrees north of its true position. He has thus represented it as being 20 times larger than it really is. This extravagant over-estimate, which had its origin in the Mythological Geography of the Indian Brâhmaṇs, and which was adopted by the islanders themselves, as well as by the Greeks, was shared also by the Arab geographers Masû'dî, Idrisi, and Abu'l-fidâ, and by such writers as Marco Polo. In consequence of these misrepresentations it came to be questioned at one time whether Ceylon or

Sumatra was the Taprobanê of the Greeks, and Kant undertook to prove that it was Madagascar (Tennent's *Ceylon*; vol. I, p. 10 and n.). Ptolemy has so far departed from his usual practice that he gives some particulars respecting it, which lie out of the sphere of Geography, strictly so called. He is mistaken in stating that the tiger is found in Ceylon, but he has not fallen into error on any other point which he has noticed. It may be remarked that the natives still wear their hair in the effeminate manner which he has noticed. In describing the island geographically he begins at its northern extremity, proceeds southward down the western coast, and returns along the east coast to Point Pedro. "In his map he has laid down the position of eight promontories, the mouths of five rivers and four bays and harbours, and in the interior he had ascertained that there were thirteen provincial divisions, and nineteen towns, besides two emporia on the coast, five great estuaries, which he terms lakes, two bays and two chains of mountains, one of them surrounding Adam's Peak, which he designates as Malaia, the name by which the hills that environ it are known in the *Mahawánso*." Tennent, from whom the foregoing summary has been quoted, observes in a foot-note (vol. I, p. 535) that Ptolemy distinguishes those indentations in the coast which he describes as *bays* (κόλπος) from the estuaries, to which he gives the epithet of *lakes*, (λιμὴν);[34] of the former he particularises two, Pati

[34] Tennent here seems to have confounded λιμὴν, a *haven* or creek, with λίμνη, *a lake*. The words are, however, etymologically connected.

and Prasôdês, the position of which would nearly correspond with the Bay of Trinkônamalai and the harbour of Colombo—of the latter he enumerates five, and from their position they seem to represent the peculiar estuaries formed by the conjoint influence of the rivers and the current, and known to the Arabs by the name of "gobbs."

Ceylon is watered by numerous streams, some of which are of considerable size. The most important is the Mahâweligangâ, which has its sources in the vicinity of Adam's Peak, and which, after separating into several branches, enters the ocean near Trinkônamalai. Ptolemy calls it the Ganges. He mentions four other rivers, the Soana, Azanos, Barakês and Phasis, which Tennent identifies with the Dedera-Oya, the Bentote, the Kambukgam and the Kangarayen respectively. Lassen, however (*Ind. Alt.*, vol. III, p. 21), identifies the Azanos with the Kâlagangâ which enters the sea a little farther north than the river of Bentote, and is a larger stream.

The mountains named by Ptolemy are the Galiba in the north-west of the island, and the Malaia, by which he designates the mountain groups which occupy the interior of the island towards the south. He has correctly located the plains or feeding grounds of the elephants to the south-east of these mountains; *malai* is the Tamil word for "mountain."

The places which he has named along the coast and in the interior have been identified, though in most cases doubtfully, by Tennent in his map of Taprobanê according to Ptolemy and Pliny, in vol. I. of his work, as follows :—

On the West Coast beginning from the north:—
Margana with Mantote.
Iôgana with Aripo.
Anarismoundou Cape with Kudramali Point, but Mannert with Kalpantyn (further south).
Sindo Kanda with Chilau (Chilau from Salâbhana—the Diving, *i. e.* Pearl Fishery.)
Port of Priapis[35] with Negombo.
Cape of Zeus at Colombo.
Prasôdês Bay, with Colombo Bay.
Noubartha with Barberyn.
Odoka with Hikkode.
Cape Orneôn (of Birds) with Point de Galle.
On the South Coast:—
Dagana with Dondra Head.
Korkobara with Tangalle.
On the East Coast:
Cape of Dionysos, with Hambangtote.
Cape Kêtaion (Whale cape) with Elephant Rock, (Bokana Yule identifies with Kambugam).
Haven of Mardos with Arukgam Bay.
Abaratha with Karativoe (but Yule with Aparatote, which is better).
Haven of the Sun with Batticalao.
Rizala Haven with Vendeloos Bay.
Oxeia Cape (Sharp point) with Foul Point.
Spatana Haven with an indentation in Trinkônamalai Bay.
Nagadiba or Nagadina with a site near the Bay.
Pati Bay with Trinkônamalai Bay.
Anoubingara with Kuchiavelli.
Modouttou with Kokelay.

---

[35] This was no doubt a name given by the Greeks.

On the North Coast:—

Mouth of the Phasis.

Talakôry or Aakotê, with Tondi Manaar. Yule places both Nagadiba and Modouttou on the north-west coast, identifying the latter with Mantote.

With respect to places in the interior of the island Tennent says (vol. I, p. 536, n. 2): "His (Ptolemy's) M a a g r a m m o n would appear on a first glance to be Mahâgâm, but as he calls it the metropolis, and places it beside the great river, it is evidently Bintenne, whose ancient name was "Mahâyangana" or "Mahâwelligâm." His A n u r o g r a m m u m, which he calls βασίλειον "the royal residence," is obviously Anurâdhapura, the city founded by Anurâdha 500 years before Ptolemy (*Mahawânso*, pp. 50-65). The province of the M o u d o u t t o i in Ptolemy's list has a close resemblance in name, though not in position, to Mantote; the people of Reyagamkorle still occupy the country assigned by him to the R h o g a n d a n o i—his N a g a d i b o i are identical with the Nâgadiva of the *Mahawânso*; and the islet to which he has given the name of B a s s a, occupies nearly the position of the Basses, which it has been the custom to believe were so-called by the Portuguese,—"Baxos" or "Baixos" "Sunken Rocks." The R h o g a n d a n o i were located in the south-west of the island. The sea, which stretched thence towards Malaka, appears to have at one time borne their name, as it was called by the Arab navigators "the sea of Horkand." The group of islands lying before Ceylon is no doubt that of the Maldives.

KLAUDIOS PTOLEMY'S GEOGRAPHY OF CENTRAL ASIA.

Having now examined in detail the whole of Ptolemy's Indian Geography, I annex as a suitable Appendix his description of the countries adjacent to India. The reader will thus be presented with his Geography in its entirety of Central and South-Eastern Asia. In the notes I have adverted only to the more salient points.

Book VI, Cap. 9.

Position of Hyrkania.

[*Map of Asia*, 7.]

1. Hyrkania is bounded on the north by that part of the Hyrkanian sea which extends from the extreme point of the boundary line with Mêdia as far as the mouth of the river Ôxos which lies in .... 100° 43° 5′

2. In which division occur these towns :—

Saramannê, a town............	94° 15′	40° 30′
Mouth of the Maxêra..........	97° 20′	41° 30′
The sources of this river ...	98°	38° 20′
Mouth of the Sokanda	97° 20′	42°
Mouth of the river Ôxos ...	100°	43° 5′

3. On the west by the part of Mêdia already mentioned as far as Mount Korônos [in which part of Mêdia is Saramannê..................... 94° 15′ 40° 30′]

4. on the south by Parthia, along the side of it described as passing through the range of Korônos, and on the east by Margianê

through the mountainous region which connects the extremities referred to.

5. The maritime ports of Hyrkania are inhabited by the Maxêrai, and the Astabênoi and below the Maxêrai by the Khrêndoi, after whom comes the country adjacent to the Korônos range, Arsîtis, and below the Astabênoi is the country called Sirakônê.

6. The cities in the interior are said to be these :—

Barangê	99°	42°
Adrapsa	98° 30′	41° 30′
Kasapê...........................	99° 30′	40° 30′
Abarbina.........................	97°	40° 10′
Sorba	98°	40° 30′
7. Sinaka	100°	39° 40′
Amarousa	96°	39° 55′
Hyrkania, the metropolis....	98° 50′	40°
Sakê (or Salê)..................	94° 15′	39° 30′
Asmourna	97° 30′	39° 30′
Maisoka (or Mausoka)	99°	39° 30′
8. And an island in the sea near it called Talka	95°	42°

The name of Hyrkania is preserved to this day in that of Gurkan or Jorjan, a town lying to the east of Asterâbâd. Its boundaries have varied at different periods of history. Speaking generally, it corresponds with the modern Mazanderan and Asterâbâd. Its northern frontier was formed by the Kaspian, which was sometimes called after it—the Hyrkanian Sea. The river Ôxos, which is called by the natives on its banks the Amu-daryâ, and by Persian writers the Jihun, falls now into the Sea of Aral, but as we learn from our author as well as from other ancient writers it was in former times an affluent of the Kaspian, a fact confirmed by modern explorations. Mount Korônos was the eastern portion of the lofty mountain chain called the Elburz, which runs along the southern shores of the Kaspian. The River Maxêra is mentioned by Pliny (lib. VI, c. xiv, sec. 18) who calls it the Maxeras. It has been variously identified, as with the Tejin, the Gurgan, the Atrek and others. The metropolis of Hyrkania is called by Ammianus Marcellinus (c. xxiii, sec. 6) Hyrkana, which is probably the Gurkan already mentioned.

Cap. 10.

Position of Margianê.

[*Map of Asia* 7.]

Margianê is bounded on the west by Hyrkania, along the side which has been already traced, and on the north by a part of Skythia extending from the mouths of the river Ôxos as far as the division towards Baktrianê, which lies in 103°—43°, and on the south by part of Areia along the parallel of latitude running from the boundary towards Hyrkania and Parthia through the Sariphi range, as far as the extreme point lying 109°—39°, and on the east by Baktrianê along the mountainous region which connects the

said extremities. A considerable stream, the Margos, flows through the country, and its sources lie in ....105° 39°
while it falls into the Ôxos in 102° 43° 30′.

2. The parts of it towards the river Ôxos are possessed by the Derbikkai, called also the Derkeboi, and below them the Massagetai, after whom the Parnoi and the Dâai, below whom occurs the desert of Margiana, and more to the east than are the Tapouroi.

3. The cities of it are—

Ariaka	103°	43°
Sina (or Sêna)................	102° 30′	42° 20′
Aratha	103° 30′	42° 30′
Argadina	101° 20′	41° 40′
Iasonion	103° 30′	41° 30′

4. There unites with the River Margos, another stream flowing from the Sariphi range

of which the sources lie......	103°	39°
Rhêa..........................	102°	40° 50′
Antiokheia Margianê.........	106°	40° 20′
Gourianê	104°	40°
Nisaia or Nigaia	105°	39° 10′

"In early periods," says Wilson (*Ariana Antiqua*, p. 148), "Margiana seems to have been unknown as a distinct province, and was, no doubt, in part at least, comprised within the limits of Parthia. In the days of the later geographers, it had undergone the very reverse relation, and had, to all appearance, extended its boundaries so as to include great part of the original Parthia. It is evident from Strabo's notice of the latter (lib. XI, c. ix) that there was left little of it except the name; and in Ptolemy no part of Parthia appears above the mountains." Strabo says of it (lib. XI, c. x) "Antiokhos Sôtêr admired its fertility, he enclosed a circle of 1,500 stadia with a wall, and founded a city, Antiokheia. The soil is well adapted to vines. They say that a vine stem has been frequently seen there which would require two men to girth it, and bunches of grapes two cubits in size." Pliny writes somewhat to the same effect. He says (lib. VI, c. xvi): "Next comes Margianê, noted for its sunny skies; it is the only vine-bearing district in all these parts, and it is shut in on all sides by pleasant hills. It has a circuit of 1,500 stadia, and is difficult of approach on account of sandy deserts, which extend for 120 miles. It lies confronting a tract of country in Parthia, in which Alexander had built Alexandria, a city, which after its destruction by the barbarians, Antiokhos, the son of Seleucus, rebuilt on the same site. The river Margus which amalgamates with the Zothale, flows through its midst. It was named Syriana, but Antiokhos preferred to have it called Antiokheia. It is 80 stadia in circumference. To this place Orodes conducted the Romans who were taken prisoners when Crassus was defeated." This ancient city is represented now by Merv. The river Margus is that now called the Murgh-âb or Meru-rûd. It rises in the mountains of the Hazâras (which are a spur of the Paropanisos and the Sariphi montes of our author), and loses itself in the sands about 50 miles north-west of the city, though in ancient times it appears to have poured its waters into the Ôxos.

The tribes that peopled Hyrkania and Margiana and the other regions that lay to the eastward of the Kaspian were for the most part of Skythian origin, and some of them were nomadic. They are described by the ancient writers as brave and hardy warriors, but of repulsive aspect and manners, and addicted to inhuman practices. Ptolemy names five as belonging to Margiana—the Derbikkai, Massagetai, Parnoi, Däai and Tapouroi.

The Derbikes are mentioned by Strabo (lib. XI, c. xi, sec. 7), who gives this account of them. "The Derbikes worship the earth. They neither sacrifice nor eat the female of any animal. Persons who attain the age of above 70 years are put to death by them, and their nearest relations eat their flesh. Old women are strangled and then buried. Those who die under 70 years of age are not eaten, but are only buried."

The Massagetai are referred to afterwards (c. xiii, sec. 3) as a tribe of nomadic Sakai, belonging to the neighbourhood of the river Askatangkas. They are mentioned by Herodotos (lib. I, c. cciv.) who says that they inhabited a great portion of the vast plain that extended eastward from the Kaspian. He then relates how Cyrus lost his life in a bloody fight against them and their queen Tomyris. Alexander came into collision with their wandering hordes during the campaign of Sogdiana as Arrian relates (*Anab.* lib. IV, cc. xvi, xvii).

As regards the origin of their name it is referred by Beal (*J. R. A. S.*, N.S., vol. XVI, pp. 257, 279) to *maiza*—'greater' (in Moeso-Gothic) and Yue-ti (or chi). He thus reverts to the old theory of Rémusat and Klaproth, that the Yue-ti were Getae, and this notwithstanding the objection of Saint-Martin stated in *Les Huns Blancs*, p. 37, n. 1. The old sound of *Yue* he observes was *Get*, correspondent with the Greek form *Getai*. In calling attention to the Moeso-Gothic words *maiza* (greater) and *minniza* (less) he suggests that "we have here the origin of the names Massagetae, and the Mins, the Ta Yue-chi (great Yue-chi) and the Sian Yue-chi (little Yue-chi)."

The Parnoi, according to Strabo, were a branch of the Dahai (lib. XI, c. vii, sec. 1) called by Herodotos (lib. I, c. lii) the Däoi, and by our author and Stephanos of Byzantium the Däai. Strabo (lib. XI, c. viii, 2) says of them: "Most of the Skythians beginning from the Kaspian Sea, are called Dahai Skythai, and those situated more towards the east, Massagetai and Sakai, the rest have the common appellation of Skythians, but each separate tribe has its peculiar name. All, or the greater part of them, are nomadic." Virgil (*Aen.* lib. VIII, l. 728) applies to the Dahae the epithet *indomiti*. It is all but certain that they have left traces of their name in the province of Dahestân, adjoining to Asterâbâd, as this position was within the limits of their migratory range. In the name Dâae, Dahae or Ta-hia (the Chinese form) it is commonly inferred that we have the term Tajik, that is Persian, for there is good reason to place Persians even in Trans-oxiana long before the barbarous tribes of the Kaspian plains were heard of (See Wilson's *Arian. Antiq.*, p. 141).

The Tapouroi appear to be the same as the Tapyroi mentioned by Strabo as occupying the country between the Hyrkanoi and the Areioi. Their position, however, varied at various times.

Nisaia or Nigaia (the Nesaia of Strabo) has been identified by Wilson (*Arian. Antiq.*, pp. 142, 148) with the modern Nissa, a small town or village on the north of the Elburz mountains, between Asterâbâd and Meshd.

Cap. 11.

Position of Baktrianê.

1. Baktrianê is bounded on the west by Margianê along the side already described, on the north and east by Sogdianê, along the rest of the course of the River Ôxos, and on the south by the rest of Areia, extending from the extreme point towards Margianê—
the position of which is...... 109° 39°
and by the Paropanisadai along the parallel thence prolonged, through where the range of Paropanisos diverges towards the sources of the Ôxos which lie in .... 119° 30′ 39°

2. The following rivers which fall into the Ôxos flow through Baktrianê :—
The river Ôkhos, whose sources lie .... 110° 39°

and the Dargamanês, whose sources lie	116° 30′	36° 20′
and the Zariaspis, whose sources lie	113°	39°
and the Artamis, whose sources lie	114°	39°
and the Dargoidos, whose sources lie	116°	39°
and the point where this joins the Ôxos lies in......	117° 30′	44°

3. Of the other tributaries the Artamis and

the Zariaspis unite in.........	113°	40° 40′
before falling into the Ôxos in	112° 30′	44°

4. The Dargamanês and the Ôkhos also

unite in	109°	40° 30′
before falling into the Ôxos in	109°	44°

5. Of the Paropanisos range, the western

part is situated in	111° 30′	39°
and [the Eastern] in	119° 30′	39°

6. The parts of Baktrianê in the north and towards the River Ôxos are inhabited by the Salaterai and the Zariaspai, and to the south of these up towards the Salaterai the Khomaroi, and below these the Kômoi, then the Akinakai, then the Tambyzoi, and below the Zariaspai the Tokharoi, a great people, and below them the Marykaioi, and the Skordai, and the Ouarnoi

(Varnoi), and still below those the Sabadioi, and the Oreisitoi, and the Amareis.

7. The towns of Baktrianê towards the river Ôxos are the following :—

Kharakharta	111°	44°
Zari(a)spa or Kharispa	115°	44°
Khoana..........................	117°	42°
Sourogana	117° 30′	40° 30′
Phratou	119°	39° 20′

8. And near the other rivers these :—

Alikhorda........................	107°	43° 30′
Khomara	106° 30′	43° 30′
Kouriandra	109° 30′	42° 10′
Kauaris	111° 20°	43°
Astakana........................	112°	42° 20′
Ebousmouanassa or Tosmouanassa	108° 30′	41° 20′
Menapia	113°	41° 20′
Eukratidia	115°	42°
9. Baktra, the king's residence (Balkh)	116°	41°
Estobara	109° 30′	45° 20′
Marakanda (Samarkand) ...	112°	39° 15′
Marakodra	115° 20′	39° 20′

The boundaries of Baktra or Baktriana varied at different periods of history, and were never perhaps at any time fixed with much precision. According to Strabo it was the principal part of Ariana, and was separated from Sogdiana on the east and north-east by the Ôxos, from Areia on the south by the chain of Paropanisos, and on the west from Margiana by a desert region. A description of Baktriana, which Burnes, in his work on Bokhara, corroborates as very accurate, is given by Curtius (lib. VII, c. iv) and is to this effect: "The nature of the Baktrian territory is varied, and presents striking contrasts. In one place it is well-wooded, and bears vines which yield grapes of great size and sweetness. The soil is rich and well-watered—and where such a genial soil is found corn is grown, while lands with an inferior soil are used for the pasturage of cattle. To this fertile tract succeeds another much more extensive, which is nothing but a wild waste of sand parched with drought, alike without inhabitant and without herbage. The winds, moreover, which blow hither from the Pontic Sea, sweep before them the sand that covers the plain, and this, when it gathers into heaps, looks, when seen from a distance, like a collection of great hills; whereby all traces of the road that formerly existed are completely obliterated. Those, therefore, who cross these plains, watch the stars by night as sailors do at sea, and direct their course by their guidance. In fact they almost see better under the shadow of night than in the glare of sunshine. They are, consequently, unable to find their way in the day-time, since there is no track visible which they can follow, for the brightness of the luminaries above is shrouded in darkness. Should now the wind which rises from the sea overtake them, the sands with which it is laden would completely overwhelm them. Nevertheless in all the more favoured localities the number of men and of horses that are there generated is exceedingly great. Baktra itself, the capital city of that region, is situated under mount Paropanisos. The river Bactrus passes by its walls: and gave the city and the region their name." This description is in agreement with the general character of the country from Balkh to Bokhara, in which oases of the most productive soil alternate with wastes of sand.

Baktra figures very early in history. Its capital indeed, Baktra (now Balkh) is one of the oldest cities in the world. The Baktrian Walls is one of the places which Euripides (*Bakkhai,* l. 15) represents Dionysos to have visited in the course of his eastern peregrinations. Ninus, as we learn through Ktêsias, marched into Baktriana with a vast army and, with the assistance of Semiramis, took its capital. In the time of Darius it was a satrapy of the Persian empire and paid a tribute of 360 talents. Alexander the Great, when marching in pursuit of Bessus, passed through Baktria and, crossing the Ôxos, proceeded as far as Marakanda (Samarkand). Having subjugated the regions lying in that direction, he returned to Baktra and there spent the winter before starting to invade India. Some years after the conqueror's death Seleukos reduced Baktria, and annexed it to his other dominions. It was wrested, however, from the hands of the third prince of his line about the year 256 B.C. or perhaps later, by Antiokhos Theos or Theodotos, who made Baktria an independent kingdom. His successors were ambitious and enterprising, and appear to have extended their authority along the downward course of the

Indus even to the ocean, and southward along the coast as far as the mouth of the Narmadâ. The names of these kings have been recovered from their coins found in great numbers both in India and in Afghanistan. This Graeko-Baktrian empire, after having subsisted for about two centuries and a half, was finally overthrown by the invasion of different hordes of the Sakai, named, as Strabo informs us, the Asioi, Pasianoi, Tokharoi and Sakarauloi.[36] These Sakai yielded in their turn to barbarians of their own kindred or at least of their own type, the Skythians, who gave their name to the Indus valley and the regions adjoining the Gulf of Khambhât. Among the most notable Indo-Skythian kings were Kadphises and Kanerkes who reigned at the end of the first and the beginning of the second century of our æra and, therefore, not very long before the time of Ptolemy. Between the Indo-Skythian and Muhammadan periods was interposed the predominancy of Persia in the regions of which we have been speaking.

Ptolemy mentions five rivers which fall into the Ôxos: the Ôkhos, Dargamanês, Zariaspis, Artamis, and Dargoidos, of which the Zariaspis and Artamis unite before reaching the Ôxos. Ptolemy's account cannot be reconciled with the existing hydrography of the country. The Dargamanês is called by Ammianus (lib. XXIII, c. vi) the Orga-

[36] The Wu-sun (of Chinese history) are apparently to be identified with the Asii or Asiani, who, according to Strabo occupied the upper waters of the Iaxartes, and who are classed as nomades with the Tokhâri and Sakarauli (? Sara-Kauli, *i.e.*, Sarikulis).—Kingsmill, in *J. R. A. S.*, N. S., vol. XIV, p. 79.

menes. The Artamis, Wilson thinks, may be the river now called the Dakash (*Ariana Antiqua*, p. 162) and the Dargamanês, the present river of Ghori or Kunduz which is a tributary of the Ôkhos and not of the Ôxos as in Ptolemy. The Ôkhos itself has not been identified with certainty. According to Kinneir it is the Tezen or Tejend which, rising in Sarâkhs, and receiving many confluents, falls into the Kaspian in N. L. 38° 41′. According to Elphinstone it is the river of Herat, either now lost in the sand or going to the Ôxos (*Ariana Antiqua*, p. 146). Bunbury (vol. II, p. 284) points out that in Strabo the Ôkhos is an independent river, emptying into the Kaspian. The Ôkhos of Artemidoros, he says, may be certainly identified with the Attrek, whose course, till lately, was very imperfectly known.

Ptolemy gives a list of thirteen tribes which inhabited Baktrianê. Their names are obscure, and are scarcely mentioned elsewhere.[37]

In the list of towns few known names occur. The most notable are Baktra, Marakanda, Eukratidia and Zariaspa. Baktra, as has been already stated, is the modern Balkh. Heeren (*Asiatic Nations*, 2nd edit., vol. I, p. 424), writes of it in these terms: "The city of Baktra must be regarded as the commercial entrepôt of Eastern Asia: its name belongs to a people who never cease to afford

---

[37] Prof. Beal (*J. R. A. S.*, N. S., Vol. XVI, p. 253), connects the name of the Tokharoi with Tu-ho-lo the name of a country or kingdom Tukhârâ, frequently mentioned by Hiuen Tsiang. The middle symbol *ho*, he says, represents the rough aspirate, and we should thus get Tahra or Tuxra, from which would come the Greek Tokharoi.

matter for historical details, from the time they are first mentioned. Not only does Baktra constantly appear as a city of wealth and importance in every age of the Persian empire, but it is continually interwoven in the traditions of the East with the accounts of Semiramis and other conquerors. It stood on the borders of the gold country, 'in the road of the confluence of nations,' according to an expression of the *Zend-avesta;* and the conjecture that in this part of the world the human race made its first advance in civilisation, seems highly probable." The name of Balkh is from the Sanskrit name of the people of Baktra, the Bahlikas. Marakanda is Samarkand. It was the capital of Sogdiana, but Ptolemy places it in Baktrianê, and considerably to the south of Baktra, although its actual latitude is almost 3 degrees to the north. It was one of the cities of Sogdiana which Alexander destroyed. Its circumference was estimated at 64 stadia, or about 7 miles. The name has been interpreted to mean "warlike province." Eukratidia received its name from the Graeko-Baktrian king, Eukratidês, by whom it was founded. Its site cannot be identified. Pliny makes Zariaspa the same as Baktra, but this must be a mistake. No satisfactory site has been as yet assigned to it.

Cap. 12.

Position of the Sogdianoi.

The Sogdianoi are bounded on the west by that part of Skythia which extends from the section of the Ôxos which is towards Baktrianê and Margianê through the Oxeian mountains as far as the section of the river Iaxartes, which lies in 110° E. 49° N.; on the north likewise by a part of Skythia along the section of the Iaxartes extended thence as far as the limit where its course bends, which lies in 120° E. 48° 30′ N. On the east by the Sakai along the (bending) of the Iaxartes as far as the sources of the bending which lie in 125° E. 43° N., and by the line prolonged from the Sakai to an extreme point which lies in 125° E. 38° 30′ N., and on the east and the south and again on the west by Baktrianê along the section of the Ôxos already mentioned and by the Kaukasian mountains especially so-called, and the adjoining line and the limits as stated, and the sources of the Ôxos.

2. The mountains called the Sogdian extend between the two rivers, and have their extremities lying in .... 111° 47°
and .... 122° 46° 30′

3. From these mountains a good many nameless rivers flow in contrary directions to meet these *two rivers*, and of these nameless rivers one forms the Oxeian Lake, the middle of which lies in 111° E. 45° N., and other two streams descend from the same hilly regions as the Iaxartes—the regions in question are called the Highlands of the Kômêdai. Each of these streams falls into the Iaxartes; one of them is called Dêmos and
its sources lie in .... 124° 43°

Its junction with the river Iaxartes occurs in	123°	47°
The other is the Baskatis whose sources lie in	123°	43°
Its junction with the river Iaxartes occurs in	121°	47° 30′

4. The country towards the Oxeian mountains is possessed by the Paskai, and the parts towards the most northern section of the Iaxartes by the Iatioi, and the Tokharoi, below whom are the Augaloi; then along the Sogdian mountains the Oxydrângkai and the Drybaktai, and the Kandaroi, and below the mountains the Mardyênoi, and along the Ôxos the Ôxeianoi and the Khôrasmioi, and farther east than these the Drepsianoi, and adjoining both the rivers, and still further east than the above the Anieseis along the Iaxartes, and the Kirrhâdai (or Kirrhodeeis) along the Ôxos, and between the Kaukasos Range and Imaos the country called Ouandabanda.

5. Towns of the Sogdianoi in the highlands along the Iaxartes are these:—

Kyreskhata	124°	43° 40′
Along the Ôxos:—		
Oxeiana	117° 30′	44° 20′
Marouka	117° 15′	43° 40′
Kholbêsina	121°	43°

6. Between the rivers and higher up—

Trybaktra........................	112° 15′	
Alexandreia Oxeianê	113°	44° 20′
Indikomordana	115°	44° 20′
Drepsa (or Rhepsa) the Metropolis	120°	45°
Alexandreia Eskhatê (*i.e.* Ultima)	122°	41°

Sogdiana was divided from Baktriana by the river Ôxos and extended northward from thence to the river Iaxartes. The Sakai lay along the eastern frontier and Skythic tribes along the western. The name exists to this day, being preserved in Soghd which designates the country lying along the river Kohik from Bokhara eastward to Samarkand. The records of Alexander's expedition give much information regarding this country, for the Makedonian troops were engaged for the better part of three years in effecting its subjugation.

In connexion with Sogdiana, Ptolemy mentions four mountain ranges—the Kaukasian, the Sogdian, the mountain district of the Kômêdai, and Imaos. Kaukasos was the general name applied by the Makedonians to the great chain which extended along the northern frontiers of Afghanistan, and which was regarded as a prolongation of the real Kaukasos. Ptolemy uses it here in a specific sense to designate that part of the chain which formed the eastern continuation of the Paropanisos towards Imaos. Imaos is the meridian chain which intersects the Kaukasos, and is now called Bolor Tâgh. Ptolemy places it about 8 degrees too far eastward. The

Sogdian Mountains, placed by Ptolemy between the Iaxartes and Ôxos, towards their sources, are the Thian Shan. The Kômêdai, who gave their name to the third range, were, according to Ptolemy, the inhabitants of the hill-country which lay to the east of Baktriana and up whose valley lay the route of the caravans from Baktra, bound for Sêrika across Imaus or the Thsung-lung. Cunningham has identified them with the Kiu-mi-tho (Kumidha) of Hiuen Tsiang. Their mountain district is that called Muz-tâgh.

The rivers mentioned in connexion with Sogdiana are the Ôxos, and the Iaxartes, with its two tributaries, the Baskatis and the Dêmos. The Ôxos takes its rise in the Pamîr[38] Lake, called the Sari-Kul (or Yellow Lake), at a distance of fully 300 miles to the south of the Iaxartes. It is fed on its north bank by many smaller streams which run due south from the Pamîr uplands, breaking the S.W. face of that region into a series of valleys, which, though rugged, are of exuberant fertility. Its course then lies for

---

[38] The Pamîr plateau between Badakshan and Yarkand connects several chains of mountains, viz. the Hindu Kush in the S.W. the Kuen-luen in the E., the Karar Korum in the Bolor, the Thian-shân chain in the north, which runs from Tirak Dawan and Ming-yol to the Western Farghana Pass. This plateau is called *Bâm-i-dunyâ* or *Roof of the World*. With regard to the name Pamîr Sir H. Rawlinson says: "My own conjecture is that the name of Pamîr, or Fâmir, as it is always written by the Arabs, is derived from the Fani (φαυνοι), who, according to Strabo bounded the Greek kingdom of Baktria to the E. (XI. 14) and whose name is also preserved in Fân-tâû, the Fân-Lake, &c. Fâmîr for Fân-mîr would then be a compound like Kashmir, Aj-mir, Jessel-mir, &c. signifying 'the lake country of the Fâ-ni.'' (*J. R. G. S.* XLII. p 189, n.).'

hundreds of miles through arid and saline steppes till before reaching the sea of Aral it is dissipated into a network of canals, both natural and artificial. Its delta, which would otherwise have remained a desert, has thus been converted into a fruitful garden, capable of supporting a teeming population, and it was one of the very earliest seats of civilization.[39] The deflexion of the waters of the Ôxos into the Aral, as Sir H. Rawlinson points out, has been caused in modern times not by any upheaval of the surface of the Turcoman desert, but by the simple accidents of fluvial action in an alluvial soil. The name of the river is in Sanskrit *Vakshu,* Mongolian, Bakshu, Tibetan *Pakshu* Chinese *Po-thsu,* Arabic and Persian *Vakhsh-an* or *âb*—from Persian *vah* = 'pure,' or Sanskrit *Vah* = 'to flow.' The region embracing the head-waters of the Ôxos appears to have been the scene of the primæval Aryan Paradise. The four rivers thereof, as named by the Brahmans, were the Sita, the Alakananda, the Vakshu, and the Bhadro = respectively, according to Wilson to the Hoang-ho, the Ganges, the Ôxos, and the Oby. According to the Buddhists the rivers were the Ganges, the Indus, the Ôxos, and the Sita, all of which they derived from a great central lake in the plateau of Pamîr, called A-neou-ta = Kara-kul or Sarik-kul Lake.

The Iaxartes is now called the Syr-darya or

[39] "Abu Rihan says that the Solar Calendar of Khwârasm was the most perfect scheme for the measurement of time with which he was acquainted. Also that the Khwârasmians dated originally from an epoch anterior by 980 years to the aera of the Seleucidae=134 B.C." (See *Quarterly Review*, No. 240, Art. on *Central Asia*).

Yellow River. The ancients sometimes called it the Araxes, but, according to D'Anville, this is but an appellative common to it with the Amu or Ôxos, the Armenian Aras and the Rha or Volga. The name Iaxartes was not properly a Greek word but was borrowed from the barbarians by whom, as Arrian states (*Anab.* lib. III. c. xxx), it was called the Orxantes. It was probably derived from the Sanskrit root *kshar*, "to flow" with a semitic feminine ending, and this etymology would explain the modern form of *Sirr.* See *J. R. G. S.* XLII. p. 492, n. The Iaxartes rises in the high plateau south of Lake Issyk-kul in the Thian Shan. Its course is first to westward through the valley of Khokan, where it receives numerous tributaries. It then bifurcates, the more northern branch retaining the name of Syr-darya. This flows towards the north-west, and after a course of 1150 miles from its source enters the Sea of Aral. Ptolemy however, like all the other classical writers, makes it enter the Kaspian sea. Humboldt accounts for this apparent error by adducing facts which go to show that the tract between the Aral and the Kaspian was once the bed of an united and continuous sea, and that the Kaspian of the present day is the small residue of a once mighty Aralo-Kaspian Sea. Ammianus Marcellinus (lib. XXIII, c. vi), describing Central Asia in the upper course of the Iaxartes which falls into the Kaspian, speaks of two rivers, the Araxates and Dymas (probably the Dêmos of Ptolemy) which, rushing impetuously down from the mountains and passing into a level plain, form therein what is called the Oxian lake, which is spread over a vast area. This is the earliest intimation of the Sea of Aral. (See Smith's *Dict. of Anc. Geog.* s. v.). Bunbury, however, says (vol. II, pp. 641-2): "Nothing but the unwillingness of modern writers to admit that the ancients were unacquainted with so important a feature in the geography of Central Asia as the Sea of Aral could have led them to suppose it represented by the Oxiana Palus of Ptolemy. While that author distinctly describes both the Jaxartes and the Oxus as flowing into the Caspian Sea, he speaks of a range of mountains called the Sogdian Mountains, which extend between the two rivers, from which flow several nameless streams into those two, one of which forms the Oxian lake. This statement exactly tallies with the fact that the Polytimetos or river of Soghd, which rises in the mountains in question, does not flow into the Oxus, but forms a small stagnant lake called Kara-kul or Denghiz; and there seems no doubt this was the lake meant by Ptolemy. It is true that Ammianus Marcellinus, in his description of these regions, which is very vague and inaccurate, but is based for the most part upon Ptolemy, terms it a large and widespread lake, but this is probably nothing more than a rhetorical flourish." The Iaxartes was regarded as the boundary towards the east of the Persian Empire, which it separated from the nomadic Skythians. The soldiers of Alexander believed it to be the same as the Tanais or Don.

In the list of the tribes of Sogdiana some names occur which are very like Indian, the Kandaroi, who may be the Gandhâras, the Mardyênoi, the Madras, the Takhoroi, the Takurs, and the

Kirrhadai (or Kirrhodeeis) the Kirâta. The name of the Khorasmioi has been preserved to the present day in that of Khwârazm, one of the designations of the Khanate of Khiva. The position of the Khorasmioi may be therefore assigned to the regions south of the Sea of Aral, which is sometimes called after them the Sea of Khwârazm. The Drepsianoi had their seats on the borders of Baktria, as Drepsa, one of their cities and the capital of the country, may be identified with Andarâb, which was a Baktrian town. It is called by Strabo Adrapsa and Darapsa—(lib. XI, c. xi, 2, and lib. XV, c. ii, 10) and Drapsaka by Arrian—(*Anab.* lib. III, c. 39). Bunbury (vol. I, p. 427, n. 3) remarks: "The Drepsa of Ptolemy, though doubtless the same *name*, cannot be the same place (as the Drapsaka of Arrian, *Anab.* lib. III, c. xxix.) as that author places it in Sogdiana, considerably to the north of Marakanda." Ptolemy, however, as I have already pointed out, places Marakanda to the south of Baktra. Kingsmill (*J. R. A. S.*, N. S., vol. XIV, p. 82) identifies Darapsa with the Lam-shi-ch'eng of the Chinese historians. It was the capital of their Ta-hia (Tokhâra—Baktria) which was situated about 2000 li south-west of Ta-wan (Yarkand), to the south of the Kwai-shui (Ôxos). The original form of the name was probably, he says, Darampsa. In Ta-wan he finds the Phrynoi of Strabo. The region between Kaukasos and Imaös, Ptolemy calls Vandabanda, a name of which, as Wilson conjectures, traces are to be found in the name of Badakshân.

With regard to the towns Mr. Vaux remarks,

(Smith's *Dict.* s. v. Sogdiana): "The historians of Alexander's march leave us to suppose that Sogdiana abounded with large towns, but many of these, as Prof. Wilson has remarked, were probably little more than forts erected along the lines of the great rivers to defend the country from the incursions of the barbarous tribes to its N. and E. Yet these writers must have had good opportunity of estimating the force of these places, as Alexander appears to have been the best part of three years in this and the adjoining province of Baktriana. The principal towns, of which the names have been handed down to us, were Kyreskhata or Kyropolis on the Iaxartes (Steph. Byz. *s. v.*; Curt. lib. VI, c. vi) Gaza (Ghaz or Ghazni, Ibn Haukal, p. 270); Alexandreia Ultima (Arrian, lib. III, c. xxx; Curt. *l. c.*; Am. Marc., lib. XXIII, c. vi) doubtless in the neighbourhood, if not on the site of the present Khojend; Alexandreia Oxiana (Steph. Byz. *s. v.*); Nautaka (Arrian, *An.* lib. III, c. xxviii; lib. IV, c. xviii) in the neighbourhood of Karshi or Naksheb. Brankhidae, a place traditionally said to have been colonized by a Greek population; and Marginia (Curt., lib. VII, c. x, 15) probably the present Marghinan."

Cap. 13.

Position of the Sakai.

[*Map of Asia* 7.]

1. The Sakai are bounded on the west by the Sogdianoi along their eastern side already described, on the north by Skythia along the

line parallel to the river Iaxartes as far as the limit *of the country* which lies in 130° E. 49° N. on the east in like manner by Skythia along the meridian lines prolonged from thence and through the adjacent range of mountains called Askatangkas as far as the station at Mount Imaös, whence traders start on their journey to Sêra which lies in 140° E. 43° N., and through Mount Imaös as it ascends to the north as far as the limit *of the country* which lies in 143° E. 35° N., and on the south by Imaös itself along the line adjoining the limits that have been stated.

2. The country of the Sakai is inhabited by nomads. They have no towns, but dwell in woods and caves. Among the Sakai is the mountain district, already mentioned, of the Kômêdai, of which the ascent from the

Sogdianoi lies in	125°	43°
And the parts towards the valley of the Kômêdai lie in.........	130°	39°
And the so-called Stone Tower lies in	135°	43°

3. The tribes of the Sakai, along the Iaxartes, are the Karatai and the Komaroi, and the people who have all the mountain region are the Kômêdai, and the people along the range of Askatangka the Massagetai; and the people between are the Grynaioi Skythai and the Toörnai, below whom, along Mount Imaös, are the Byltai.

In the name of the mountain range on the east of the Sakai, A s k a-t a n g k-a s, the middle syllable represents the Turkish word *tágh*—'mountain.' The tribe of the K a r a t a i, which was seated along the banks of the Iaxartes, bears a name of common application, chiefly to members of the Mongol family—that of Karait. The name of the Massagetai, Latham has suggested, may have arisen out of the common name *Mustágh*, but Beal, as already stated, refers it to the Moeso-gothic "*maiza*" and "Yue-chi—Getæ." The B y l t a i are the people of what is now called Little Tibet and also Baltistân.

Cap. 14.

Position of Skythia within Imaös.

[*Map of Asia* 7.]

1. S k y t h i a within Imaös is bounded on the west by Sarmatia in Asia along the side already traced, on the north by an unknown land, on the east by Mount Imaös ascending to the north pretty nearly along the meridian of the starting-place already mentioned as far as the unknown land .... 140° 63°, on the south and also on the east by the Sakai and the Sogdianoi and by Marginê along their meridians already mentioned as far as the Hyrkanian Sea at the mouth of the Ôxos, and also by the part of the Hyrkanian Sea lying between the north of the Ôxos and the river Rhâ according to such an outline.

2. The bend of the River Rhâ which marks the boundary of Sarmatia and

Skythia........................	85°	54°
with the mouth of the river Rhâ which lies in	87° 30′	48° 50′
Mouth of the river Rhymmos	91°	48° 45′
Mouth of the river Daïx ...	94°	48° 45′
Mouth of the river Iaxartes	97°	48°
Mouth of the river Iästos ...	100°	47° 20′
Mouth of the river Polytimêtos	103°	45° 30′
Aspabôta, a town	102°	44°

after which comes the mouth of the Ôxos.

3. The mountains of Skythia within Imaös are the more eastern parts of the Hyperborean hills and the mountains called

Alana, whose extremities lie	105°	59°
and	118°	59° 30′

4. And the Rymmik mountains whose ex-

tremities lie	90°	54°
and	99°	47° 30′

from which flow the Rymmos and some other streams that discharge into the River Rhâ, uniting with the Daïx river.

5. And the Norosson range, of which the

extremities lie..................	97°	53° 30′
and	106°	52° 30′

and from this range flow the Daïx and some other tributaries of the Iaxartes.

6. And the range of mountains called Aspisia whose extremities lie 111° 55° 30′
and .... 117° 52° 30′
and from these some streams flow into the River Iaxartes.

7. And the mountains called Tapoura whose extremities lie .... 120° 56°
and .... 125° 49°
from which also some streams flow into the Iaxartes.

8. In addition to these in the depth of the region of the streams are the Syêba mountains whose extremities lie .... 121° 58°
and .... 132° 62°
and the mountains called the Anarea whose extremities lie .... 130° 56°
and.............................. 137° 50°
after which is the bend in the direction of Imaŏs continuing it towards the north.

9. All the territory of this Skythia in the north, adjoining the unknown regions, is inhabited by the people commonly called the Alanoi Skythai and the Souobênoi and the Alanorsoi, and the country below these by the Saitianoi and the Massaioi and the Syêboi, and along Imaös on the outer side the Tektosakes, and near the most eastern sources of the river Rhâ the Rhoboskoi below whom the Asmanoi.

10. Then the Paniardoi, below whom, more towards the river, the country of Kano-

dipsa, and below it the Koraxoi, then the Orgasoi, after whom as far as the sea the Erymmoi, to east of whom are the Asiôtai, then the Aorsoi, after whom are the Iaxartai, a great race seated along their homonymous river as far as to where it bends towards the Tapoura Mountains, and again below the Saitanioi are the Mologênoi, below whom, as far as the Rymmik range, are the Samnîtai.

11. And below the Massaioi and the Alana Mountains are the Zaratai and the Sasones, and further east than the Rymmik Mountains are the Tybiakai, after whom, below the Zaratai, are the Tabiênoi and the Iâstai and the Makhaitêgoi along the range of Norosson, after whom are the Norosbeis and the Norossoi, and below these the Kakhagai Skythai along *the country of* the Iaxartai.

12. Further west than the Aspisia range are the Aspisioi Skythai, and further east the Galaktophagoi Skythai, and in like manner the parts farther east than the Tapoura and Syêba ranges are inhabited by the Tapoureoi.

13. The slopes and summits of the Anarea Mountains and Mount Askatangkas are inhabited by the homonymous Anareoi Skythai below the Alanorsoi, and the Askatangkai

S k y t h a i further east than the Tapoureoi, and as far as Mount Imaös.

14. But the parts between the Tapoura Mountains and the slope towards the mouth of the Iaxartes and the seacoast between the two rivers are possessed by the A r i a k a i, along the Iaxartes and below these the N a m o s t a i, then the S a g a r a u k a i, and along the river Ôxos the R h i b i o i, who have a town

Dauaba ....104° 45°.

The country of the Skyths is spread over a vast area in the east of Europe and in Western and Central Asia. The knowledge of the Skyths by the Greeks dates from the earliest period of their literature, for in Homer (*Iliad,* lib. XIII, l. 4) we find mention made of the Galaktophagoi (milk-eaters) and the Hippemologoi (mare-milkers) which must have been Skythic tribes, since the milking of mares is a practice distinctive of the Skyths. Ptolemy's division of Skythia into within and beyond Imaös is peculiar to himself, and may have been suggested by his division of India into within and beyond the Ganges. Imaös, as has already been pointed out is the Bolor chain, which has been for ages the boundary between Turkistân and China. Ptolemy, however, placed Imaös too far to the east, 8° further than the meridian of the principal source of the Ganges. The cause of this mistake, as a writer in Smith's *Dictionary* points out, arose from the circumstance that the data upon which Ptolemy came to his conclusion were selected from two different sources. The Greeks first became acquainted with the

Kômêdorum Montes when they passed the Indian Kaukasos between Kâbul and Balkh, and advanced over the plateau of Bâmiyân along the west slopes of Bolor, where Alexander found in the tribe of the Sibae the descendants of Hêraklês, just as Marco Polo and Burnes met with people who boasted that they had sprung from the Makedonian conquerors. The north of Bolor was known from the route of the traffic of the Sêres. The combination of notations obtained from such different sources was imperfectly made, and hence the error in longitude. This section of Skythia comprised Khiva, the country of the Kosaks, Ferghâna, Tashkend, and the parts about the Balkash.

The rivers mentioned in connexion with Skythia within Imaös are the Ôxos, Iaxartes, Rhâ, Rhymmos, Daïx, Iästos and Polytimêtos. The Rhâ is the Volga, which is sometimes called the Rhau by the Russians who live in its neighbourhood. Ptolemy appears to be the first Greek writer who mentions it. The Rhymmos is a small stream between the Rhâ and the Ural river called the Narynchara. The Daïx is the Isik or Ural river. The Iästos was identified by Humboldt with the Kizil-darya, which disappeared in the course of last century, but the dry bed of which can be traced in the barren wastes of Kizil-koum in W. Turkestân. With regard to the Polytimêtos, Wilson says (*Arian. Antiq.* p. 168); "There can be no hesitation in recognizing the identity of the Polytimêtês and the Zarafshân, or river of Samarkand, called also the Kohik, or more correctly the river of *the* Kohak; being so termed from its passing by a rising ground, a Koh-ak, a 'little hill' or 'hillock,' which lies to the east of the city. According to Strabo, this river traversed Sogdiana and was lost in the sands. Curtius describes it as entering a cavern and continuing its course underground. The river actually terminates in a small lake to the south of Bokhara, the Dangiz, but in the dry weather the supply of water is too scanty to force its way to the lake, and it is dispersed and evaporated in the sands. What the original appellation may have been does not appear, but the denominations given by the Greeks and Persians 'the much-honoured' or 'the gold-shedding' stream convey the same idea, and intimate the benefits it confers upon the region which it waters." Ptolemy is wide astray in making it enter the Kaspian.

The mountains enumerated are the Alana, Rhymmika, Norosson, Aspisia, Tapoura, Syêba, and Anarea. By the Alana Mountains, which lay to the east of the Hyperboreans, it has been supposed that Ptolemy designated the northern part of the Ural Chain. If so, he has erroneously given their direction as from west to east. The Rhymmik mountains were probably another branch of that great meridian chain which consists of several ranges which run nearly parallel. The Norosson may be taken as Ptolemy's designation for the southern portion of this chain. The Aspisia and Tapoura mountains lay to the north of the Iaxartes. The latter, which are placed three degrees further east than the Aspisia, may be the western part of the Altai. The Syêba stretched still farther eastward with an inclination northward. To the southward of them were the A n a r e a, which may be placed near the sources of the Obi and the Irtish, forming one of the western branches of the Altai. Ptolemy erroneously prolongs the chain of Imaös to these high latitudes.

Ptolemy has named no fewer than 38 tribes belonging to this division of Skythia. Of these the best known are the A l a n i, who belonged also to Europe, where they occupied a great portion of Southern Russia. At the time when Arrian the historian was Governor of Kappadokia under Hadrian, the Asiatic Alani attacked his province, but were repelled. He subsequently wrote a work on the tactics to be observed against the Alani (ἔκταξις κατ 'Αλανῶν) of which some fragments remain. The seats of the Alani were in the north of Skythia and adjacent to *the unknown land*, which may be taken to mean the regions stretching northward beyond Lake Balkash. The position of the different tribes is fixed with sufficient clearness in the text. These tribes were essentially nomadic, pastoral and migratory—hence in Ptolemy's description of their country towns are singularly conspicuous by their absence.

CAP. 15.

THE POSITION OF S K Y T H I A BEYOND I M A Ö S.

[*Map of Asia*, 8.]

1. S k y t h i a beyond Mount Imaös is bounded on the west by Skythia within Imaös, and the Sakai along the whole curvature of the

mountains towards the north, and on the north by the unknown land, and on the east by Serikê in a straight line whereof the extremities

lie in .... 150° 63°

and .... 160° 35°

and on the south by a part of India beyond the Ganges along the parallel of latitude which cuts the southern extremity of the line just mentioned.

2. In this division is situated the western part of the Auxakian Mountains, of which the

extremities lie .... 149° 49°

and .... 165° 54°

and the western part of the mountains

called Kasia, whose extremities lie in 152° 41°

and.................................. 162° 44°

and also the western portion of Emôdos,

whose extremities lie in .... 153° 36°

and.................................. 165° 36°

and towards the Auxakians, the source

of the River Oikhardês lying in...... 153° 51°

3. The northern parts of this Skythia are possessed by the Abioi Skythai, and the parts below them by the Hippophagoi Skythai, after whom the territory of Auxakîtis extends onward, and below this again, at the starting place already mentioned, the Kasian land, below which are the Khatai Skythai, and then succeeds the Akhasa land, and below it along the Emôda the Kharaunaioi Skythai.

4. The towns in this division are these:—

Auxakia	143°	49° 40′
Issêdôn Skythikê	150°	49° 30′
Khaurana................................	150°	37° 15′
Soita..	145°	35° 20′

Skythia beyond Imaös embraced Ladakh, Tibet, Chinese Tartary and Mongolia. Its mountains were the Auxakian and Kasian chains, both of which extended into Sêrikê, and Emôdos. The Auxakians may have formed a part of the Altai, and the Kasians, which Ptolemy places five degrees further south, are certainly the mountains of Kâshgar. The Emôdos are the Himalayas.

The only river named in this division is the Oikhardês, which has its sources in three different ranges, the Auxakian, the Asmiraean and the Kasian. According to a writer in Smith's *Dictionary* the Oikhardês "may be considered to represent the river formed by the union of the streams of Khotan, Yarkand, Kashgar and Ushi, and which flows close to the hills at the base of the Thian-shan. Saint-Martin again inclines to think Œchardês may be a designation of the Indus, while still flowing northward from its sources among the Himalayas. "Skardo," he says, (*Étude*, p. 420) "the capital of the Balti, bears to the name of the Oikhardês (Chardi in Amm. Marc. 2) a resemblance with which one is struck. If the identification is well founded, the river Oichardês will be the portion of the Indus which traverses Balti and washes the walls of Skardo."

In the north of the division Ptolemy places the Abioi Skythai. Homer, along with the Galaktophagoi and Hippêmolgoi, mentions the Abioi. Some think that the term in the passage designates a distinct tribe of Skythians, but others take it to be a common adjective, characterizing the Skythians in general as very scantily supplied with the means of subsistence. On the latter supposition the general term must in the course of time have become a specific appellation. Of the four towns which Ptolemy assigns to the division, one bears a well-known name, Issêdôn, which he calls Skythikê, to distinguish it from Issêdôn in Serikê. The name of the Issêdônes occurs very early in Greek literature, as they are referred to by the Spartan poet Alkman, who flourished between 671 and 631 B. C. He calls them Assedones *Frag.* 94, ed. Welcker). They are mentioned also by Hekataios of Miletos. In very remote times they were driven from the steppes over which they wandered by the Arimaspians. They then drove out the Skythians, who in turn drove out the Kimmerians. Traces of these migrations are found in the poem of Aristeas of Prokonnesos, who is fabled to have made a pilgrimage to the land of the Issêdones. Their position has been assigned to the east of Ichin, in the steppe of the central horde of the Kirghiz, and that of the Arimaspi on the northern declivity of the Altai. (Smith's *Dict.* s. v.) This position is not in accordance with Ptolemy's indications. Herodotos, while rejecting the story of the Arimaspians and the griffins that guarded their gold, admits at the same time that by far the greatest quantity of gold came from the north of Europe, in which he included the tracts along the Ural, and Altai ranges. The abundance of gold among the Skythians on the Euxine is attested by the contents of their tombs, which have been opened in modern times. (See Bunbury, vol. I, p. 200.)

Regarding Ptolemy's Skythian geography, Bunbury says (vol. II. p. 597): "It must be admitted that Ptolemy's knowledge of the regions on either side of the Imaos was of the vaguest possible character. Eastward of the Rhâ (Volga), which he regarded as the limit between Asiatic Sarmatia and Skythia, and north of the Iaxartes, which he describes like all previous writers as falling into the Kaspian—he had, properly speaking, no geographical knowledge whatever. Nothing had reached him beyond the names of tribes reported at second-hand, and frequently derived from different authorities, who would apply different appellations to the same tribe, or extend the same name to one or more of the wandering hordes, who were thinly dispersed over this vast extent of territory. Among the names thus accumulated, a compilation that is probably as worthless as that of Pliny, notwithstanding its greater pretensions to geographical accuracy, we find some that undoubtedly represent populations really existing in Ptolemy's time, such as the Alani, the Aorsi, &c., associated with others that were merely poetical or traditional, such as the Abii, Galaktophagi and Hippophagi, while the Issêdones, who were placed by Herodotos immediately east of the Tanais, are strangely transferred by Ptolemy to the far East, on the very borders of Serika; and he has even the name of a *town* which he calls Issedon Serika, and to which he assigns a position in longitude 22° east of Mount Imaös, and not less than 46° east of Baktra. In one essential point, as has been already pointed out, Ptolemy's conception of Skythia differed from that of all preceding geographers, that instead of regarding it as bounded on the north and east by the sea, and consequently of comparatively limited extent, he considered it as extending without limit in both directions, and bounded only by 'the unknown land,' or, in other words, limited only by his own knowledge."

CAP. 16.

POSITION OF SERIKÊ.

[*Map of Asia*, 8].

Serikê is bounded on the west by Skythia, beyond Mount Imaös, along the line already mentioned, on the north by the unknown land along the same parallel as that through Thulê, and on the east, likewise by the unknown land along the meridian of which the extremities lie..180° 63°
and ....180° 55°
and on the south by the rest of India beyond the Ganges through the same parallel as far as the extremity lying ....173° 55°
and also by the Sinai, through the line prolonged till it reaches the already mentioned extremity towards the unknown land.

2. Serikê is girdled by the mountains called Anniba, whose extremities lie ...153° 60°
and ....171° 56°

and by the eastern part of the Auxakians, of which the extremity lies ......165° 54°
and by the mountains called the Asmiraia whose extremities lie ....167° 47° 30′
and ....174° 47° 30′
and by the eastern part of the Kasia range, whose extremities lie ....162° 44°
and ....171° 40°
and by Mount Thagouron whose centre lies..............................170° 43°
and also by the eastern portion of the mountains called Emôda and Sêrika, whose extremity lies..............................165° 36°
and by the range called Ottorokorrhas, whose extremities lie ....169° 36°
and ....176° 38°

3. There flow through the far greatest portion of Sêrikê two rivers, the Oikhardês, one of whose sources is placed with the Auxakioi, and the other which is placed in the Asmiraian mountains lies in.....................174° 47° 30′
and where it bends towards the Kasia range ....160° 48° 30′
but the source in them lies.........161° 44° 15′
and the other river is called the Bautisos, and this has one of its sources in the Kasia range in ....160° 43°
another in Ottorokorrha...........176° 39°
and it bends towards the Emôda in168° 39°
and its source in these lies.........160° 37°

4. The most northern parts of Sêrike are

inhabited by tribes of cannibals, below whom is the nation of the Anniboi, who occupy the slopes and summits of the homonymous mountains. Between these and the Auxakioi is the nation of the Syzyges, below whom are the Dâmnai, then as far as the river Oikhardes the Pialai (or Piaddai), and below the river the homonymous Oikhardai.

5. And again farther east than the Anniboi are the Garinaioi and the Rhabannai or Rhabbanaioi, and below the country of Asmiraia, above the homonymous mountains. Beyond these mountains as far as the Kasia range the Issêdones, a great race, and further east than these the Throanoi, and below these the Ithagouroi, to the east of the homonymous mountains, below the Issêdones, the Aspakârai, and still below those the Bâtai, and farthest south along the Emôda and Sêrika ranges the Ottorokorrhai.

6. The cities in Sêrikê are thus named:—

Damna	156°	51° 20′
Piala (or Piadda)	160°	49° 40′
Asmiraia	170°	48°
Throana	174° 40′	47° 40′
7. Issêdôn Serikê	162°	45°
Aspakara (or Aspakaia)	162° 30′	41° 40′
Drôsakhê (or Rhosakla)	167° 40′	42° 30′
Paliana	162° 30′	41°
Abragana	163° 30′	39° 30′

8. Thogara	171° 20′	39° 40′
Daxata	174°	39° 30′
Orosana	162°	37° 30′
Ottorokorrha	165°	37° 15′
Solana	169°	37° 30′
Sôra metropolis	177°	38° 35′

The chapter which Ptolemy has devoted to S ê r i k ê has given rise to more abortive theories and unprofitable controversies than any other part of his work on Geography. The position of Serikê itself has been very variously determined, having been found by different writers in one or other of the many countries that intervene between Eastern Turkistan in the north and the province of Pegu in the south. It is now however generally admitted that by Sêrikê was meant the more northern parts of China, or those which travellers and traders reached by land. At the same time it is not to be supposed that the names which Ptolemy in his map has spread over that vast region were in reality names of places whose real positions were to be found so very far eastward. On the contrary, most of the names are traceable to Sanskrit sources and applicable to places either in Kaśmîr or in the regions immediately adjoining. This view was first advanced by Saint-Martin, in his dissertation on the Serikê of Ptolemy (*Étude*, pp. 411 ff.) where he has discussed the subject with all his wonted acuteness and fulness of learning. I may translate here his remarks on the points that are most prominent: "All the nomenclature," he says (p. 414), "except some names at the extreme points north and east, is certainly of Sanskrit origin. . . . . . To the south of the mountains, in the Panjâb, Ptolemy indicates under the general name of Kaspiraei an extension genuinely historical of the Kaśmîrian empire, with a detailed nomenclature which ought to rest upon informations of the 1st century of our æra; whilst to the north of the great chain we have nothing more than names thrown at hazard in an immense space where our means of actual comparison show us prodigious displacements. This difference is explained by the very nature of the case. The Brâhmans, who had alone been able to furnish the greater part of the information carried from India by the Greeks regarding this remotest of all countries, had not themselves, as one can see from their books, anything but the most imperfect notions. Some names of tribes, of rivers, and of mountains, without details or relative positions—this is all the Sanskrit poems contain respecting these high valleys of the North. It is also all that the tables of Ptolemy give, with the exception of the purely arbitrary addition of graduations. It is but recently that we ourselves have become a little better acquainted with these countries which are so difficult of access. We must not require from the ancients information which they could not have had, and it is of importance also that we should guard against a natural propensity which disposes us to attribute to all that antiquity has transmitted to us an authority that we do not accord without check to our best explorers. If the meagre nomenclature inscribed by Ptolemy on his map, of the countries situated beyond (that is to the east) of Imaös, cannot lead to a regular correspondence with our existing notions, that which one can recognize, suffices nevertheless to determine and circumscribe its general position. Without wishing to carry into this more precision than is consistent with the nature of the indications, we may say, that the indications, taken collectively, place us in the midst of the Alpine region, whence radiate in different directions the Himâlaya, the Hindu-Kôh and the Bolor chain—enormous elevations enveloped in an immense girdle of eternal snows, and whose cold valleys belong to different families of pastoral tribes. Kaśmîr, a privileged oasis amidst these rugged mountains, appertains itself to this region which traverses more to the north the Tibetan portion of the Indus (above the point where the ancients placed the sources of the Indus) and whence run to the west the Ôxos and Iaxartes. With Ptolemy the name of Imaös (the Greek transcription of the usual form of the name of Himâlaya) is applied to the central chain from the region of the sources of the Ganges (where rise also the Indus and its greatest affluent, the Śatadru or Satlaj) to beyond the sources of the Iaxartes. The general direction of this great axis is from south to north, saving a bend to the south-east from Kaśmîr to the sources of the Ganges; it is only on parting from this last point that the Himâlaya runs directly to the east, and it is there also that with Ptolemy the name of Emôdos begins, which designates the Eastern Himâlaya. Now it is on Imaos itself or in the vicinity of this grand system of mountains to the north of our Panjâb and to the east of the valleys of the Hindu-Kôh and of the upper Ôxos that there come to be placed, in a space from 6 to 7 degrees at most from south to north, and less perhaps than that in the matter of the longitudes, all the names which can be identified on the map where Ptolemy has wished to represent, in giving them an extension of nearly 40 degrees from west to east, the region which he calls Skythia beyond Imaös and Serika. One designation is there immediately recognizable among all the others—that of Kasia. Ptolemy indicates the situation of the country of Kasia towards the bending of Imaös to the east above the sources of the Ôxos, although he carries his Montes Kasii very far away from that towards the east; but we are sufficiently aware beforehand that here, more than in any other part of the Tables, we have only to attend to the nomenclature, and to leave the notations altogether out of account. The name of the Khaśa has been from time immemorial one of the appellations the most spread through all the Himâlayan range. To keep to the western parts of the chain, where the indication of Ptolemy places us, we there find Khaśa mentioned from the heroic ages of India, not only in the *Itihâsas* or legendary stories of the *Mahâbhârata*, but also in the law book of Manu, where their name is read by the side of that of the Darada, another people well known, which borders in fact on the Khaśa of the north. The Khaśa figure also in the Buddhist Chronicles of Ceylon, among the people subdued by Aśôka in the upper Panjâb, and we find them mentioned in more than 40 places of the Kaśmîr *Chronicle* among the chief mountain tribes that border on Kaśmîr. Baber knows also that a people of the name of Khas is indigenous to the high valleys in the neighbourhood of the Eastern Hindu-Kôh; and, with every reason, we attach to this indigenous people the origin of the name of Kâshgar, which is twice reproduced in the geography of these high regions. Khaśagiri in Sanskrit, or, according to a form more approaching the Zend, Khaśaghaïri, signifies properly the mountains of the Khaśa. The Akhasa Khôra, near the Kasia regio, is surely connected with the same nationality. The Aspakârai, with a place of the same name (Aspakara) near the Kasii Montes, have no correspondence actually known in these high valleys, but the form of the name connects it with the Sanskrit or Iranian nomenclature. Beside the Aspakarai, the Batai are found in the Bâutta of the *Rájataraṅgiṇi*. . . . . . In the 10th century of our æra, the Chief of Ghilghit took the title of Bhâtshâh or Shah of the Bhât. The Balti, that we next name, recall a people, mentioned by Ptolemy in this high region, the Byltai. The accounts possessed by Ptolemy had made him well acquainted with the general situation of the Byltai in the neighbourhood of the Imaös, but he is either ill informed or has ill applied his information as to their exact position, which he indicates as being to the west of the great chain of Bolor and not to the east of it, where they were really to be found. The Ramana and the Daśamana, two people of the north, which the *Mahábhárata* and the Pauranik lists mention along with the Chîna, appear to us not to differ from the Rhabannae and the Damnai of Ptolemy's table." Saint-Martin gives in the sequel a few other identifications—that of the Throanoi (whose name should be read Phrounoi, or rather Phaunoi as in Strabo) with the Phuna of the *Lalitavistara* (p. 122)—of the Kharaunaioi with the Kajana, whose language proves them to be Daradas, and of the Ithagouroi with the Dangors, Dhagars or Dakhars, who must at one time have been the predominant tribe of the Daradas. The country called Asmiraia he takes, without hesitation, to be Kaśmîr itself. As regards the name Ottorokorrha, applied by Ptolemy to a town and a people and a range of mountains, it is traced without difficulty to the Sanskrit—Uttara-kuru, *i.e.*, the Kuru of the north which figures in Indian mythology as an earthly paradise sheltered on every side by an encircling rampart of lofty mountains, and remarkable for the longevity of its inhabitants, who lived to be 1000 and 10,000 years old. Ptolemy was not aware that this was but an imaginary region, and so gave it a place within the domain of real geography. The land of the Hyperboreans is a western repetition of the Uttarakúru of Kaśmîr.

Cap. 17.

Position of Areia.

[*Map of Asia* 9.]

Areia is bounded on the north by Marganê and by a part of Baktrianê along its southern side, as already exhibited. On the west by

Parthia and by the Karmanian desert along their eastern meridians that have been defined, on the south by Drangianê along the line which, beginning from the said extremity towards Karmania, and curving towards the north, turns through Mount Bagôos towards the east on to the extreme point which lies ....111° 34°
the position where the mountain curves is ....105° 32°
The boundary on the east is formed by the Paropanisadai along the line adjoining the extremities already mentioned through the western parts of Paropanisos; the position may be indicated at three different points, the southern ....111° 36°
the northern ....111° 30′ 39°
and the most eastern ....119° 30′ 39°

2. A notable river flows through this country called the Areias, of which the sources that are in Paropanisos, lie ....111° 38° 15′
and those that are in the Sariphoi..118° 33° 20′
The part along the lake called Areia, which is below these mountains, lies in ...108° 40′ 36°

3. The northern parts of Areia are possessed by the Nisaioi and the Astauênoi or Astabênoi, but those along the frontier of Parthia and the Karmanian desert by the Masdôranoi or Mazôranoi, and those along the frontier of Drangianê by the Kaseirôtai, and those along the Paropanisadai by the Parautoi, below whom are the Obareis

and intermediately the Drakhamai, below whom the Aitymandroi, then the Borgoi, below whom is the country called Skorpiophoros.

4. The towns and villages in Areia are these :—

Dista	102° 30′	38° 15′
Nabaris	105° 40′	38° 20′
Taua	109°	38° 45′
Augara	102°	38°
Bitaxa	103° 40′	38°
Sarmagana	105° 20′	38° 10′
Spharê	107° 15′	38° 15′
Rhaugara	109° 30′	38° 10′
5. Zamoukhana	102°	37°
Ambrôdax	103° 30′	37° 30′
Bogadia	104° 15′	37° 40′
Ouarpna (Varpna)	105° 30′	37°
Godana	110° 30′	37° 30′
Phoraua	110°	37°
Khatriskhê	103°	36° 20′
Khaurina	104°	36° 20′
6. Orthiana	105° 15′	36′ 20′
Taukiana	106° 10′	36°
Astauda	107° 40′	36°
Artikaudna	109° 20′	36° 10′
Alexandreia of the Areians	110°	36°
Babarsana or Kabarsana	103° 20′	35° 20′
Kapoutana	104° 30′	35° 30′

7. Areia, a city	105°	35°
Kaskê	107° 20′	35° 20′
Sôteira	108° 40′	35° 30′
Ortikanô	109° 20′	35° 30′
Nisibis	111°	35° 20′
Parakanakê	105° 30′	34° 20′
Sariga	106° 40′	34° 40′
8. Darkama	111°	34° 20′
Kotakê	107° 30′	33° 40′
Tribazina	106°	33°
Astasana	105°	33°
Zimyra	102° 30′	33° 15′

Areia was a small province included in Ariana, a district of wide extent, which comprehended nearly the whole of ancient Persia. The smaller district has sometimes been confounded with the larger, of which it formed a part. The names of both are connected with the well-known Indian word *ârya*, 'noble' or 'excellent.' According to Strabo, Aria was 2,000 stadia in length and only 300 stadia in breadth. "If," says Wilson (*Ariana Antiq.*, p. 150) "these measurements be correct, we must contract the limits of Aria much more than has been usually done; and Aria will be restricted to the tract from about Meshd to the neighbourhood of Herat, a position well enough reconcilable with much that Strabo relates of Aria, its similarity to Margiana in character and productions, its mountains and well-watered valleys in which the vine flourished, its position as much to the north as to the south of the chain of Taurus or Alburz, and its being bounded by Hyrkania,

Margiana, and Baktriana on the north, and Drangiana on the south."

Mount Bagôos, on its south-east border, has been identified with the Ghûr mountains. The Montes Sariphi are the Hazâras. The river Areias, by which Aria is traversed, is the Hari Rûd or river of Herat which, rising at Oba in the Paropanisan mountains, and having run westerly past Herat, is at no great distance lost in the sands. That it was so lost is stated both by Strabo and Arrian. Ptolemy makes it terminate in a lake; and hence, Rennell carried it south into the Lake of Seistân, called by Ptolemy the Areian lake. It receives the Ferrah-Rûd, a stream which passes Ferrah or Farah, a town which has been identified with much probability with the Phra mentioned by Isidôros in his *Mans. Parth.*, sec. 16. It receives also the Etymander (now the Helmand) which gave its name to one of the Areian tribes named by Ptolemy.

He has enumerated no fewer than 35 towns belonging to this small province, a long list which it is not possible to verify, but a number of small towns, as Wilson points out, occur on the road from Meshd to Herat and thence towards Qandahâr or Kâbul, and some of these may be represented in the Table under forms more or less altered. The capital of Areia, according to Strabo and Arrian, was Artakoana (v. ll. Artakakna, Artakana) and this is no doubt the Artikaudna of Ptolemy, which he places on the banks of the Areian lake about two-thirds of a degree north-west of his Alexandreia of the Areians. The identification of this Alexandreia is uncertain; most probably it was Herat, or some place in its neighbourhood. Herat is called by oriental writers Hera, a form under which the Areia of the ancients is readily to be recognized. Ptolemy has a city of this name, and Wilson (*Ariana Antiqua*, p. 152), is of opinion that "Artakoana, Alexandria and Aria are aggregated in Herat." With reference to Alexandria he quotes a memorial verse current among the inhabitants of Herat: "It is said that Hari was founded by Lohrasp, extended by Gushtasp, improved by Bahman and completed by Alexander." The name of Sôteira indicates that its founder was Antiokhos Sôtêr.

Cap. 18.

Position of the Paropanisadai.

[*Map of Asia* 9.]

1. The Paropanisadai are bounded on the west by Areia along the aforesaid side, on the north by the part of Baktrianê as described, on the east by a part of India along the meridian line prolonged from the sources of the river Ôxos, through the Kaukasian mountains as far as a terminating point which
lies in ....119° 30′ 39°
and on the south by Arakhôsia along the line connecting the extreme points already determined.

2. The following rivers enter the country—the Dargamanês, which belongs to Baktrianê, the position of the sources of which has

been already stated; and the river which falls into the Kôa, of which the sources lie........................115° 34° 30′.

3. The northern parts are possessed by the Bôlitai, and the western by the Aristophyloi, and below them the Parsioi, and the southern parts by the Parsyêtai, and the eastern by the Ambautai.

4. The towns and villages of the Paropanisadai are these:—

Parsiana	118° 30′	38° 45′
Barzaura	114°	37° 30′
Artoarta	116° 30′	37° 30′
Baborana	118°	37° 10′
Katisa	118° 40′	37° 30′
Niphanda	119°	37°
Drastoka	116°	36° 30′
Gazaka or Gaudzaka	118° 30′	36° 15′
5. Naulibis	117°	35° 30′
Parsia	113° 30′	35°
Lokharna	118°	34°
Daroakana	118° 30′	34° 20′
Karoura, called also Ortospana	118°	35°
Tarbakana	114° 20′	33° 40′
Bagarda	116° 40′	33° 40′
Argouda	118° 45′	33° 30′

The tribes for which Paropanisadai was a collective name were located along the southern and eastern sides of the Hindu-Kush, which Ptolemy calls the Kaukasos, and of which his Paropanisos formed a part. In the tribe which he calls the

Bôlitai we may perhaps have the Kabolitae, or people of Kabul, and in the Ambautai the Ambashṭha of Sanskrit. The Parsyêtai have also a Sanskrit name—'mountaineers,' from *parvata*, 'a mountain,' so also the Parautoi of Areia. The principal cities of the Paropanisadai were Naulibis and Karoura or Ortospana. Karoura is also written as Kaboura and in this form makes a near approach to Kabul, with which it has been identified. With regard to the other name of this place, Ortospana, Cunningham (*Anc. Geog. of Ind.*, p. 35) says: "I would identify it with Kâbul itself, with its Bala Hisâr, or 'high fort,' which I take to be a Persian translation of Ortospana or Urddhasthâna, that is, high place or lofty city." Ptolemy mentions two rivers that crossed the country of the Paropanisadai—the Dargamanês from Baktriana that flowed northward to join the Ôxos, which Wilson (*Ariana Antiqua*, p. 160) takes to be either the Dehas or the Gori river. If it was the Dehas, then the other river which Ptolemydoes not name, but which he makes to be a tributary of the Kôa, may be the Sarkhâb or Gori river, which, however, does not join the Kôa but flows northward to join the Ôxos. Pâṇini mentions Parśusthâna, the country of the Parśus, a warlike tribe in this reign, which may correspond to Ptolemy's Parsioi or Parsyetai.[40] The following places have been identified:—

Parsiana with Pañjshir; Barzaura with Bazârak; Baborana with Parwân; Drastoka with Istargaṛh; Parsia (capital of the

[40] See Beal's *Bud. Rec. of Wn. Count.* vol. II, p. 285n.

Parsii) with Farzah, and Lokharna with Lôgaṛh south of Kâbul.

Cap. 19.

Position of Drangianê.

[*Map of Asia* 9.]

Drangianê is bounded on the west and north by Areia along the line already described as passing through Mount Bagôos, and on the east by Arakhôsia along the meridian line drawn from an extreme point lying in the country of the Areioi and that of the Paropanisadai to another extreme point, of which the position is in ....111° 30′ 28′
and on the south by a part of Gedrôsia along the line joining the extreme points already determined, passing through the Baitian mountains.

2. There flows through the country a river which branches off from the Arabis of which the sources lie ....109° 32° 30′

3. The parts towards Areia are possessed by the Darandai, and those towards Arakhôsia by the Baktrioi, the country intermediate is called Tatakênê.

4. The towns and villages of Drangianê are said to be these :—

Prophthasia ....110° 32° 20′
Rhouda ....106° 30′ 31° 30′

Inna	109°	31° 30′
Arikada	110° 20′	31° 20′
5. Asta	117° 30′	30° 40′
Xarxiarê	106° 20′	29° 15′
Nostana	108°	29° 40′
Pharazana	110°	30°
Bigis	111°	29° 40′
Ariaspê	108° 40′	28° 40′
Arana	111°	28° 15′

Drangianê corresponds in general position and extent with the province now called Seistân. The inhabitants were called Drangai, Zarangae, Zarangoi, Zarangaioi and Sarangai. The name, according to Burnouf, was derived from the Zend word, *zarayo*, 'a lake,' a word which is retained in the name by which Ptolemy's Areian lake is now known—Lake Zarah. The district was mountainous towards Arakhôsia, which formed its eastern frontier, but in the west, towards Karmania, it consisted chiefly of sandy wastes. On the south it was separated from Gedrôsia by the Baitian mountains, those now called the Washati. Ptolemy says it was watered by a river derived from the Arabis, but this is a gross error, for the Arabis, which is now called the Purali, flows from the Baitian mountains in an opposite direction from Drangiana. Ptolemy has probably confounded the Arabis with the Etymander or Helmand river which, as has already been noticed, falls into Lake Zarah.

Ptolemy has portioned out the province among three tribes, the Darandai (Drangai?) on the north, the Baktrioi to the south-east, and the people of Tatakênê between them.

The capital was P r o p h t h a s i a which was distant, according to Eratosthenes, 1500 or 1600 stadia from Alexandria Areiôn (Herat). Wilson therefore fixes its site at a place called Peshawarun, which is distant from Herat 183 miles, and where there were relics found of a very large city. This place lies between Dushak and Phra, *i.e.* Farah, a little to the north of the lake. These ruins are not, however, of ancient date, and it is better therefore to identify Prophthasia with Farah which represents Phra or Phrada, and Phrada, according to Stephanos of Byzantium, was the name of the city which was called by Alexander Prophthasia (Bunbury, vol. I, p. 488). Dashak, the actual capital of Seistân, is probably the Zarang of the early Muhammadan writers which was evidently by its name connected with Drangiana. In the Persian cuneiform inscription at Behistun the country is called Zasaka, as Rawlinson has pointed out (see Smith's *Dictionary*, s. v. Drangiana). The place of next importance to the capital was A r i a s p ê, which Arrian places on the Etymander (*Anab.*, lib. IV, c. vii). The people were called Ariaspai at first, or Agriaspai, but afterwards Euergetai,—a title which they had earned by assisting Cyrus at a time when he had been reduced to great straits.

Cap. 20.

Position of Arakhôsia.

A r a k h ô s i a is bounded on the west by Drangianê, on the north by the Paropanisadai, along the sides already determined, on the east by the part of India lying along the meridian

line extended from the boundary towards the Paropanisadai as far as an extreme point lying ....119° 28°
and on the south by the rest of Gedrôsia along the line joining the extreme points already determined through the Baitian range.

2. A river enters this country which branches off from the Indus of which the sources lie in ....114° 32° 30′
and the divarication (ἐκτροπή)
in ....121° 30′ 27° 30′
and the part at the lake formed by it which is called Arakhôtos Krênê (fountain)—
lies in...........................115° 28° 40′

3. The people possessing the north parts of the country are the Parsyêtai, and those below them the Sydroi, after whom are the Rhôploutai and the Eôrîtai.

4 The towns and villages of Arakhôsia are said to be these :—

Ozola (or Axola)	114° 15′	32° 15′
Phôklis	118° 15′	32° 10′
Arikaka	113°	31° 20′
Alexandreia	114°	31° 20′
Rhizana	115°	31° 30′
Arbaka	118°	31° 20′
Sigara	113° 15′	30°
Khoaspa	115° 15′	30° 10′
5. Arakhôtos	118°	30° 20′
Asiakê	112° 20′	29° 20′
Gammakê	116° 20′	29° 20′

Malianê ....118° 29° 20′
Dammana ....113° 28° 20′

Arakhôsia comprised a considerable portion of Eastern Afghanistan. It extended westward beyond the meridian of Qandahâr and its eastern frontier was skirted by the Indus. On the north it stretched to the mountains of Ghûr, the western section of the Hindu-Kush, and on the south to Gedrôsia from which it was separated by the Baitian mountains, a branch of the Brahui range. The name has been derived from Haraqiati, the Persian form of the Sanskrit Sarasvatî, a name frequently given to rivers (being a compound of *saras*, 'flowing water,' and the affix *vatî*) and applied among others to the river of Arakhôsia. The province was rich and populous, and what added greatly to its importance, it was traversed by one of the main routes by which Persia communicated with India. The principal river was that now called the Helmand which, rising near the Koh-i-bâbâ range west of Kâbul, pursues a course with a general direction to the south-west, and which, after receiving from the neighbourhood of Qandahâr the Argand-âb with its affluents, the Tarnak and the Arghasan, flows into the lake of Zarah. Ptolemy mentions only one river of Arakhôsia and this, in his map, is represented as rising in the Paryêtai mountains (the Hazâras) and flowing into a lake from which it issues to fall into the Indus about 3½ degrees below its junction with the combined rivers of the Panjâb. This lake, which, he says, is called Arakhotos Krênê, he places at a distance of not less than 7 degrees from his Areian lake. In the text he says that the river is an arm of the Indus, a statement for which it is difficult to find a reason.

The capital of Arakhôsia was Arakhôṭos, said by Stephanos of Byzantium to have been founded by Semiramis. Regarding its identification Mr. Vaux (Smith's *Dictionary*, s. v.) says: "Some difference of opinion has existed as to the exact position of this town, and what modern city or ruins can be identified with the ancient capital? M. Court has identified some ruins on the Arghasan river, 4 parasangs from Qandahâr, on the road to Shikarpur, with those of Arakhôtos, but these Prof. Wilson considers to be too much to the S.E. Rawlinson (*Jour. Geog. Soc.*, vol. XII, p. 113) thinks that he has found them at a place now called Ulân Robât. He states that the most ancient name of the city, Kophen, mentioned by Stephanos and Pliny, has given rise to the territorial designation of Kipin, applied by the Chinese to the surrounding country. The ruins are of a very remarkable character, and the measurements of Strabo, Pliny, and Ptolemy are, he considers, decisive as to the identity of the site. Stephanos has apparently contrasted two cities—Arakhôsia, which he says is not far from the Massagetae, and Arakhôtas, which he calls a town of India. Sir H. Rawlinson believes the contiguity of the Massagetae and Arakhôsia, may be explained by the supposition that by Massagetae, Stephanos meant the Sakai, who colonized the Hazâra mountains on their way from the Hindu-Kush to Sakastân or Seistân." Another account of the origin of the name Seistân is that it is a corruption of the word Saghistân, *i. e.*, the country of the *saghis*, a kind of wood which abounds in the province and is used as fuel. Arakhôsia, according to Isidoros of Kharax, was called by the Parthians "White India."

Cap. 21.

Position of Gedrôsia.

Gedrôsia is bounded on the west by Karmania along the meridian line, already determined as far as the sea, and on the north by Drangianê and Arakhôsia along the separate meridian lines passing through these countries, and on the east by part of India along the river Indus following the line prolonged from the boundary towards Arakhôsia to its termination at the sea in .... 109° 20°
and on the south by a part of the Indian Ocean. It is thus described through its circuit.

2. After the extremity towards Karmania

the mouth of the River Arabis	105°	20° 15′
the sources of the river	110°	27° 30′
the divarication of the river entering Drangianê	107° 30′	25°
Rhagiraua, a city	106°	20°
Women's Haven (Gynaikôn limên)	107°	20° 15′
Koiamba	108°	20°
Rhizana	108° 20′	20° 15′
After which the extreme point at the sea already mentioned...........................	109°	20°

3. Through Gedrôsia run the mountains called the Arbita, whose extreme points lie in..........................160° (107 ?) 22°
and ....113° 26° 30′
from these mountains some rivers join the Indus and the source of one of these lies ....111° 25° 30′
and also there are some streams flowing through Gedrôsia, that descend from the Baitian range.

4. The maritime parts are possessed by the villages of the Arbitai, and the parts along Karamania by the Parsidai (or Parsirai), and the parts along Arakhôsia by the Mausarnaioi, all the interior of the country is called Paradônê, and below it Parisiênô, after which the parts towards the Indus river are possessed by the Rhamnaî.

5. The towns and villages of Gedrôsia are accounted to be these:—

Kouni	110°	27°
Badara	113°	27°
Mousarna	115°	27° 30′
Kottobara	118°	27° 30′
Soxestra or Sôkstra	118° 30′	25° 45′
Oskana	115°	26°
Parsis, the Metropolis	106° 30′	23° 30′
Omiza	110°	23° 30′
Arbis, a city	105°	22° 30′

6. The islands adjacent to Gedrôsia are—

Asthaia	105°	18°
Kodanê	(107 ?) 160° 30′	17°

Gedrôsia corresponds to the modern Baluchistân. Its coast line extended from the mouth of the Indus to Cape Jask near the Straits, which open into the Persian Gulf. Ptolemy however assigned the greater portion of this coast to Karmania which according to his view must have begun somewhere near Cape Passence. Arrian restricted the name of Gedrôsia to the interior of the country, and assigned the maritime districts beginning from the Indus to the Arabies, the Oreitai and the Ikhthyophagoi in succession. The ancient and the modern names of the province, Major Mockler tries to identify in his paper in the *Jour. R. As. Soc.*, N. S., vol. XI. pp. 129-154.

The people that possessed the maritime region immediately adjoining the Indus were called the Arbitai or Arabies. In one of their harbours the fleet of Nearkhos at the outset of his memorable voyage was detained for 24 days waiting till the monsoon should subside. This harbour was found to be both safe and commodious, and was called by Nearkhos the Port of Alexander. It is now Karâchi, the great emporium for the commerce of the Indus. The name of the people was applied also to a chain of mountains and to a river, the Arabis, now called the Purali, which falls into the Bay of Sonmiyâni. Ptolemy's Arabis, however, lay nearer Karmania, and may be taken to be the Bhasul, which demarcated the western frontier of the Oreitai, and to the east of which the district is still known by the name of Arbu. Ptolemy does not mention the Oreitai, but seems to have included their territory in that of the Arbitai.

The Rhamnai are placed in Ptolemy's map in the northern part of the province and towards the river Indus. This race appears to have been one that was widely diffused, and one of its branches, as has been stated, was located among the Vindhyas.

The Parsidai, who bordered on Karmania, are mentioned in the *Periplûs* (c. xxxvii) and also in Arrian's *Indika* (c. xxvi) where they are called Pasireës. They gave their names to a range of mountains which Ptolemy makes the boundary between Gedrôsia and Karmania, and also to a town, Parsis, which formed the capital of the whole province.

Of the other towns enumerated only one is mentioned in Arrian's *Indika*, Gynaikôn Limên, or women's haven, the port of Morontobara, near Cape Monze, the last point of the Pab range of mountains. The haven was so named because the district around had, like Carthage, a woman for its first sovereign.

The names of the two towns Badara and Mousarna occur twice in Ptolemy, here as inland towns of Gedrôsia, and elsewhere as seaport towns of Karmania. Major Mockler, who personally examined the Makrân coast from Gwadar to Cape Jâsk, and has thereby been enabled to correct some of the current identifications, has shown that Gwadar and Badara are identical. Badara appears in the *Indika* of Arrian as Barna.

I here subjoin, for comparison, a passage from Ammianus Marcellinus which traverses the ground covered by Ptolemy's description of Central and Eastern Asia. Ammíanus wrote about the middle of the fourth century of our æra, and was a well informed writer, and careful in his statement of facts. The extract is from the 23rd Book of his *History*:—

AMMIANUS MARCELLINUS—Book XXIII.

"If you advance from Karmania into the interior (*of Asia*) you reach the Hyrkanians, who border on the sea which bears their name. Here, as the poorness of the soil kills the seeds committed to it, the inhabitants care but little for agriculture. They live by hunting game, which is beyond measure varied and abundant. Tigers show themselves here in thousands, and many other wild beasts besides. I bear in mind that I have already described the nature of the contrivances by which these animals are caught. It must not be supposed, however, that the people never put hands to the plough, for where the soil is found richer than usual the fields are covered with crops. In places, moreover, that are adapted for being planted-out, gardens of fruit-trees are not wanting, and the sea also supplies many with the means of livelihood. Two rivers flow through the country whose names are familiar to all, the Oxus and Maxera. Tigers at times, when pressed by hunger on their own side of these rivers, swim over to the opposite side and, before the alarm can be raised, ravage all the neighbourhood where they land. Amidst the smaller townships there exist also cities of great power, two on the sea-board, Socunda and Saramanna, and the others inland—Azmorna and Solen, and Hyrkana, which rank above the others. The country next to this people on the north is said to be inhabited by the

Abii, a most pious race of men, accustomed to despise all things mortal, and whom Jupiter (as Homer with his over-fondness for fable sings) looks down upon from the summits of Mount Ida. The seats immediately beyond the Hyrkanians form the dominions of the Margiani, who are nearly on all sides round hemmed in by high hills, and consequently shut out from the sea. Though their territory is for the most part sterile, from the deficiency of water, they have nevertheless some towns, and of these the more notable are Jasonion and Antiochia and Nisæa. The adjoining region belongs to the Baktriani, a nation hitherto addicted to war and very powerful, and always troublesome to their neighbours, the Persians, before that people had reduced all the surrounding states to submission, and absorbed them into their own name and nationality. In old times, however, even Arsakes himself found the kings who ruled in Baktriana formidable foes to contend with. Most parts of the country are, like Margiana, far distant from the sea, but the soil is productive, and the cattle that are pastured on the plains and hill-sides, are compact of structure, with limbs both stout and strong, as may be judged from the camels which were brought from thence by Mithridates and seen by the Romans during the siege of Cyzicus, when they saw this species of animal for the first time. A great many tribes, among which the Tochari are the most distinguished, obey the Baktrians. Their country is watered, like Italy, by numerous rivers, and of these the Artemis and Zariaspes after their union, and in like manner th combined Ochus and Orchomanes, swell with their confluent waters the vast stream of the Ôxos. Here also cities are to be found, and these are laved by different rivers. The more important of them are Chatra and Charte and Alicodra and Astacia and Menapila, and Baktra itself, which is both the capital and the name of the nation. The people, who live at the very foot of the mountains, are called the Sogdii, through whose country flow two rivers of great navigable capacity, the Araxates and Dymas, which rushing impetuously down from the mountains and passing into a level plain, form a lake of vast extent, called the Oxian. Here, among other towns, Alexandria, and Kyreschata, and Drepsa the Metropolis, are well known to fame. Contiguous to the Sogdians are the Sacae, an uncivilized people, inhabiting rugged tracts that yield nothing beyond pasture for cattle, and that are, therefore, unadorned with cities. They lie under Mounts Askanimia and Komedus. Beyond the valleys at the foot of these mountains and the village which they call Lithinon Pyrgon (Stone Tower) lies the very long road by which traders pursue their journey who start from this point to reach the Sêres. In the parts around are the declivities by which the mountains called Imaus and the Tapourian range, sink down to the level of the plains. The Skythians are located within the Persian territories, being conterminous with the Asiatic Sarmatians, and touching the furthest frontier of the Alani. They live, as it were, a sort of secluded life, and are reared in solitude, being scattered over districts that lie far apart, and that yield for the sustenance of life a mean and scanty fare. The tribes which inhabit these tracts are various, but it would be superfluous for me to enumerate them, hastening as I am to a different subject. One fact must, however, be stated, that there are in these communities which are almost shut out from the rest of mankind by the inhospitable nature of their country, some men gentle and pious, as for instance, the Jaxartes and the Galaktophagi, mentioned by the poet Homer in this verse:

Γλακτοφάγων ἀβίωντε δικαιοτάτων ἀνθρώπων.

"Among the many rivers of Skythia which either fall naturally into larger ones, or glide onward to reach at last the sea, the Roemnus is of renown, and the Jaxartes and the Talicus, but of cities they are not known to have more than but three, Aspabota and Chauriana and Saga.

"Beyond these places in the two Skythias and on their eastern side lie the Sêres, who are girt in by a continuous circle of lofty mountain-peaks, and whose territory is noted for its vast extent and fertility. On the west they have the Skythians for their next neighbours, and on the north and east they adjoin solitudes covered over with snow, and on the south extend as far as India and the Ganges. The mountains referred to are called Anniva and Nazavicium and Asmira and Emodon and Opurocara. Through this plain which, as we have said, is cinctured on all sides by steep declivities, and through regions of vast extent, flow two famous rivers, the Œchardes and the Bautisus, with a slower current. The country is diversified in its character, here expanding into open plains, and there rising in gentle undulations. Hence it is marvellously fruitful and well-wooded, and teeming with cattle. Various tribes inhabit the most fertile districts, and of these the Alitrophagi and Annibi and Sizyges and Chardi are exposed to blasts from the north and to frosts, while the Rabannae and Asmirae and Essedones, who outshine all the other tribes, look towards the rising sun. Next to these, on their western side, are the Athagorae and the Aspacarae. The Betae, again, are situated towards the lofty mountains fringing the south, and are famed for their cities which, though few in number are distinguished for their size and wealth; the largest of them being Asmira, and Essedon and Asparata and Sera, which are beautiful cities and of great celebrity. The Sêres themselves lead tranquil lives, and are averse to arms and war, and since people whose temper is thus sedate and peaceful relish their ease, they give no trouble to any of their neighbours. They enjoy a climate at once agreeable and salubrious; the sky is clear and the prevailing winds are wonderfully mild and genial. The country is well-shaded with woods, and from the trees the inhabitants gather a product which they make into what may be called fleeces by repeatedly besprinkling it with water. The material thus formed by saturating the soft down with moisture is exquisitely fine, and when combed out and spun into woof is woven into silk, an article of dress formerly worn only by the great, but now without any distinction even by the very poorest.[41]

[41] It was a notion long prevalent that silk was combed from the leaves of trees. Thus Virgil (*Georg.* II, 121)

The Sêres themselves live in the most frugal manner, more so indeed than any other people in the world. They seek after a life as free as possible from all disquiet, and shun intercourse with the rest of mankind. So when strangers cross the river into their country to buy their silks or other commodities, they exchange no words with them, but merely intimate by their looks the value of the goods offered for sale; and so abstemious are they that they buy not any foreign products. Beyond the Sêres live the Ariani, exposed to the blasts of the north wind. Through their country flows a navigable river called the Arias, which forms a vast lake bearing the same name. This same Aria has numerous towns, among which Bitana Sarmatina, and Sotera and Nisibis and Alexandria are the most notable. If you sail from Alexandria *down the river* to the Caspian Sea the distance is 1,500 stadia.

Immediately adjoining these places are the Paropanisatae, who look on the east towards the Indians and on the west towards Caucasus, lying themselves towards the slopes of the mountains. The River Ortogordomaris, which is larger than any of the others, and rises among the Baktriani, flows through their territory. They too, have some towns, of which the more celebrated are Agazaca and Naulibus and Ortopana, from which the navi-

---

"Velleraque ut foliis depectant tenuia Seres." Strabo (XV, i, 20) describes silk as carded off the bark of certain trees. Pausanias, who wrote about 180 A.D. is the first classical author who writes with some degree of correctness about silk and the silk-worm. Conf. P. Mela, i, 2, 3; iii, 7, 1; Pliny, VI, 17, 20; Prop. i, 14, 22; Sol. 50; Isid. Orig. xix, 17, 6; ib. 27, 5.

gation along the coast to the borders of Media in the immediate neighbourhood of the Caspian Gates extends to 2,200 stadia. Contiguous to the Paropanisatae just named are the Drangiani, seated quite close to the hills and watered by a river called the Arabian, because it rises in Arabia. Among their other towns they have two to boast of in particular, Prophthasia and Ariaspe, which are both opulent and famous. After these, and directly confronting them, Arachosia comes into view, which on its right side faces the Indians. It is watered by a stream of copious volume derived from the Indus, that greatest of rivers, after which the adjacent regions have been named. This stream, which is less than the Indus, forms the lake called Arachotoscrene. The province, among other important cities, has Alexandria and Arbaca and Choaspa. In the very interior of Persia is Gedrosia, which on the right touches the Indian frontier. It is watered by several streams, of which the Artabius is the most considerable. Where it is inhabited by the Barbitani the mountains sink down to the plains. A number of rivers issue from their very base to join the Indus, and these all lose their names when absorbed into that mightier stream. Here too, besides the islands there are cities, of which Sedratyra and Gunaikon Limen (Women's haven) are considered to be superior to the others. But we must bring this description here to an end, lest in entering into a minute account of the seaboard on the extremities of Persia we should stray too far from the proper argument."

APPENDIX OF ADDITIONAL NOTES.

1. On the latitude of Byzantium and of Tâsh-Kurghân—(p. 14).

Ptolemy, like Hipparkhos and all the ancients except Strabo, erroneously took the latitude of Byzantium (41° 1′) to be the same as that of Marseilles (43° 18′). The latitude of Tâsh-kurghân in the Pâmîr is 37° 46′ and its longitude 75° 10′ E.; the latitude of Tashkend is 42° 58′, and that of Och or Ush (near which there is a monument called at this day the Takht-i-Suleiman, 'Throne of Soliman,' which Heeren took to be the veritable stone tower of Ptolemy) is 40° 19′.

2. On Kouroula—(pp. 22, 63, and 64).

Lieut.-Colonel Branfill (*Names of Places in Tanjore*, p. 8), thinks this may be represented by Kurla or Koralai-gorla on the East Coast. "There is," he points out, "*Gorla*pâlem near Nizâmpaṭṭanam. (*Cf. Vingorla*, South Concan, Malabar Coast)."

3. Argaric Gulf and Argeirou (pp. 22, 59, and 60).

Branfill in the work cited (pp. 8 and 9) says:—"Âṛṛaṅkarai (pronounced nowadays Âtraṅkarai), at the mouth of the Vagai looks very like the ancient '*Argari*,' and '*Sinus Argalicus*' (Yule), the Argaric Gulf . . Αγχείρου looks like Aṇaikarai, the ancient name of Adam's Bridge, so called by the Tamils as being the bridge or causeway *par excellence* . . . . . In the middle ages, before Pâmban was separated from the mainland by the storm that breached the famous causeway, there is said to have been a great city, remains of which are still to be seen on the spit of sand opposite to Pâmban." Αγχείρου in Nobbe's edition appears as Αργείρου.

4. On Thelkheir—(pp. 63 and 64).

Branfill (p. 12), would identify this with Chidambaram—" the town between the Veḷḷâṛ and Kolladam (*Coleroon*) rivers, ... from *chit* = wisdom, and *ambara*, horizon, sky; = *Heaven of Wisdom*. Tillai, or Tillaivanam is the former name of this place, and it is familiarly known as Tillai even now amongst the natives. May not this be the ancient Thellyr and Θελχείρ of Ptolemy and the ancient geographers? But perhaps Teḷḷûr (near Vandavasi) may be it." Tillai, he points out (p. 30), is a tree with milky sap.

5. On Orthoura—(pp. 64 and 184).

Branfill (pp. 7 and 8), identifying this, says :—" Orattûr (pronounced Oratthûru) is found repeatedly in this (Kavêri Delta) and the adjacent districts, and may represent the 'Orthura' of ancient geographers, for which Colonel Yule's Map of Ancient India gives *Ureiyour*, and Professor Lassen's Wadiur."

6. On Arkatos—(p. 64).

Branfill, who takes this to designate a place and not a king, says (p. 11) :—" Âṛkâḍ or Âṛukâḍu = six forests; the abode of six Rishis in old times. There are several places of this name in Tanjore and S. Arcot, besides the town of 'Arcot' near 'Vellore' (Αρκατοῦ βασίλειον Σῶρα). One of these would correspond better than that with Harkâtu of Ibn Batuta, who reached it the first evening of his march inland after landing from Ceylon, apparently on the shallow coast of Madura or Tanjore (fourteenth century)."

7. On the River Adamas—(p. 71).

Professor V. Ball, in his Presidential Address to the Royal Geological Society of Ireland (read March 19, 1883), says:—"The *Adamas* River of Ptolemy, according to Lassen's analysis of the data, was not identical with the Mahanadi, as I have suggested in my 'Economic Geology' (p. 30), but with the Subanrikha, which is, however, so far as we know, not a diamond-bearing river, nor does it at any part of its course traverse rocks of the age of those which contain the matrix in other parts of India. This *Adamas* River was separated from the Mahanadi by the Tyndis and Dosaron; the latter, according to Lassen, taking its rise in the country of Kokkonaga (*i.e.* Chutia Nâgpur), and to which the chief town Dosara (the modern Doesa) gave its name. But, according to this view, the Dosaron must have been identical with the modern Brahmini, which in that portion of its course called the Sunk (or Koel), included a diamond locality. I cannot regard this identification as satisfactory, as it does not account for the Tyndis intervening between the Dosaron and Mahanadi, since, as a matter of fact, the Brahmini and Mahanadi are confluent at their mouths. Lassen, however, identifies the Dosaron with the Baiturnee, and the Tyndis with the Brahmini. This destroys the force of his remark, as to the origin of the name of the former, since at its nearest point it is many miles distant from Doesa."

8. On Mount Sardônyx—(p. 77).

Professor Ball in the address above cited, says:—"The sardonyx mines of Ptolemy are probably identical with the famous carnelian and agate mines of Rajpipla, or, rather, as it should be called, Ratanpur."

9. On Talara—(p. 90).

Branfill suggests the identification of this with Tellâr or Tillârampaṭṭu (p. 8).

10. On Pounnata—(p. 180).

"Punâḍu, Punnâḍu, or Punnâta, as it is variously written, seems also to be indicated by the Pannuta in Lassen's Map of Ancient India according to Ptolemy, and by the Paunata of Colonel Yule's Map of Ancient India, *ubi beryllus*." This place is about 70 miles to the south-east of Seringapatam.

11. On Arembour—(pp. 180, 182).

Branfill—(p. 8), identifies this with Arambaûr.

12. On Abour—(p. 184).

Branfill (p. 11), identifies this with "Âvûr, *cow-villa*, a decayed town, 5 miles S.W. of Kambakôṇam, with a temple and a long legend about a cow(â). May not this be the ancient Abur of the Map of Ancient India in Smith's Classical Atlas? Colonel Yule suggests Amboor, but this Âvûr seems nearer, and if not this there are several places in S. Arcot named Amûr."

13. On Argyrê—(p. 196).

Professor Ball says:—"There are no silver mines in Arakan, and considering the geological structure of the country, it is almost certain there never were any. I have been recently informed by General Sir A. Phayre that Argyrê is probably a transliteration of an ancient Burmese name for Arakan. It seems likely therefore that it was from putting a Greek interpretation to this name that the story of the silver-mines owed its origin."

14. On the Golden Khersonese—(p. 197).

"Gold," says Mr. Colquhoun (*Amongst the Shans*, p. 2), "has been for centuries washed from the beds of the Irrawadi, Sitang, Salween, Mékong, and Yang-tsi-kiang rivers." The gold-reefs of Southern India which have of late attracted so much notice, are, he points out, but outcrops of the formation which extends on the surface for thousands of square miles in the Golden Peninsula.

15. On the Loadstone rocks (p. 242).

Professor Ball thinks these rocks may possibly be identified with certain hill-ranges in Southern India which mainly consist of magnetic iron (*Economic Geology of India*, p. 37).

16. On the sandy deserts of Baktria (p. 270).

In the *Proceedings* of the Royal Geographical Society for April last will be found a description of the Kara-kum sands, by M. Paul Lessar, who divides them into three classes. The *burkans* which form his 3rd class are of the nature described by Curtius. "The sand is wholly of a drifting nature; the slightest puff of wind effaces the fresh track of a caravan." He notices a place in the Khanate of Bokhara where whole caravans have been buried.

17. On the river Ôchos (p. 273).

"What hitherto has been taken for the dry bed of the Ochus is not the bed of a river, but merely a natural furrow between sand-hills. Thus the bed of the Ochus has still to be discovered." *Proceedings* of the Royal Geog. Socy. for April 1885.

18. On the Avestic names of rivers, &c. in Afghanistan—(pp. 305-19).

In the 1st chapter of the Vendîdâd the names are given of the sixteen lands said to have been created by Ahura Mazda. Of these the following nine have been thus identified by Darmesteter in his translation of the Zend-Avesta, *Sacred Books of the East*, Vol. IV. p. 2):—

Zend name.	Old Persian.	Greek.	Modern.
Sughdha	Suguda	Sogdianê	(Samarkand)
Môuru	Margu	Margianê	Merv
Bâkhdhi	Bakhtri	Baktra	Balkh
Harôyu	Haraiva	Areia	Hari-Rûd
Vehrkâna	Varkâna	Hyrkania	Jorjân
Harahvaiti	Harauvati	Arakhôtos	Harût
Haêtuma*n*t		Etymandros	Helmend
Ragha	Raga	Rhagai	Raï
Hapta hi*n*du	Hi*n*davas	Indoi	(Pañjâb)

Some of these and other names are examined in an article in *The Academy* (May 16, 1885, No. 680), signed by Auriel Stein, from which the following particulars are gathered: "We recognize the 'powerful, faithful *Mourva*' as the modern Merv, the 'beautiful *Bâkhdhi*' as Balkh, *Haraêva* as Herât, the mountain *Vâitigaesa* as the Bâdhgês of recent notoriety. The river *Harahvaiti* (Sansk. Sarasvatî) has been known in successive ages as Arakhôtos and Arghand-âb; but more important for Avestic geography is the large stream of which it is a tributary, the 'bountiful, glorious *Haêtumañt*,' the Etymandros and Hermandus of classic authors, the modern *Helmand.*'" A passage is quoted from the Avesta where eight additional rivers seem to be named. "At its foot (the mountain Ushidao's, *i.e.* the Koh-i-Baba and Siâh-Kôh's) gushes and flows forth the *Hvâstra* and the Hvaśpa, the *Fradatha* and the beautiful *Hvarenanhaiti* and *Ustavaiti* the mighty, and *Urvadha*, rich of pastures, and the *Erezi* and *Zarenumaiti*." The *Hvaśtra* Stein thinks may be the Khâsh-Rûd, and the *Hvaśpa* the Khuspâs-Rûd, both of which come from the south slope of the Siâh-Kôh and reach the eastern basin of the lagune where the lower course of the Helmand is lost. "In Khuspâs," he adds, "a place on the upper course of the Khuspâs-Rûd, we may recognize the town *Khoaspa* mentioned by Ptolemy in Arakhôsia. The name *hvaśpa* means "having good horses," and seems to have been a favourite designation for rivers in Irân. Besides the famous Khoaspês near Susa, we hear of another Khoaspes, a tributary of the Kabûl River." In Kâsh, a town on the Khâsh-Rûd may be recognized the station called Cosata by the Anonymous Ravennas. The *Fradatha* is Pliny's Ophradus (*i.e.* ὁ Φράδος of the Greek original) and now the Farâh-Rûd. The *Prophthasia* of Ptolemy and Stephanos of Byzantium is a literal rendering of *fradatha*, which in common use as neuter means (literally "proficiency"), "progress," "increase." The *Havrenanihaiti* is the *Pharnacotis* of Pliny and now the Harrût-Rûd, which like the Farâh-Rûd enters the lake of Seistân. *Farnakvati* has been suggested as the original and native form of Pharnacotis.

19. On the Griffins or Gryphons—(p. 295).

Professor Ball in a paper published in the *Proceedings* of the Royal Irish Academy, 2nd Ser., Vol. II. No. 6, pp. 312-13 (Pol. Lit. and Antiq.) says: "In the account which Photios gives of the Griffins, if we exclude from it the word *birds*, and for feathers read hair, we have a tolerably accurate description of the hairy black-and-tan coloured Thibetan mastiffs, which are now, as they were doubtless formerly, the custodians of the dwellings of the Thibetans, those of gold-miners, as well as of others. They attracted the special attention of Marco Polo, as well as of many other travellers in Thibet, and for a recent account of them reference may be made to Capt. Gill's '*River of Golden Sand.*'"

ERRATA.

Page 8, n. 5, for *Noble* read *Nobbe*.

„ 14, n. 12, after Tâsh-Kurghân insert its Lat. 37° 46′ (long. 75° 4′).

„ 20, n. for [IXXXIXI] read [|XXIX|].

„ 25, for *censure* in last line but one read *use*.

„ 51, l. 20, for Kandionoi read Pandionoi.

„ 63, l. 16, for *outlet* read *outset*.

„ 64, l. 13, omit *the* before Kolkhoi.

„ 68, l. 15, for *Gûdrû* read Gûdûr.

„ 70, l. 27, Katikardama should begin the line after.

„ 71, l. 18, after Dôsarôn instead of the dash insert the sign of equality (=) and so after "Adamas" in the next line, and after "Ganges" in line 21.

„ 75, Section 21 should have been *immediately* followed by the next 4 sections which appear on p. 78.

„ 76, l. 16, for '*punishment*' of the '*gods*' read '*punishment of the gods*.'

„ 80, l. 21, for Ṛikshavant read Ṛikshavat.

„ 81, l. 29, for Bidasis read Bibasis.

„ 87, l. 7, for the *comma* after the bracket put *period*.

„ 88, l. 26, for Rhonadis read Rhouadis.

Page 124, The sections 47-50 should have been placed after the notice of Iomousa on p. 126.

" 140, l. 29, after 'second group' insert (*sections* 57 *and* 58).

" 140, last line, after 'fourth group,' insert (*section* 61).

" 141, l. 15, after 'sixth group' insert (*section* 64.)

INDEX.

B

H

L

GENERAL INDEX.

BOMBAY: PRINTED AT THE EDUCATION SOCIETY'S PRESS, BYCULLA.

BRITISH
12SE85
MUSEUM

TRÜBNER'S
Oriental & Linguistic Publications.

A CATALOGUE

OF

BOOKS, PERIODICALS, AND SERIALS,

ON THE

History, Languages, Religions, Antiquities, Literature, and Geography of the East,

AND KINDRED SUBJECTS.

PUBLISHED BY

TRÜBNER & CO.

LONDON:
TRÜBNER & CO., 57 AND 59, LUDGATE HILL.

1885.

CONTENTS.

TRÜBNER'S ORIENTAL SERIES.

"A knowledge of the commonplace, at least, of Oriental literature, philosophy, and religion is as necessary to the general reader of the present day as an acquaintance with the Latin and Greek classics was a generation or so ago. Immense strides have been made within the present century in these branches of learning; Sanscrit has been brought within the range of accurate philology, and its invaluable ancient literature thoroughly investigated; the language and sacred books of the Zoroastrians have been laid bare; Egyptian, Assyrian, and other records of the remote past have been deciphered, and a group of scholars speak of still more recondite Accadian and Hittite monuments; but the results of all the scholarship that has been devoted to these subjects have been almost inaccessible to the public because they were contained for the most part in learned or expensive works, or scattered throughout the numbers of scientific periodicals. Messrs. TRÜBNER & Co., in a spirit of enterprise which does them infinite credit, have determined to supply the constantly-increasing want, and to give in a popular, or, at least, a comprehensive form, all this mass of knowledge to the world."—*Times.*

THE FOLLOWING WORKS ARE NOW READY.

Post 8vo. cloth, uniformly bound.

ESSAYS ON THE SACRED LANGUAGE, WRITINGS, AND RELIGION of THE PARSIS. By MARTIN HAUG, Ph.D. late Professor of Sanskrit and Comparative Philology at the University of Munich. Edited and enlarged by Dr. E. W. WEST. To which is also added a Biographical Memoir of the late Dr. Haug, by Prof. Evans. Third Edition, pp. xlviii. and 428. 1884. 16*s.*

TEXTS FROM THE BUDDHIST CANON, commonly known as Dhammapada. With accompanying Narratives. Translated from the Chinese by S. BEAL, B.A., Professor of Chinese, University College, London. pp. viii. and 176. 1878. 7*s.* 6*d.*

THE HISTORY OF INDIAN LITERATURE. By ALBRECHT WEBER. Translated from the German by J. MANN, M.A., and T. ZACHARIAE, Ph.D., with the sanction of the Author. Second Edition, pp. xxiii. 360. 1882. 10*s.* 6*d.*

A SKETCH OF THE MODERN LANGUAGES OF THE EAST INDIES. By ROBERT CUST. Accompanied by Two Language Maps. pp. xii. and 198. 1878. 12*s.*

THE BIRTH OF THE WAR GOD. A Poem by KÁLIDÁSA. Translated from the Sanskrit into English Verse. By RALPH T. H. GRIFFITH, M.A., Principal of Benares College. Second Edition. pp. xii.-116. 1879. 5*s.*

A CLASSICAL DICTIONARY OF HINDU MYTHOLOGY AND HISTORY, GEOGRAPHY AND LITERATURE. By JOHN DOWSON, M.R.A.S., late Professor in the Staff College. pp. xix. and 412. 1879. 16*s.*

SELECTIONS FROM THE KORAN. With a COMMENTARY. Translated by the late EDWARD WILLIAM LANE, Author of an "Arabic-English Lexicon," etc. A New Edition, Revised, with an Introduction on the History and Development of Islam, especially with reference to India. By STANLEY LANE POOLE. pp. cxii. and 176. 1879. 9*s.*

METRICAL TRANSLATIONS FROM SANSKRIT WRITERS. With an Introduction, many Prose Versions, and Parallel Passages from Classical Authors. By J. MUIR, C.I.E., D.C.L. pp. xliv. and 376. 1879. 14*s.*

MODERN INDIA AND THE INDIANS. Being a Series of Impressions, Notes, and Essays. By MONIER WILLIAMS, D.C.L., Boden Professor of Sanskrit in the University of Oxford. Third Revised Edition, pp. 366. With map. 1879. 14*s.*

MISCELLANEOUS ESSAYS RELATING TO INDIAN SUBJECTS. By BRIAN HOUGHTON HODGSON, F.R.S., late of the Bengal Civil Service, etc., etc 2 vols. pp. viii. and 408, and viii. and 348. 1880. 28*s.*

THE LIFE OR LEGEND OF GAUDAMA, the Buddha of the Burmese. With Annotations, The Ways to Neibban, and Notice on the Phongyies or Burmese Monks. By the Right Reverend P. BIGANDET, Bishop of Ramatha, Vicar Apostolic of Ava and Pegu. Third Edition 2 vols. pp. xx. and 268, and viii. and 326. 1880. 21*s.*

THE GULISTAN; or, Rose Garden of Shekh Mushliu'd-din Sadi of Shiraz. Translated for the first time into Prose and Verse, with a Preface, and a Life of the Author, from the Ātish Kadah, by E. B. EASTWICK, F.R.S., M.R.A.S., etc. Second Edition, pp. xxvi. and 244. 1880. 10*s.* 6*d.*

CHINESE BUDDHISM. A Volume of Sketches, Historical and Critical. By J. EDKINS, D.D., pp. xxvi. and 454. 1880. 18*s.*

THE HISTORY OF ESARHADDON (SON OF SENNACHERIB) KING OF ASSYRIA, B.C. 681–668. Translated from the Cuneiform Inscriptions upon Cylinders and Tablets in the British Museum Collection. With the Original Texts, a Grammatical Analysis of each Word, Explanations of the Ideographs by Extracts from the Bi-Lingual Syllabaries, and list of Eponyms, etc. By E. A. BUDGE, B.A., etc. pp. xii. and 164. 1880. 10*s.* 6*d.*

A TALMUDIC MISCELLANY; or, One Thousand and One Extracts from the Talmud, the Midrashim, and the Kabbalah. Compiled and Translated by P. J. Hershon. With a Preface by the Rev. F. W. FARRAR, D.D., Canon of Westminster. With Notes and Copious Indexes. pp. xxviii. and 362. 1880. 14*s.*

BUDDHIST BIRTH STORIES; or, Jātaka Tales. The oldest collection of Folk-lore extant: being the Jātakatthavannanā, for the first time edited in the original Pali, by V. FAUSBÖLL, and translated by T. W. Rhys Davids. Translation. Vol. I. pp. cxvi. and 348. 1880. 18*s.*

THE CLASSICAL POETRY OF THE JAPANESE. By BASIL CHAMBERLAIN, Author of "Yeigio Henkaku, Ichiran," pp. xii. and 228. 1880. 7*s.* 6*d.*

LINGUISTIC AND ORIENTAL ESSAYS. Written from the year 1846–1878. By R. CUST. pp. xii. and 484. 1880. 18*s.*

THE MESNEVĪ. (Usually known as the Mesnevīyi Sherif, or Holy Mesnevī) of Mevlānā (our Lord) Jelālu'd-Din Muhammed er-Rūmī. Book I. With a Life of the Author. Illustrated by a Selection of Characteristic Anecdotes, by Mevlānā Shemsu'd-Din Ahmed el Eflākī, el 'Ārifī. Translated and the Poetry Versified in English. By J. W. REDHOUSE, M.R.A.S. pp. xv. and 135, v. and 290. 1881. 21*s.*

EASTERN PROVERBS AND EMBLEMS, Illustrating Old Truths. By the Rev. J. LONG, M.R.A.S., F.R.G.S. pp. xvi. and 280. 1881. 6*s.*

INDIAN POETRY. Containing "The Indian Song of Songs," from the Sanskrit of the "Gita Govinda" of Jayadeva; Two Books from "the Iliad of India" (Mahabharata); and other Oriental Poems. Third Edition. By EDWIN ARNOLD, M.A., C.S.I. pp. viii. and 270. 1884. 7*s.* 6*d.*

HINDU PHILOSOPHY. The Sankhya Karika of Iswara Krishna. An Exposition of the System of Kapila. With an Appendix on the Nyaya and Vaiseshika Systems. By J. DAVIES, M.A. pp. viii. and 152. 1881. 6*s.*

THE RELIGIONS OF INDIA. By A. BARTH. Authorised Translation by Rev. J. WOOD. pp. 336. 1881. 16*s.*

A Manual of Hindu Pantheism. The Vedantasara. Translated with Copious Annotations, by Major G. A. Jacob, B.S.C. With Preface by E. B. Cowell, M.A., Prof. of Sanskrit in Cambridge University. pp. x. and 129. 1881. 6*s.*

The Quatrains of Omar Khayyám. Translated by E. H. Whinfield, M.A., late of H.M. Bengal Civil Service. pp. 96. 1881. 5*s.*

The Mind of Mencius; or, Political Economy founded upon Moral Philosophy. A Systematic Digest of the Doctrine of the Chinese Philosopher Mencius. Translated from the Original Text, and Classified with Comments and Explanations. By the Rev. Ernst Faber, Rhenish Mission Society. Translated from the German with Additional Notes, by the Rev. A. B. Hutchinson, C.M.S., Hong-Kong. pp. xvi. and 294. 1881. 10*s.* 6*d.*

Tsuni-||Goam, the Supreme Being of the Khoi-Khoi. By Theophilus Hahn, Ph.D., Custodian of the Grey Collection, Cape Town, etc. pp. xii. and 154. 1881. 7*s.* 6*d.*

Yusef and Zulaikha. A Poem by Jámi. Translated from the Persian into English Verse. By R. T. H. Griffith. pp. xiv. and 304. 1882. 8*s.* 6*d.*

The Indian Empire: its History, People, and Products. By W. W. Hunter, C.I E., LL.D. pp. 568. With Map. 1882. 16*s.*

A Comprehensive Commentary to the Quran: comprising Sale's Translation and Preliminary Discourse, with Additional Notes and Emendations. With a complete Index to the Text, Preliminary Discourse, and Notes. By Rev. E. M. Wherry, M.A., Lodiana. Vol. I. pp. xii. and 392. 1882. 12*s.* 6*d.* Vol. II. pp. xii.—408. 1884. 12*s.* 6*d.*

Comparative History of the Egyptian and Mesopotamian Religions. By C. P. Tiele. Egypt, Babel-Assur, Yemen, Harran, Phœnicia, Israel. Vol. I. History of the Egyptian Religion. Translated from the Dutch, with the co-operation of the Author, by James Ballingal. pp. xxiv.-230, 1882. 7*s.* 6*d.*

The Sarva-Darsana-Samgraha; or Review of the different Systems of Hindu Philosophy. By Madhava Acharya. Translated by E. B. Cowell M.A., Cambridge; and A. E. Gough, M.A., Calcutta. pp. xii.-282. 1882. 10*s.* 6*d.*

Tibetan Tales, Derived from Indian Sources. Translated from the Tibetan of the Kah-Gyur. By F. Anton von Schiefner. Done into English from the German, with an Introduction, by W. R. S. Ralston, M.A. pp. lxvi.-368. 1882. 14*s.*

Linguistic Essays. By Carl Abel, Ph.Dr. pp. viii.-266. 1882. 9*s.*

Contents.—Language as the Expression of National Modes of Thought—The Conception of Love in some Ancient and Modern Languages—The English Verbs of Command—The discrimination of Synonyms—Philological Methods—The Connection between Dictionary and Grammar—The Possibility of a Common Literary Language for the Slave Nations Coptic Intensification—The Origin of Language—The Order and Position of Words in the Latin Sentence.

Hindū Philosophy. The Bhagavad Gītā or the Sacred Lay. A Sanskrit Philosophical Poem. Translated, with Notes, by John Davies, M.A. (Cantab.) M.R.A.S. pp. vi.-208. 1882. 8*s.* 6*d.*

The Philosophy of the Upanishads and Ancient Indian Metaphysics. By A. E. Gough, M.A. Calcutta. Pp. xxiv.-268. 1882. 9*s.*

Udanavarga: A Collection of Verses from the Buddhist Canon. Compiled by Dharmatrata. The Northern Buddhist Version of Dhammapada. Translated from the Tibetan of Bkah hgyur, Notes and Extracts from the Commentary of Pradjnavarman, by W. W. Rockhill. Pp. xvi.-224. 1883. 9*s.*

A History of Burma. Including Burma Proper, Pegu, Taungu, Tenasserim, and Arakan. From the Earliest Time to the End of the First War with British India. By Lieut.-General Sir A. P. Phayre, G.C.M.G., K.C.S.I., &c. pp. xii. and 312, with Maps and Plan. 1883. 14*s*.

The Quatrains of Omar Khayyám. The Persian Text, with an English Verse Translation. By E. H. Whinfield, M.A., late of the Bengal Civil Service. pp. xxxii. and 336. 1883. 10*s*. 6*d*.

A Sketch of the Modern Languages of Africa. By R. N. Cust. Accompanied by a Language Map. By E. G. Ravenstein. Two Vols. pp. xvi.-288, viii.-278, with Thirty-one Autotype Portraits. 1883. 25*s*.

Outlines of the History of Religion to the Spread of the Universal Religions. By Prof. C. P. Tiele. Translated from the Dutch by J. E. Carpenter, M.A., with the Author's assistance. Third Edition, pp. xx. and 250. 1884. 7*s*. 6*d*.

Religion in China; containing a brief Account of the Three Religions of the Chinese; with Observations on the Prospects of Christian Conversion amongst that People. By Joseph Edkins, D.D., Peking. Third Edition, pp. xvi. and 260. 1884. 7*s*. 6*d*.

The Life of the Buddha and the Early History of his Order. Derived from Tibetan Works in the Bkah-hgyur and Bstan-hgyur. Followed by notices on the Early History of Tibet and Khoten. Translated by W. W. Rockhill, Second Secretary U.S. Legation in China. pp. x.—274, cloth. 1884. 9*s*.

Buddhist Records of the Western World. Translated from the Chinese of Hiuen Tsiang (A.D. 629). By S. Beal. Dedicated by permission to H.R.H. the Prince of Wales. 2 volumes, pp. cviii.—242, and viii.—370, cloth. 1884. 24*s*.

The Sankhya Aphorisms of Kapila. With Illustrative Extracts from the Commentaries. Translated by J. R. Ballantyne, LL.D., late Principal of Benares College. Edited by Fitzedward Hall. Third Edition. pp. viii.—464, cloth. 1884. 16*s*.

The Ordinances of Manu. Translated from the Sanskrit. With an Introduction by the late A. C. Burnell, Ph.D., C.I.E. Completed and Edited by E. W. Hopkins, Ph.D., Columbia College, New York. pp. xlviii.—398, cloth. 1884. 12*s*.

THE FOLLOWING WORKS ARE IN PREPARATION:—

The Six Jewels of the Law. With Pali Texts and English Translation. By R. Morris, LL.D.

Oriental Religions in their Relation to Universal Religion. By Samuel Johnson. Second Section—China. In Two Volumes.

A Comparative History of the Egyptian and Mesopotamian Religions. By Dr. C. P. Tiele. In two Volumes. Vol. II. History of the Assyrian Religion. Translated from the Dutch, with the Assistance of the Author, by James Ballingal.

SERIALS AND PERIODICALS.

Asiatic Society of Great Britain and Ireland.—JOURNAL OF THE ROYAL ASIATIC SOCIETY OF GREAT BRITAIN AND IRELAND, from the Commencement to 1863. First Series, complete in 20 Vols. 8vo., with many Plates, Price £10; or, in Single Numbers, as follows:—Nos. 1 to 14, 6*s.* each; No. 15, 2 Parts, 4*s.* each; No. 16, 2 Parts, 4*s.* each; No. 17, 2 Parts, 4*s.* each; No. 18, 6*s.* These 18 Numbers form Vols. I. to IX.—Vol. X., Part 1, o.p.; Part 2, 5*s.*; Part 3, 5*s.*—Vol. XI., Part 1, 6*s.*; Part 2 not published.—Vol. XII., 2 Parts, 6*s.* each.—Vol. XIII., 2 Parts, 6*s.* each.—Vol. XIV., Part 1. 5*s.*; Part 2 not published.—Vol. XV., Part 1, 6*s.*; Part 2, with 3 Maps, £2 2*s.* —Vol. XVI., 2 Parts, 6*s.* each.—Vol. XVII., 2 Parts, 6*s.* each.—Vol. XVIII., 2 Parts, 6*s.* each.—Vol. XIX., Parts 1 to 4, 16*s.*—Vol. XX., Parts 1 and 2, 4*s.* each. Part 3, 7*s.* 6*d.*

Asiatic Society.—JOURNAL OF THE ROYAL ASIATIC SOCIETY OF GREAT BRITAIN AND IRELAND. *New Series.* Vol. I. In Two Parts. pp. iv. and 490, sewed. 1864–5. 16*s.*

CONTENTS.—I. Vajra-chhedikâ, the "Kin Kong King," or Diamond Sûtra. Translated from the Chinese by the Rev. S. Beal.—II. The Páramitá-hridaya Sûtra, or, in Chinese, "Mo-ho-pô-ye-po-lo-mih-to-sin-king," *i.e.* "The Great Páramitá Heart Sûtra." Translated from the Chinese by the Rev. S. Beal.—III. On the Preservation of National Literature in the East. By Col. F. J. Goldsmid.—IV. On the Agricultural, Commercial, Financial, and Military Statistics of Ceylon. By E. R. Power.—V. Contributions to a Knowledge of the Vedic Theogony and Mythology. By J. Muir, D.C.L.—VI. A Tabular List of Original Works and Translations, published by the late Dutch Government of Ceylon at their Printing Press at Colombo. Compiled by Mr. M. P. J. Ondaatje.—VII. Assyrian and Hebrew Chronology compared, with a view of showing the extent to which the Hebrew Chronology of Ussher must be modified, in conformity with the Assyrian Canon. By J. W. Bosanquet.—VIII. On the existing Dictionaries of the Malay Language. By Dr. H. N. van der Tuuk.—IX. Bilingual Readings: Cuneiform and Phœnician. Notes on some Tablets in the British Museum, containing Bilingual Legends (Assyrian and Phœnician). By Major-Gen. Sir H. Rawlinson, K.C.B.—X. Translations of Three Copper-plate Inscriptions of the Fourth Century A.D., and Notices of the Châlukya and Gurjjara Dynasties. By Prof. J. Dowson, Staff College, Sandhurst.—XI. Yama and the Doctrine of a Future Life, according to the Rig-Yajur-, and Atharva-Vedas. By J. Muir, D.C.L.—XII. On the Jyotisha Observation of the Place of the Colures, and the Date derivable from it. By W. D. Whitney, Prof. of Sanskrit, Yale College, U.S.A.—Note on the preceding Article. By Sir E. Colebrooke, Bart., M.P.—XIII. Progress of the Vedic Religion towards Abstract Conceptions of the Deity. By J. Muir, D.C.L.—XIV. Brief Notes on the Age and Authenticity of the Work of Aryabhata, Varâhamihira, Brahmagupta, Bhattotpala, and Bhâskarâchârya. By Dr. Bhâu Dâjî.—XV. Outlines of a Grammar of the Malagasy Language. By H. N. Van der Tuuk.—XVI. On the Identity of Xandrames and Krananda. By E. Thomas, Esq.

Vol. II. In Two Parts. pp. 522, sewed. 1866–7. 16*s.*

CONTENTS.—I. Contributions to a Knowledge of Vedic Theogony and Mythology. No. 2. By J. Muir.—II. Miscellaneous Hymns from the Rig- and Atharva-Vedas. By J. Muir.—III. Five hundred questions on the Social Condition of the Natives of Bengal. By the Rev. J. Long.—IV. Short account of the Malay Manuscripts belonging to the Royal Asiatic Society. By Dr. H. N. van der Tuuk.—V. Translation of the Amitâbha Sûtra from the Chinese. By the Rev. S. Beal.—VI. The initial coinage of Bengal. By E. Thomas.—VII. Specimens of an Assyrian Dictionary. By E. Norris.—VIII. On the Relations of the Priests to the other classes of Indian Society in the Vedic age By J. Muir.—IX. On the Interpretation of the Veda. By the same.—X. An attempt to Translate from the Chinese a work known as the Confessional Services of the great compassionate Kwan Yin, possessing 1000 hands and 1000 eyes. By the Rev. S. Beal.—XI. The Hymns of the Gaupâyanas and the Legend of King Asamâti. By Prof. Max Müller.—XII. Specimen Chapters of an Assyrian Grammar. By the Rev. E. Hincks, D.D.

Vol. III. In Two Parts. pp. 516, sewed. With Photograph. 1868. 22*s.*

CONTENTS.—I. Contributions towards a Glossary of the Assyrian Language. By H. F. Talbot.—II. Remarks on the Indo-Chinese Alphabets. By Dr. A. Bastian.—III. The poetry of Mohamed Rabadan, Arragonese. By the Hon. H. E. J. Stanley.—IV. Catalogue of the Oriental Manuscripts in the Library of King's College, Cambridge. By E. H. Palmer, B.A.—V. Description of the Amravati Tope in Guntur. By J. Fergusson, F.R.S.—VI. Remarks on Prof. Brockhaus' edition of the Kathâsarit-sâgara, Lambaka IX. XVIII. By Dr. H. Kern, Prof. of Sanskrit, University of Leyden.—VII. The source of Colebrooke's Essay "On the Duties of a Faithful Hindu Widow." By Fitzedward Hall, D.C.L. Supplement: Further detail of proofs that Colebrooke's Essay, "On the Duties of a Faithful Hindu Widow," was not indebted to the Vivâdabhangârnava. By F. Hall.—VIII. The Sixth Hymn of the First Book of the Rig Veda. By Prof. Max Müller.—IX. Sassanian Inscriptions. By E. Thomas.—X. Account of an Embassy from Morocco to Spain in 1690 and 1691. By the Hon. H. E. J. Stanley.—XI. The Poetry of Mohamed Rabadan, of Arragon. By the same.—XII. Materials for the History of

India for the Six Hundred Years of Mohammadan rule, previous to the Foundation of the British Indian Empire. By Major W. Nassau Lees, LL.D.—XIII. A Few Words concerning the Hill people inhabiting the Forests of the Cochin State. By Capt. G. E. Fryer, M.S.C.—XIV. Notes on the Bhojpurí Dialect of Hindí, spoken in Western Behar. By J. Beames, B.C.S.

Vol. IV. In Two Parts. pp. 521, sewed. 1869–70. 16*s.*

Contents.—I. Contribution towards a Glossary of the Assyrian Language. By H. F. Talbot. Part II.—II. On Indian Chronology. By J. Fergusson, F.R.S.—III. The Poetry of Mohamed Rabadan of Arragon. By the Hon. H. E. J. Stanley.—IV. On the Magar Language of Nepal. By J. Beames, B.C.S.—V. Contributions to the Knowledge of Parsee Literature. By E. Sachau, Ph.D.—VI. Illustrations of the Lamaist System in Tibet, drawn from Chinese Sources. By W. F. Mayers, of H.B.M. Consular Service, China.—VII. Khuddaka Pátha, a Páli Text, with a Translation and Notes. By R. C. Childers, late Ceylon C.S.—VIII. An Endeavour to elucidate Rashiduddin's Geographical Notices of India. By Col. H. Yule, C.B.—IX. Sassanian Inscriptions explained by the Pahlavi of the Pârsis. By E. W. West.—X. Some Account of the Senbyû Pagoda at Mengún, near the Burmese Capital, in a Memorandum by Capt. E. H. Sladan, Political Agent at Mandalé; with Remarks on the Subject by Col. H. Yule, C.B.—XI. The Brhat-Sanhitâ; or, Complete System of Natural Astrology of Varâha-Mihira. Translated from Sanskrit into English by Dr. H. Kern.—XII. The Mohammedan Law of Evidence, and its influence on the Administration of Justice in India. By N. B. E. Baillie.—XIII. The Mohammedan Law of Evidence in connection with the Administration of Justice to Foreigners. By the same.—XIV. A Translation of a Bactrian Páli Inscription. By Prof. J. Dowson.—XV. Indo-Parthian Coins By E. Thomas.

Vol. V. In Two Parts. pp. 463, sewed. With 10 full-page and folding Plates. 1871–2. 18*s.* 6*d.*

Contents.—I. Two Játakas. The original Páli Text, with an English Translation. By V. Fausböll.—II. On an Ancient Buddhist Inscription at Keu-yung kwan, in North China. By A. Wylie.—III. The Brhat Sanhitâ; or, Complete System of Natural Astrology of Varâha-Mihira Translated from Sanskrit into English by Dr. H. Kern.—IV. The Pongol Festival in Southern India. By C. E. Gover.—V. The Poetry of Mohamed Rabadan, of Arragon. By the Right Hon. Lord Stanley of Alderley.—VI. Essay on the Creed and Customs of the Jangams. By C. P. Brown.—VII. On Malabar, Coromandel, Quilon, etc. By C. P. Brown.—VIII. On the Treatment of the Nexus in the Neo-Aryan Languages of India. By J. Beames, B.C.S.—IX. Some Remarks on the Great Tope at Sânchi. By the Rev. S. Beal.—X. Ancient Inscriptions from Mathura. Translated by Prof. J. Dowson.—Note to the Mathura Inscriptions. By Major-Gen. A. Cunningham.—XI. Specimen of a Translation of the Adi Granth. By Dr. E. Trumpp.—XII. Notes on Dhammapada, with Special Reference to the Question of Nirvâna. By R. C. Childers, late Ceylon C.S—XIII. The Brhat-Sanhitâ; or, Complete System of Natural Astrology of Varâha-mihira. Translated from Sanskrit into English by Dr. H. Kern.—XIV. On the Origin of the Buddhist Arthakathâs. By the Mudliar L. Comrilla Vijasinha, Government Interpreter to the Ratnapura Court, Ceylon. With Introduction by R. C. Childers, late Ceylon C.S.—XV. The Poetry of Mohamed Rabadan, of Arragon. By the Right Hon. Lord Stanley of Alderley.—XVI. Proverbia Communia Syriaca. By Capt. R. F. Burton.—XVII. Notes on an Ancient Indian Vase, with an Account of the Engraving thereupon. By C. Horne, late B.C.S.—XVIII. The Bhar Tribe. By the Rev. M. A. Sherring, LL.D., Benares. Communicated by C. Horne, late B.C.S.—XIX. Of *Jihad* in Mohammedan Law, and its application to British India. By N. B. E. Baillie.—XX. Comments on Recent Pehlvi Decipherments. With an Incidental Sketch of the Derivation of Aryan Alphabets. And Contributions to the Early History and Geography of Tabaristán. Illustrated by Coins. By E. Thomas, F.R.S.

Vol. VI., Part 1, pp. 212, sewed, with two plates and a map. 1872. 8*s.*

Contents.—The Ishmaelites, and the Arabic Tribes who Conquered their Country. By A. Sprenger.—A Brief Account of Four Arabic Works on the History and Geography of Arabia. By Captain S. B. Miles.—On the Methods of Disposing of the Dead at Llassa, Thibet, etc. By Charles Horne, late B.C.S. The Brhat-Sanhitâ; or, Complete System of Natural Astrology of Varâha-mihira, Translated from Sanskrit into English by Dr. H. Kern.—Notes on Hwen Thsang's Account of the Principalities of Tokhâristân, in which some Previous Geographical Identifications are Reconsidered. By Colonel Yule, C.B.—The Campaign of Ælius Gallus in Arabia. By A. Sprenger.—An Account of Jerusalem, Translated for the late Sir H. M. Elliot from the Persian Text of Násir ibn Khusrû's Safanâmah by the late Major A. R. Fuller.—The Poetry of Mohamed Rabadan, of Arragon. By the Right Hon. Lord Stanley of Alderley.

Vol. VI., Part II., pp. 213 to 400 and lxxxiv., sewed. Illustrated with a Map, Plates, and Woodcuts. 1873. 8*s.*

Contents.—On Hiouen-Thsang's Journey from Patna to Ballabhi. By James Fergusson, D.C.L., F.R.S.—Northern Buddhism. [Note from Colonel H. Yule, addressed to the Secretary.] —Hwen Thsang's Account of the Principalities of Tokhâristân, etc. By Colonel H. Yule, C.B.—The Brhat-Sañhitâ; or, Complete System of Natural Astrology of Varâha-mihira. Translated from Sanskrit into English by Dr. H. Kern.—The Initial Coinage of Bengal, under the Early Muhammadan Conquerors. Part II. Embracing the preliminary period between A.H. 614-634 (A.D. 1217-1236-7). By Edward Thomas, F.R.S.—The Legend of Dipañkara Buddha. Translated from the Chinese (and intended to illustrate Plates xxix. and L., 'Tree and Serpent Worship'). By S. Beal.—Note on Art. IX., antè pp. 213-274, on Hiouen-Thsang's Journey from Patna to Ballabhi. By James Fergusson. D.C.L., F.R.S.—Contributions towards a Glossary of the Assyrian Language. By H. F. Talbot.

Vol. VII., Part I., pp. 170 and 24, sewed. With a plate. 1874. 8*s.*

CONTENTS.—The *Upasampadâ-Kammavâcâ*, being the Buddhist Manual of the Form and Manner of Ordering of Priests and Deacons. The Pâli Text, with a Translation and Notes. By J. F. Dickson, B.A.—Notes on the Megalithic Monuments of the Coimbatore District, Madras. By M. J. Walhouse, late M.C.S.—Notes on the Sinhalese Language. No. 1. On the Formation of the Plural of Neuter Nouns. By R. C. Childers, late Ceylon C.S.—The Pali Text of the *Mahâparinibbâna Sutta* and Commentary, with a Translation. By R. C. Childers, late Ceylon C.S.—The Brihat-Sanhitâ; or, Complete System of Natural Astrology of Varâha-mihira. Translated from Sanskrit into English by Dr. H. Kern.—Note on the Valley of Choombi. By Dr. A. Campbell, late Superintendent of Darjeeling.—The Name of the Twelfth Imâm on the Coinage of Egypt. By H. Sauvaire and Stanley Lane Poole.—Thre Inscriptions of Parâkrama Bâbu the Great from Pulastipura, Ceylon (date circa 1180 A.D.). By T. W. Rhys Davids.—Of the Kharâj or Muhammadan Land Tax; its Application to British India, and Effect on the Tenure of Land. By N. B. E. Baillie.—Appendix: A Specimen of a Syriac Version of the Kalilah wa-Dimnah, with an English Translation. By W. Wright.

Vol. VII., Part II., pp. 191 to 394, sewed. With seven plates and a map. 1875. 8*s*

CONTENTS.—Sîgiri, the Lion Rock, near Pulastipura, Ceylon; and the Thirty-ninth Chapter of the Mahâvamsa. By T. W. Rhys Davids.—The Northern Frontagers of China. Part I. The Origines of the Mongols. By H. H. Howorth.—Inedited Arabic Coins. By Stanley Lane-Poole.—Notice on the Dinârs of the Abbasside Dynasty. By Edward Thomas Rogers.—The Northern Frontagers of China. Part II. The Origines of the Manchus. By H. H. Howorth.—Notes on the Old Mongolian Capital of Shangtu. By S. W. Bushell, B.Sc., M.D.—Oriental Proverbs in their Relations to Folklore, History, Sociology; with Suggestions for their Collection, Interpretation, Publication. By the Rev. J. Long.—Two Old Simhalese Inscriptions. The Sahasa Malla Inscription, date 1200 A.D., and the Ruwanwæli Dagaba Inscription, date 1191 A.D. Text, Translation, and Notes. By T. W. Rhys Davids.—Notes on a Bactrian Pali Inscription and the Samvat Era. By Prof. J. Dowson.—Note on a Jade Drinking Vessel of the Emperor Jahângîr. By Edward Thomas, F.R.S.

Vol. VIII., Part I., pp. 156, sewed, with three plates and a plan. 1876. 8*s.*

CONTENTS.—Catalogue of Buddhist Sanskrit MSS. in the Possession of the R.A.S. (Hodgson Collection). By Prof. E. B. Cowell and J. Eggeling.—On the Ruins of Sîgiri in Ceylon. By T. H. Blakesley, Ceylon.—The Pâtimokkha, being theBuddhist Office of the Confession of Priests. The Pali Text, with a Translation, and Notes. By J. F. Dickson, M.A., Ceylon C.S.—Notes on the Sinhalese Language. No. 2. Proofs of the Sanskritic Origin of Sinhalese. By R. C. Childers, late of the Ceylon Civil Service.

Vol. VIII., Part II., pp. 157-308, sewed. 1876. 8*s.*

CONTENTS.—An Account of the Island of Bali By R. Friederich.—The Pali Text of the Mahâparinibbâna Sutta and Commentary, with a Translation. By R C. Childers, late Ceylon C.S.—The Northern Frontagers of China. Part III. The Kara Khitai. By H. H. Howorth.—Inedited Arabic Coins. II. By S. L. Poole.—On the Form of Government under the Native Sovereigns of Ceylon. By A. de Silva Ekanâyaka, Mudaliyar, Ceylon.

Vol. IX., Part I., pp. 156, sewed, with a plate. 1877. 8*s.*

CONTENTS.—Bactrian Coins and Indian Dates. By E. Thomas, F.R.S.—The Tenses of the Assyrian Verb. By the Rev. A. H. Sayce, M.A.—An Account of the Island of Bali. By R. Friederich (continued from Vol. VIII. N.S. p. 218).—On Ruins in Makran. By Major Mockler.—Inedited Arabic Coins. III. By Stanley Lane Poole,—Further Note on a Bactrian Pali Inscription and the Samvat Era. By Prof. J. Dowson.—Notes on Persian Belûchistan. From the Persian of Mirza Mehdy Khân. By A. H. Schindler.

Vol IX., Part II., pp. 292, sewed, with three plates. 1877. 10*s.* 6*d.*

CONTENTS.—The Early Faith of Asoka. By E. Thomas, F.R.S.—The Northern Frontagers of China. Part II. The Manchus (Supplementary Notice). Part IV. The Kin or Golden Tatars. By H. H. Howorth.—On a Treatise on Weights and Measures by Eliyâ, Archbishop of Nisîbîn. By M. H. Sauvaire.—On Imperial and other Titles. By Sir T. E. Colebrooke, Bart., M.P.—Affinities of the Dialects of the Chepang and Kusundah Tribes of Nipâl with those of the Hill Tribes of Arracan. By Capt. C. J. F. Forbes F.R.G.S., M.A.S. Bengal, etc.—Notes on Some Antiquities found in a Mound near Damghan. By A. H. Schindler.

Vol. X., Part I., pp. 156, sewed, with two plates and a map. 1878. 8*s.*

CONTENTS.—On the Non-Aryan Languages of India. By E. L. Brandreth.—A Dialogue on the Vedantic Conception of Brahma. By Pramadâ Dâsa Mittra, late Offi. Prof. of Anglo-Sanskrit, Gov. College, Benares.—An Account of the Island of Bali. By R. Friederich (continued from Vol. IX. N.S. p. 120).—Unpublished Glass Weights and Measures. By E. T. Rogers.—China viâ Tibet. By S. C. Boulger.—Notes and Recollections on Tea Cultivation in Kumaon and Garhwâl. By J. H. Batten, late B.C S.

Vol. X., Part II., pp. 146, sewed. 1878. 6*s.*

CONTENTS.—Note on Pliny's Geography of the East Coast of Arabia. By Major-Gen. S. B. Miles, B.S.C.—The Maldive Islands; with a Vocabulary taken from François Pyrard de Laval, 1602—1607. By A. Gray, late Ceylon C.S.—On Tibeto-Burman Languages. By Capt. C. J. F. S. Forbes, Burmese C.S. Commission.—Burmese Transliteration. By H. L. St. Barbe, Resident at Mandelay.—On the Connexion of the Mōns of Pegu with the Koles of Central India. By Capt. C. J. F. S. Forbes, Burmese C.C.—Studies on the Comparative Grammar of the Semitic Languages, with Special Reference to Assyrian. By P. Haupt. The Oldest Semitic Verb-Form.—Arab Metrology. II. El Djabarty. By M. H. Sauvaire.—The Migrations and Early History of the White Huns; principally from Chinese Sources. By T. W. Kingsmill.

Vol. X., Part III., pp. 204, sewed. 1878. 8*s.*

CONTENTS —On the ill Canton of Sálár,—the most Easterly Settlement of the Turk Race. By Robert B. Shaw. Geological Notes on the River Indus. By Griffin W. Vyse, Executive Engineer P.W.D. Panjab.—Educational Literature for Japanese Women. By B. H. Chamberlain.—On the Natural Phenomenon Known in the East by the Names Sub-hi-Kâzib, etc., etc. By J. W. Redhouse.—On a Chinese Version of the Sánkhya Káriká, etc., found among the Buddhist Books comprising the Tripitaka and two other works. By the Rev. S. Beal.—The Rock-cut Phrygian Inscriptions at Doganlu. By E. Thomas, F.R.S.—Index.

Vol. XI., Part. I., pp. 128, sewed, with seven illustrations. 1879. 5*s.*

CONTENTS.—On the Position of Women in the East in the Olden Time. By E. Thomas, F.R.S. —Notice of Scholars who have Contributed to our Knowledge of the Languages of British India during the last Thirty Years. By R. N. Cust.—Ancient Arabic Poetry: its Genuineness and Authenticity. By Sir W. Muir, K.C.S.I.—Note on Manrique's Mission and the Catholics in the time of Sháh Jahán. By H. G. Keene.—On Sandhi in Pali. By the late R. C. Childers.—On Arabic Amulets and Mottoes. By E. T. Rogers.

Vol. XI., Part II., pp. 256, sewed, with map and plate. 1879. 7*s.* 6*d.*

CONTENTS.—On the Identification of Places on the Makran Coast mentioned by Arrian, Ptolemy, and Marcian. By Major E. Mockler.—On the Proper Names of the Mohammadans. By Sir T. E. Colebrooke, Bart., M P.—Principles of Composition in Chinese, as deduced from the Written Characters. By the Rev. Dr. Legge. On the Identification of the Portrait of Chosroes II. among the Paintings in the Caves at Ajanta. By James Fergusson, Vice-President.—A Specimen of the Zoongee (or Zurngee) Dialect of a Tribe of Nagas, bordering on the Valley of Assam, between the Dikho and Desoi Rivers, embracing over Forty Villages. By the Rev. Mr. Clark

Vol. XI. Part III. pp. 104, cxxiv. 16, sewed. 1879. 8*s.*

CONTENTS.—The Gaurian compared with the Romance Languages. Part I. By E. L. Brandreth.—Dialects of Colloquial Arabic. By E. T. Rogers.—A Comparative Study of the Japanese and Korean Languages. By W. G. Aston.—Index.

Vol. XII. Part I. pp. 152, sewed, with Table. 1880. 5*s.*

CONTENTS.—On "The Most Comely Names," *i.e.* the Laudatory Epithets, or the Titles of Praise, bestowed on God in the Qur'ân or by Muslim Writers. By J. W. Redhouse.—Notes on a newly-discovered Clay Cylinder of Cyrus the Great. By Major-Gen. Sir H. C. Rawlinson, K.C.B.—Note on Hiouen-Thsang's Dhanakacheka. By Robert Sewell, M.C.S.—Remarks by Mr. Fergusson on Mr. Sewell's Paper.—A Treatise on Weights and Measures. By Eliyá, Archbishop of Nisíbín. By H. Sauvaire. (Supplement to Vol. IX., pp. 291-313)—On the Age of the Ajantá Caves. By Rájendralála Mitra, C.I.E.—Notes on Babu Rájendralá Mitra's Paper on the Age of the Caves at Ajantá. By J. Fergusson, F.R S.

Vol. XII. Part II. pp. 182, sewed, with map and plate. 1880. 6*s.*

CONTENTS.—On Sanskrit Texts Discovered in Japan. By Prof. Max Müller.—Extracts from Report on the Islands and Antiquities of Bahrein. By Capt. Durand. Followed by Notes by Major-Gen. Sir H. C. Rawlinson, K.C.B.—Notes on the Locality and Population of the Tribes dwelling between the Brahmaputra and Ningthi Rivers. By the late G. H. Damant, Political Officer Nága Hills.—On the Saka, Samvat, and Gupta Eras. A Supplement to his Paper on Indian Chronology. By J. Fergusson, D.C.L.—The Megha-Sûtra. By C. Bendall.—Historical and Archæological Notes on a Journey in South-Western Persia, 1877-1878. By A. Houtum-Schindler.—Identification of the "False Dawn" of the Muslims with the "Zodiacal Light" of Europeans. By J. W. Redhouse.

Vol. XII. Part III. pp. 100, sewed. 1880. 4*s.*

CONTENTS.—The Gaurian compared with the Romance Languages. Part II. By E. L. Brandreth.—The Uzbeg Epos. By Arminius Vambéry.—On the Separate Edicts at Dhauli and Jaugada. By Prof. Kern —Grammatical Sketch of the Kakhyen Language. By Rev. J. N. Cushing.—Notes on the Libyan Languages, in a Letter addressed to R. N. Cust, Esq., by Prof. F. W. Newman.

Vol. XII. Part IV. pp. 152, with 3 plates. 1880. 8*s.*

CONTENTS.—The Early History of Tibet, from Chinese Sources. By S. W. Bushell, M.D.—Notes on some Inedited Coins from a Collection made in Persia during the Years 1877-79. By Guy Le Strange, M.R.A.S.—Buddhist Nirvâna and the Noble Eightfold Path. By Oscar Frankfurter, Ph.D.—Index.—Annual Report, 1880.

Vol. XIII. Part I. pp. 120, sewed. 1881. 5*s.*

CONTENTS. -Indian Theistic Reformers. By Prof. Monier Williams, C.I.E.—Notes on the Kawi Language and Literature. By Dr. H. N. Van der Tuuk.—The Invention of the Indian Alphabet. By John Dowson. The Nirvana of the Northern Buddhists. By the Rev. J. Edkins, D.D.—An Account of the Malay "Chiri," a Sanskrit Formula. By W. E. Maxwell.

Vol. XIII. Part II. pp. 170, with Map and 2 Plates. 1881. 8*s.*

CONTENTS.—The Northern Frontagers of China. Part V. The Khitai or Khitans. By H. H. Howorth.—On the Identification of Nagarahara, with reference to the Travels of Hiouen-Thsang. By W. Simpson.—Hindu Law at Madras. By J. H. Nelson, M.C.S.—On the Proper Names of the Mohammedans. By Sir T. E. Colebrooke, Bart., M.P.—Supplement to the Paper on Indian Theistic Reformers, published in the January Number of this Journal. By Prof. Monier Williams, C.I.E.

Vol. XIII. Part III. pp. 178, with plate. 1881. 7s. 6d.

CONTENTS.—The Avâr Language. By C. Graham.—Caucasian Nationalities. By M. A. Morrison.—Translation of the Markandeya Purana. Books VII., VIII. By the Rev. B. H. Wortham.—Lettre à M. Stanley Lane Poole sur quelques monnaies orientales rares ou inédites de la Collection de M. Ch. de l'Ecluse. Par H. Sauvaire.—Aryan Mythology in Malay Traditions. By W. E. Maxwell, Colonial Civil Service.—The Koi, a Southern Tribe of the Gond. By the Rev. J. Cain, Missionary.—On the Duty which Mohammedans in British India owe, on the Principles of their own Law, to the Government of the Country. By N. B. E. Baillie.—The L-Poem of the Arabs, by Shanfara. Re-arranged and translated by J. W. Redhouse, M.R.A.S.

Vol. XIII. Part IV. pp. 130, cxxxvi. 16, with 3 plates. 1881. 10s. 6d.

CONTENTS.—The Andaman Islands and the Andamanese. By M. V. Portman.—Notes on Marco Polo's Itinerary in Southern Persia. By A. Houtum-Schindler.—Two Malay Myths: The Princess of the Foam, and the Raja of Bamboo. By W. E. Maxwell.—The Epoch of the Guptas. By E. Thomas, F.R.S.—Two Chinese-Buddhist Inscriptions found at Buddha Gaya. By the Rev. S. Beal. With 2 Plates.—A Sanskrit Ode addressed to the Congress of Orientalists at Berlin. By Rama Dasa Sena, the Zemindar of Berhampore: with a Translation by S. Krishnavarma.—Supplement to a paper, "On the Duty which Mahommedans in British India owe, on the Principles of their own Law, to the Government of the Country." By N. B. E. Baillie.—Index.

Vol. XIV. Part I. pp. 124, with 4 plates. 1882. 5s.

CONTENTS.—The Apology of Al Kindy: An Essay on its Age and Authorship. By Sir W Muir, K.C.S.I.—The Poet Pampa. By L. Rice.—On a Coin of Shams ud Dunyâ wa ud Din Mahmûd Shâh. By C. J. Rodgers, Amritsar.—Note on Pl. xxviii. fig. 1, of Mr. Fergusson's "Tree and Serpent Worship," 2nd Edition. By S. Beal, Prof of Chinese, London University.—On the present state of Mongolian Researches. By Prof. B. Julg, in a Letter to R. N Cust.—A Sculptured Tope on an Old Stone at Dras, Ladak. By W. Simpson, F.R.G S.—Sanskrit Ode addressed to the Fifth International Congress of Orientalists assembled at Berlin, September, 1881. By the Lady Pandit Rama-bai, of Silchar, Kachar, Assam; with a Translation by Prof. Monier Williams, C.I.E.—The Intercourse of China with Eastern Turkestan and the Adjacent Countries in the Second Century B.C. By T. W. Kingsmill.—Suggestions on the Formation of the Semitic Tenses. A Comparative and Critical Study. By G. Bertin.—On a Lolo MS. written on Satin. By M. T. de La Couperie.

Vol. XIV. Part II. pp. 164, with three plates. 1882. 7s. 6d.

CONTENTS.—On Tartar and Turk. By S. W. KOELLE, Ph.D.—Notice of Scholars who have Contributed to our Knowledge of the Languages of Africa. By R. N. Cust.—Grammatical Sketch of the Hausa Language. By the Rev. J. F. Schön, F.R.G.S.,—Buddhist Saint Worship. By A. Lillie.—Gleanings from the Arabic. By H. W. Freeland, M.A.—Al Kahirah and its Gates. By H. C. Kay, M.A.—How the Mahâbhârata begins. By Edwin Arnold, C.S.I.—Arab Metrology. IV. Ed-Dahaby. By M. H. Sauvaire.

Vol. XIV. Part III. pp. 208, with 8 plates. 1882. 8s.

CONTENTS.—The Vaishnava Religion, with special reference to the Sikshā-patrī of the Modern Sect called Svāmi-Nārāyana. By Monier Williams, C.I.E., D.C.L.—Further Notes on the Apology of Al-Kindy. By Sir W. Muir, K.C.S.I., D.C.L., LL.D.—The Buddhist Caves of Afghanistan By W. Simpson.—The Identification of the Sculptured Tope at Sanchi. By W. Simpson.—On the Genealogy of Modern Numerals. By Sir E. C. Bayley, K.C.S.I., C.I.E.—The Cuneiform Inscriptions of Van, deciphered and translated. By A. H. Sayce.

Vol. XIV. Part IV. pp. 330, clii. 1882. 14s.

CONTENTS.—The Cuneiform Inscriptions of Van, Deciphered and Translated. By A. H. Sayce.—Sanskrit Text of the Sikshā-Patrī of the Svāmi-Nārayana Sect. Edited and Translated by Prof. M. Williams, C.I.E.—The Successors of the Siljuks in Asia Minor. By S. L. Poole.—The Oldest Book of the Chinese (*The Yh-King*) and its Authors. By T. de la Couperie.

Vol. XV. Part I. pp. 134, with 2 plates. 1883. 6s.

CONTENTS.—The Genealogy of Modern Numerals. Part II. Simplification of the Ancient Indian Numeration. By Sir E C. Bayley, C.I.E.—Parthian and Indo-Sassanian Coins. By E. Thomas, F.R.S.—Early Historical Relations between Phrygia and Cappadocia. By W. M. Ramsay.

Vol. XV. Part II. pp. 158, with 6 tables. 1883. 5s.

CONTENTS.—The Tattva-muktavalī of Gauda-pûrnânandachakravartin. Edited and Translated by Professor E. B. Cowell.—Two Modern Sanskrit slokas. Communicated by Prof. E. B. Cowell.—Malagasy Place-Names. By the Rev. James Sibree, jun.—The Namakkâra, with Translation and Commentary. By H. L. St. Barbe.—Chinese Laws and Customs. By Christopher Gardner.—The Oldest Book of the Chinese (the *Yh-King*) and its Authors (continued). By Terrien de LaCouperie.—Gleanings from the Arabic. By H. W. Freeland.

Vol. XV. Part III. pp. 62-cxl. 1883. 6s.

CONTENTS.—Early Kamada Authors. By Lewis Rice.—On Two Questions of Japanese Archæology. By B. H. Chamberlain, M.R.A S.—Two Sites named by Hiouen-Thsang in the 10th Book of the Si-yu-ki. By the Rev. S. Beal.—Two Early Sources of Mongol History. By H. H. Howorth, F.S.A.—Proceedings of Sixtieth Anniversary of the Society, held May 21, 1883.

Vol. XV. Part IV. pp. 140-iv.-20, with plate. 1883. *5s.*

Contents.—The Rivers of the Vedas, and How the Aryans Entered India. By Edward Thomas, F.R.S.—Suggestions on the Voice-Formation of the Semitic Verb. By G Bertin, M.R.A.S. —The Buddhism of Ceylon. By Arthur Lillie, M.R.A.S.—The Northern Frontagers of China. Part VI. Hia or Tangut. By H. H. Howorth, F.S.A.—Index.—List of Members.

Vol. XVI. Part I. pp. 138, with 2 plates. 1884. *7s.*

Contents.—The Story of Devasmitâ. Translated from the Kathâ Sarit Sâgara, Tarânga 13, Sloka 54, by the Rev. B. Hale Wortham.—Pujahs in the Sutlej Valley, Himalayas. By William Simpson, F.R.G.S.—On some New Discoveries in Southern India. By R. Sewell, Madras C.S.—On the Importance to Great Britain of the Study of Arabic. By Habib A. Salmoné.—Grammatical Note on the Gwamba Language in South Africa. By P. Berthoud, Missionary of the Canton de Vaud, Switzerland, stationed at Valdézia, Spelonken, Transvaal. (Prepared at the request of R. N. Cust.)—Dialect of Tribes of the Hindu Khush, from Colonel Biddulph's Work on the subject (corrected).—Grammatical Note on the Simnûnî Dialect of the Persian Language. By the Rev. J. Bassett, American Missionary, Tabriz. (Communicated by R. N. Cust.)

Vol. XVI. Part II. pp. 184, with 1 plate. *9s.*

Contents.—Etymology of the Turkish Numerals. By S. W. Koelle, Ph.D., late Missionary of the Church Missionary Soc., Constantinople.—Grammatical Note and Vocabulary of the Kor-kū, a Kolarian Tribe in Central India. (Communicated by R. N. Cust.) The Pariah Caste in Travancore. By S. Mateer.—Some Bihārī Folk-Songs. By G. A. Grierson, B C.S., Offi. Magistrate, Patna.—Some further Gleanings from the Si-yu-ki. By the Rev. S. Beal.—On the Sites of Brahmanâbâd and Mansûrah in Sindh; with notices of others of less note in their Vicinity. By Major-Gen. M. R. Haig.—Antar and the Slave Daji. A Bedoueen Legend. By St. C. Baddeley.—The Languages of the Early Inhabitants of Mesopotamia. By G. Pinches.

Vol. XVI. Part III. pp. 74.—clx. 10*s*. 6*d*.

Contents.—On the Origin of the Indian Alphabet. By R. N. Cust.—The Yi king of the Chinese as a Book of Divination and Philosophy By Rev. Dr. Edkins—On the Arrangement of the Hymns of the Rig-veda. By F. Pincott.—Proceedings of the Sixty-first Anniversary Meeting of the Society, May 19, 1884.

Vol. XVI. Part IV. pp. 134. 8*s*.

Contents.—S'uka-sandesah. A Sanskrit Poem, by Lakshmî-dâsa. With Preface and Notes in English by H. H. Rama Varma, the Maharaja of Travancore, G.C.S.I.—The Chinese Book of the Odes, for English Readers. By C. F. R. Allen.—Note sur les Mots Sanscrits composés avec पति Par J. van den Gheyn, S.J.—Some Remarks on the Life and Labours of Csoma de Körös, delivered on the occasion when his Tibetan Books and MSS. were exhibited before the R.A.S., June 16, 1884. By Surgeon-Major T. Duka, M.D., late of the Bengal Army.—Arab Metrology. V. Ez-Zahrâwy. Translated and Annotated by M. H. Sauvaire, de l'Académie de Marseille.

Asiatic Society.—Transactions of the Royal Asiatic Society of Great Britain and Ireland. Complete in 3 vols. 4to., 80 Plates of Facsimiles, etc., cloth. London, 1827 to 1835. Published at £9 *5s.*; reduced to £5 *5s.*

The above contains contributions by Professor Wilson, G. C. Haughton, Davis, Morrison, Colebrooke, Humboldt, Dorn, Grotefend, and other eminent Oriental scholars.

Asiatic Society of Bengal.—Journal of the Asiatic Society of Bengal. Edited by the Honorary Secretaries. 8vo. 8 numbers per annum, 4*s*. each number.

Asiatic Society of Bengal.—Proceedings of the Asiatic Society of Bengal. Published Monthly. 1*s*. each number.

Asiatic Society of Bengal.—Journal of the Asiatic Society of Bengal. A Complete Set from the beginning in 1832 to the end of 1878, being Vols. 1 to 47. Proceedings of the same Society, from the commencement in 1865 to 1878. A set quite complete. Calcutta, 1832 to 1878. Extremely scarce. £100.

Asiatic Society.—Bombay Branch.—Journal of the Bombay Branch of the Royal Asiatic Society. Nos. 1 to 35 in 8vo. with many plates. A complete set. Extremely scarce. Bombay, 1844–78. £13 10*s*.

Asiatic Society of Bombay.—The Journal of the Bombay Branch of the Royal Asiatic Society. Edited by the Secretary. Nos. 1 to 35 7*s.* 6*d.* to 10*s.* 6*d.* each number. Several Numbers are out of print.
No. 36, Vol. XIV., 1879, pp. 163 and xviii., with plates. 10*s.* 6*d.*
No. 37, Vol. XIV., 1880, pp. 104 and xxiii., with plates. 10*s.* 6*d.*
No. 38, Vol. XIV., 1880, pp. 172 and vi., with plate. 7*s.* 6*d.*
No. 39, Vol. XV., 1881, pp. 150, with plate. 5*s.*
No. 40, Vol. XV., 1882, pp. 176, with plates. 9*s.*
No. 41, Vol. XVI., 1883, pp. 129. 7*s.* 6*d.*
No. 42, Vol. XVI., 1884, pp. 166—xviii., with plate. 9*s.*

Asiatic Society.—Ceylon Branch.—Journal of the Ceylon Branch of the Royal Asiatic Society (Colombo). Part for 1845. 8vo. pp. 120, sewed. Price 7*s.* 6*d.*

Contents:—On Buddhism. No. 1. By the Rev. D. J. Gogerly.—General Observations on the Translated Ceylonese Literature. By W. Knighton, Esq.—On the Elements of the Voice in reference to the Roman and Singalese Alphabets. By the Rev. J. C. Macvicar.—On the State of Crime in Ceylon.—By the Hon. J. Stark.—Account of some Ancient Coins. By S. C. Chitty, Esq.—Remarks on the Collection of Statistical Information in Ceylon. By John Capper, Esq.—On Buddhism. No 2. By the Rev. D. J. Gogerly.

1846. 8vo. pp. 176, sewed. Price 7*s.* 6*d.*

Contents:—On Buddhism. By the Rev. D. J. Gogerly.—The Sixth Chapter of the Tiruvathavur Purana, translated with Notes. By S. Casie Chitty, Esq.—The Discourse on the Minor Results of Conduct, or the Discourse Addressed to Subba. By the Rev. D. J. Gogerly.—On the State of Crime in Ceylon. By the Hon Mr. J. Stark.—The Language and Literature of the Singalese. By the Rev. S. Hardy.—The Education Establishment of the Dutch in Ceylon. By the Rev. J. D. Palm.—An Account of the Dutch Church in Ceylon. By the Rev. J. D. Palm.—Notes on some Experiments in Electro-Agriculture. By J. Capper, Esq.—Singalo Wada, translated by the Rev. D. J. Gogerly.—On Colouring Matter Discovered in the husk of the Cocoa Nut. By Dr. R. Gygax.

1847–48. 8vo. pp. 221, sewed. Price 7*s.* 6*d.*

Contents:—On the Mineralogy of Ceylon. By Dr. R. Gygax.—An Account of the Dutch Church in Ceylon. By the Rev. J. D. Palm.—On the History of Jaffna, from the Earliest Period to the Dutch Conquest. By S. C. Chitty.—The Rise and Fall of the Calany Ganga, from 1843 to 1846. By J. Capper.—The Discourse respecting Ratapala. Translated by the Rev. D. J. Gogerly.—On the Manufacture of Salt in the Chilaw and Putlam Districts. By A. O. Brodie.—A Royal Grant engraved on a Copper Plate. Translated, with Notes. By the Rev. D. J. Gogerly.—On some of the Coins. Ancient and Modern, of Ceylon. By the Hon. Mr. J. Stark.—Notes on the Climate and Salubrity of Putlam. By A. O. Brodie.—The Revenue and Expenditure of the Dutch Government in Ceylon, during the last years of their Administration. By J. Capper.—On Buddhism. By the Rev. D. J. Gogerly.

1853–55. 3 parts. 8vo. pp. 56 and 101, sewed. Price £1.

Contents of Part I.:—Buddhism: Chariya Pitaka. By the Rev. D. J. Gogerly.—The Laws of the Buddhist Priesthood. By the Rev. D. J. Gogerly. To be continued.—Statistical Account of the Districts of Chilaw and Putlam, North Western Province. By A. O. Brodie, Esq.—Rock Inscription at Gooroo Godde Wihare, in the Magool Korle, Seven Korles. By A. O. Brodie, Esq.—Catalogue of Ceylon Birds. By E. F. Kelaart, Esq., and E. L. Layard, Esq. (To be continued.)

Contents of Part II. Price 7*s.* 6*d.*

Catalogue of Ceylon Birds. By E. F. Kelaart, Esq., and E. L. Layard.—Notes on some of the Forms of Salutations and Address known among the Singalese. By the Hon. Mr. J. Stark.—Rock Inscriptions. By A. O. Brodie, Esq.—On the Veddhas of Bintenne. By the Rev. J. Gillings.—Rock Inscription at Piramanenkandel. By S C. Chitty, Esq.—Analysis of the Great Historical Poem of the Moors, entitled Surah. By S. C. Chitty, Esq. (To be continued).

Contents of Part III. 8vo. pp. 150. Price 7*s.* 6*d.*

Analysis of the Great Historical Poem of the Moors, entitled Surah. By S. C. Chitty, Esq. (Concluded).—Description of New or little known Species of Reptiles found in Ceylon. By E. F. Kelaart.—The Laws of the Buddhist Priesthood By the Rev. D. J. Gogerly. (To be continued).—Ceylon Ornithology. By E. F. Kelaart.—Some Account of the Rodiyas, with a Specimen of their Language. By S. C. Chitty, Esq.—Rock Inscriptions in the North-Western Province. By A. O. Brodie, Esq.

1865-6. 8vo. pp. xi. and 184. Price 7*s.* 6*d.*

Contents:—On Demonology and Witchcraft in Ceylon. By Dandris de Silva Gooneratne Modliar.—The First Discourse Delivered by Buddha. By the Rev. D. J. Gogerly. Pootoor Well.—On the Air Breathing Fish of Ceylon. By Barcroft Boake, B.A. (Vice President Asiatic Society, Ceylon).—On the Origin of the Sinhalese Language. By J. D'Alwis, Assistant Secretary.—A Few Remarks on the Poisonous Properties of the Calotropis Gigantea, etc. By W. C. Ondaatjie, Esq., Colonial Assistant Surgeon.—On the Crocodiles of Ceylon. By Barcroft Boake, Vice-President, Asiatic Society, Ceylon.—Native Medicinal Oils.

1867-70. Part I. 8vo. pp. 150. Price 10*s*.

CONTENTS:—On the Origin of the Sinhalese Language. By James De Alwis.—A Lecture on Buddhism. By the Rev. D. J. Gogerly.—Description of two Birds new to the recorded Fauna of Ceylon. By H. Nevil.—Description of a New Genus and Five New Species of Marine Univalves from the Southern Province, Ceylon. By G. Nevill.—A Brief Notice of Robert Knox and his Companions in Captivity in Kandy for the space of Twenty Years, discovered among the Dutch Records preserved in the Colonial Secretary's Office, Colombo. By J. R. Blake.

1867-70. Part II. 8vo. pp. xl. and 45. Price 7*s*. 6*d*.

CONTENTS:—Summary of the Contents of the First Book in the Buddhist Canon, called the Párájika Book.—By the Rev. S. Coles.—Párájika Book—No. 1.—Párájika Book—No. 2.

1871-72. 8vo. pp. 66 and xxxiv. Price 7*s*. 6*d*.

CONTENTS:—Extracts from a Memoir left by the Dutch Governor, Thomas Van Rhee, to his successor, Governor Gerris de Heer, 1697. Translated from the Dutch Records preserved in the Colonial Secretariat at Colombo. By R. A. van Cuylenberg, Government Record Keeper.—The Food Statistics of Ceylon. By J. Capper.—Specimens of Sinhalese Proverbs. By L. de Zoysa, Mudaliyar, Chief Translator of Government.—Ceylon Reptiles: being a preliminary Catalogue of the Reptiles found in, or supposed to be in Ceylon, compiled from various authorities. By W. Ferguson.—On an Inscription at Dondra. No. 2. By T. W. Rhys Davids, Esq.

1873. Part I. 8vo. pp. 79. Price 7*s*. 6*d*.

CONTENTS:—On Oath and Ordeal. By Bertram Fulke Hartshorne.—Notes on Prinochilus Vincens. By W. V. Legge.—The Sports and Games of the Singhalese. By Leopold Ludovici.—On Miracles. By J. De Alwis.—On the Occurrence of Scolopax Rusticola and Gallinago Scolopacina in Ceylon. By W. V. Legge.—Transcript and Translation of an Ancient Copper-plate Sannas. By Mudliyar Louis de Zoysa, Chief Translator to Government.

1874. Part I. 8vo. pp. 94. Price 7*s*. 6*d*.

CONTENTS:—Description of a supposed New Genus of Ceylon. Batrachians. By W. Ferguson.—Notes on the Identity of Piyadasi and Asoka. By Mudaliyar Louis de Zoysa, Chief Translator to Government.—On the Island Distribution of the Birds in the Society's Museum. By W. Vincent Legge.—Brand Marks on Cattle. By J. De Alwis.—Notes on the Occurrence of a rare Eagle new to Ceylon; and other interesting or rare birds. By S. Bligh, Esq., Kotmalé.—Extracts from the Records of the Dutch Government in Ceylon. By R. van Cuylenberg, Esq.—The Stature of Gotama Buddha. By J. De Alwis.

1879. 8vo. pp. 58. Price 5*s*.

CONTENTS.—Notes on Ancient Sinhalese Inscriptions.—On the Preparation and Mounting of Insects for the Binocular Microscope.—Notes on Neophron Puenopterus (Savigny) from Nuwara Eliya.—On the Climate of Dimbula.—Note on the supposed cause of the existence of Patanas or Grass Lands of the Mountain Zone of Ceylon.

1880. Part I. 8vo. pp. 90. Price 5*s*.

CONTENTS.—Text and Translation of the Inscription of Mahinde III. at Mihintale.—Glossary.—A Paper on the Vedic and Buddhistic Polities.—Customs and Ceremonies connected with the Paddi Cultivation.—Gramineae, or Grasses Indigenous to or Growing in Ceylon.

1880. Part II. 8vo. pp. 48. Price 5*s*.

CONTENTS.—Gramineae, or Grasses Indigenous to or Growing in Ceylon.—Translation of two Jatakas.—On the supposed Origin of Tamana, Nuwara, Tambapanni and Taprobane.—The Rocks and Minerals of Ceylon.

1881. Vol. VII. Part I. (No. 23.) 8vo. pp. 56. Price 5*s*.

CONTENTS.—Hindu Astronomy: as compared with the European Science. By S. Mervin.—Sculptures at Horana. By J. G. Smither.—Gold. By A. C. Dixon.—Specimens of Sinhalese Proverbs. By L. De Zoysa.—Ceylon Bee Culture. By S. Jayatilaka.—A Short Account of the Principal Religious Ceremonies observed by the Kandyans of Ceylon. By C. J. R. Le Mesurier.—Valentyn's Account of Adam's Peak. By A. Spense Moss.

1881. Vol. VII. Part II. (No. 24.) 8vo. pp. 162. Price 5*s*.

CONTENTS.—The Ancient Emporium of Kalah, etc., with Notes on Fa-Hian's Account of Ceylon. By H. Nevill.—The Sinhalese Observance of the Kaláwa. By L. Nell.—Note on the Origin of the Veddás, with Specimens of their Songs and Charms. By L. de Zoysa.—A Húniyam Image. By L. Nell.—Note on the Mirá Kantiri Festival of the Muhammadans. By A. T. Sham-ud-diû.—Tericulture in Ceylon. By J. L. Vanderstraaten.—Sinhalese Omens. By S. Jayatilaka.

1882. Extra Number. 8vo. pp. 60. Price 5*s*.

CONTENTS.—Ibn Batuta in the Maldives and Ceylon. Translated from the French of M. M. Defremery and Sanguinetti. By A. Gray.

Asiatic Society (North China Branch).—JOURNAL OF THE NORTH CHINA BRANCH OF THE ROYAL ASIATIC SOCIETY. Old Series, 4 numbers, and New Series. Parts 1 to 12. The following numbers are sold separately: OLD SERIES—No. II. May, 1859, pp. 145 to 256. No. III. December, 1859, pp. 257 to 368. 7*s.* 6*d.* each. Vol. II. No. I. September, 1860, pp. 128. 7*s.*6*d.* NEW SERIES—No. I. December, 1864, pp. 174. 7*s.* 6*d.* No. II. December, 1865, pp. 187, with maps. 7*s.* 6*d.* No. III. December, 1866, pp. 121. 9*s.* No. IV. December, 1867, pp 266. 10*s.* 6*d.* No. VI. for 1869 and 1870, pp. xv. and 200. 7*s.* 6*d.* No. VII. for 1871 and 1872, pp. ix. and 260. 10*s.* No. VIII. pp. xii. and 187. 10*s.* 6*d.* No. IX. pp. xxxiii. and 219. 10*s.* 6*d.* No. X. pp. xii. and 324 and 279. £1 1*s.* No. XI. (1877) pp. xvi. and 184. 10*s.* 6*d.* No. XII. (1878) pp. 337, with many maps. £1 1*s.* No. XIII. (1879) pp. vi. and 132, with plates, 10*s.* 6*d.* No. XIV. (1879) pp. xvi.-64, with plates, 4*s.* No. XV. (1880) pp. xliii. and 316, with plates, 15*s.* No. XVI. (1881) pp. 248. 12*s.* 6*d.* No. XVII. (1882) pp. 246 with plates. 12*s.* 6*d.*

Asiatic Society of Japan.—TRANSACTIONS OF THE ASIATIC SOCIETY OF JAPAN. Vol. I. From 30th October, 1872, to 9th October, 1873. 8vo. pp. 110, with plates. 1874. Vol. II. From 22nd October, 1873, to 15th July, 1874. 8vo. pp. 249. 1874. Vol. III. Part I. From 16th July, 1874, to December, 1874, 1875. Vol. III. Part II. From 13th January, 1875, to 30th June, 1875. Vol. IV. From 20th October, 1875, to 12th July, 1876. Vol. V. Part I. From 25th October, 1876, to 27th June, 1877. Vol. V. Part II. (A Summary of the Japanese Penal Codes. By J. H. Longford.) Vol. VI. Part I. pp. 190. Vol. VI. Part II. From 9th February, 1878, to 27th April, 1878. Vol. VI. Part III. From 25th May, 1878, to 22nd May, 1879. 7*s.* 6*d.* each Part.—Vol. VII. Part I. (Milne's Journey across Europe and Asia.) 5*s.*—Vol. VII. Part II. March, 1879. 5*s.*—Vol. VII. Part III. June, 1879. 7*s.* 6*d.* Vol. VII. Part IV. November, 1879. 10*s.* 6*d.* Vol. VIII. Part I. February, 1880. 7*s* 6*d.* Vol. VIII. Part II. May, 1880. 7*s.* 6*d.* Vol. VIII. Part. III. October, 1880. 10*s.* 6*d.* Vol. VIII. Part IV. December, 1880. 5*s.* Vol. IX. Part I. February, 1881. 7*s.* 6*d.* Vol. IX. Part II. August, 1881. 7*s.* 6*d.* Vol. IX. Part III. December, 1881. 5*s.* Vol. X. Part I. May, 1882. 10*s.* Vol. X. Part II. October, 1882. 7*s.* 6*d.* Vol. X. Supplement, 1883. £1. Vol. XI. Part I. April, 1883. 7*s.* 6*d.* Vol. XI. Part II. September, 1883. 7*s.* 6*d.* Vol. XII. Part I. November, 1883. 5*s.* Vol. XII. Part II. May, 1884. 5*s.*

Asiatic Society.—Straits Branch.—JOURNAL OF THE STRAITS BRANCH OF THE ROYAL ASIATIC SOCIETY. No. 1. 8vo. pp. 130, sewed, 3 folded Maps and 1 Plate. July, 1878. Price 9*s.*

CONTENTS.—Inaugural Address of the President. By the Ven. Archdeacon Hose, M.A.—Distribution of Minerals in Sarawak. By A. Hart Everett.—Breeding Pearls. By N. B. Dennys, Ph.D.—Dialects of the Melanesian Tribes of the Malay Peninsula. By M. de Mikluho-Maclay.—Malay Spelling in English. Report of Government Committee (reprinted).—Geography of the Malay Peninsula. Part I. By A. M. Skinner.—Chinese Secret Societies. Part I. By W. A. Pickering.—Malay Proverbs. Part. I. By W. E. Maxwell.—The Snake-eating Hamadryad. By N. B. Dennys, Ph.D.—Gutta Percha. By H. J. Murton.—Miscellaneous Notices.

No. 2. 8vo. pp. 130, 2 Plates, sewed. December, 1878. Price 9*s.*

CONTENTS:—The Song of the Dyak Head-feast. By Rev. J. Perham.—Malay Proverbs. Part II. By E. W. Maxwell.—A Malay Nautch. By F. A. Swettenham.—Pidgin English. By N. B. Dennys, Ph.D.—The Founding of Singapore. By Sir T. S. Raffles.—Notes on Two Perak Manuscripts. By W. E. Maxwell.—The Metalliferous Formation of the Peninsula. By D. D. Daly.—Suggestions regarding a new Malay Dictionary. By the Hon. C. J. Irving.—Ethnological Excursions in the Malay Peninsula. By N. von Mikluho-Maclay.—Miscellaneous Notices.

No. 3. 8vo. pp. iv. and 146, sewed. July, 1879. Price 9*s.*

CONTENTS:—Chinese Secret Societies, by W. A. Pickering.—Malay Proverbs, Part III., by W. E. Maxwell.—Notes on Gutta Percha, by F. W. Burbidge, W. H. Treacher, H. J. Murton.—The Maritime Code of the Malays, reprinted from a translation by Sir S. Raffles.—A Trip to Gunong Bumut, by D. F. A. Hervey.—Caves at Sungei Batu in Selangor, by D. D. Daly.—Geography of Aching, translated from the German by Dr. Beiber.—Account of a Naturalist's Visit to Selangor, by A. J. Hornady.—Miscellaneous Notices: Geographical Notes, Routes from Selangor to Pahang, Mr. Deane's Survey Report, A Tiger's Wake, Breeding Pearls, The Maritime Code, and Sir F. Raffles' Meteorological Returns.

No. 4. 8vo. pp. xxv. and 65, sewed. December, 1879. Price 9*s.*

Contents.—List of Members.—Proceedings, General Meeting.—Annual Meeting.—Council's Annual Report for 1879.—Treasurer's Report for 1879.—President's Address.—Reception of Professor Nordenskjold.—The Marine Code. By Sir S. Raffles.—About Kinta. By H. W. C. Leech.—About Shin and Bernam. By H. W. Leech.—The Aboriginal Tribes of Perak. By W. E. Maxwell.—The Vernacular Press in the Straits. By E. W. Birch.—On the Guliga of Borneo. By A. H. Everett.—On the name "Sumatra."—A Correction.

No. 5. 8vo. pp. 160, sewed. July, 1879. Price 9*s.*

Contents.—Selesilah (Book of the Descent) of the Rajas of Bruni. By H. Low.—Notes to Ditto.—History of the Sultins of Bruni.—List of the Mahomedan Sovereigns of Bruni.—Historic Tablet.—Acheh. By G. P. Talson.—From Perak to Shin and down the Shin and Bernam Rivers. By F. A. Swettenham.—A Contribution to Malayan Bibliography. By N. B. Dennys.—Comparative Vocabulary of some of the Wild Tribes inhabiting the Malayan Peninsula, Borneo, etc.—The Tiger in Borneo. By A. H. Everett.

No. 6. 8vo. pp. 133, with 7 Photographic Plates, sewed. December, 1880. Price 9*s.*

Contents.—Some Account of the Independent Native States of the Malay Peninsula. Part I. By F. A. Swettenham.—The Ruins of Boro Burdur in Java. By the Ven. Archdeacon G. F. Hose. A Contribution to Malayan Bibliography. By N. B. Dennys.—Report on the Exploration of the Caves of Borneo. By A. H. Everett.—Introductory Remarks. By J. Evans.—Notes on the Report.—Notes on the Collection of Bones. By G. Bush.—A Sea-Dyak Tradition of the Deluge and Consequent Events. By the Rev. J. Perham.—The Comparative Vocabulary.

No. 7. 8vo. pp. xvi. and 92. With a Map, sewed. June, 1881. Price 9*s.*

Contents.—Some account of the Mining Districts of Lower Perah. By J. Errington de la Croix.—Folklore of the Malays. By W. E. Maxwell —Notes on the Rainfall of Singapore. By J. J. L Wheatley.—Journal of a Voyage through the Straits of Malacca on an Expedition to the Molucca Islands. By Captain W. C. Lennon.

No. 8. 8vo. pp. 56. With a Map. sewed. December, 1881. Price 9*s.*

Contents.—The Endau and its Tributaries. By D. F A. Hervey.—Itinerary from Singapore to the Source of the Sembrong and up the Madek.—Petara, or Sea Dyak Gods. By the Rev. J. Perham.—Klouwang and its Caves, West Coast of Atchin. Translated by D. F. A. Hervey.—Miscellaneous Notes: Varieties of "Getah" and "Rotan."—The "Ipoh" Tree, Perak.—Comparative Vocabulary.

No. 9. 8vo. pp. xxii. and 172. With three Col. Plates, sd. June, 1882. Price 12*s.*

Contents.—Journey on Foot to the Patani Frontier in 1876. By W. E. Maxwell.—Probable Origin of the Hill Tribes of Formosa. By John Dodd —History of Perak from Native Sources. By W. E Maxwell.—Malayan Ornithology. By Captain H. R. Kelham.—On the Transliteration of Malay in the Roman Character. By W. E. Maxwell.—Kota Glanggi, Pahang. By W. Cameron.—Natural History Notes. By N. B. Dennys.—Statement of Haji of the Madek Ali.—Pantang Kapur of the Madek Jakun.—Stone from Batu Pahat.—Rainfall at Lankat, Sumatra.

No. 10. 8vo. pp. xv. and 117, sewed. December, 1882. Price 9*s.*

Contents.—Journal of a Trip from Sarawak to Meri. By N. Denison.—The Mentra Traditions. By the Hon. D. F. A. Hervey.—Probable Origin of the Hill Tribes of Formosa. By J. Dodd.—Sea Dyak Religion. By the Rev. J. Perham.—The Dutch in Perak. By W. E. Maxwell.—Outline History of the British Connection with Malaya. By the Hon. A. M. Skinner.—Extracts from Journals of the Société de Geographie of Paris.—Memorandum on Malay Transliteration.—The Chiri.—Register of Rainfall.

No. 11. 8vo. pp. 170. With a Map, sewed. June, 1883. Price 9*s.*

Contents.—Malayan Ornithology. By Captain H. R. Kelham.—Malay Proverbs. By the Hon. W. E. Maxwell.—The Pigmies. Translated by J. Errington de la Croix.—On the Patani, By W. Cameron.—Latah. By H. A. O'Brien.—The Java System. By the Hon. A. M. Skinner.—Bâtu Kôdok.—Prigi Acheh.—Dutch Occupation of the Dindings, etc.

No. 12. 8vo. pp. xxxii-116, sewed. December, 1883. Price 9*s.*

American Oriental Society.—Journal of the American Oriental Society. Vols. I. to X. and Vol. XII. (all published). 8vo. Boston and New Haven, 1849 to 1881. A complete set. Very rare. £14.

Volumes 2 to 5 and 8 to 10 and 12 may be had separately at £1 1*s.* each.

Anthropological Society of London, Memoirs read before the, 1863-1864. 8vo., pp. 542, cloth. 21*s.*

Anthropological Society of London, Memoirs read before the, 1865-1866. Vol. II. 8vo., pp. x. 464, cloth. 21*s.*

Anthropological Institute of Great Britain and Ireland (The Journal of the). Published Quarterly. 8vo. sewed.

Biblical Archæology, Society of.—TRANSACTIONS OF THE. 8vo. Vol. I. Part. I., 12*s*. 6*d*. Vol. I., Part II., 12*s*. 6*d*. (this part cannot be sold separately, or otherwise than with the complete sets). Vols. II. and III., 2 parts, 10*s*. 6*d*. each. Vol. IV., 2 parts, 12*s*. 6*d*. each. Vol. V., Part. I., 15*s*.; Part. II., 12*s*. 6*d*. Vol. VI., 2 parts, 12*s*. 6*d*. each. Vol. VII. Part I. 10*s*. 6*d*. Parts II. and III. 12*s*. 6*d*. each.

Bibliotheca Indica. A Collection of Oriental Works published by the Asiatic Society of Bengal. Old Series. Fasc. 1 to 247. New Series. Fasc. 1 to 493. (Special List of Contents and prices to be had on application.)

Browning Society's Papers (THE).—1881–4. Part I. 8vo. pp. 116, wrapper. 10*s*.

Browning.—Bibliography of Robert Browning from 1833-81. Part II. pp. 142. 10*s*. Part III. pp. 168. 10*s*. Part IV. pp. 148. 10*s*.

Calcutta Review (THE).—Published Quarterly. Price 8*s*. 6*d*. per number.

Calcutta Review.—A COMPLETE SET FROM THE COMMENCEMENT IN 1844 to 1882. Vols 1. to 75, or Numbers 1 to 140. A fine clean copy. Calcutta, 1844–82. Index to the first fifty volumes of the Calcutta Review, 2 parts. (Calcutta, 1873). Nos. 39 and 40 have never been published. £66. Complete sets are of great rarity.

Calcutta Review (Selections from the).—Crown 8vo. sewed. Nos. 1. to 45. 5*s*. each.

Cambridge Philological Society (Transactions of the).—Vol. I. From 1872 to 1880. 8vo. pp. xvi. and 420, wrapper. 1881. 15*s*.

CONTENTS.—Preface.—The Work of a Philological Society. J. P. Postgate.—Transactions of the Cambridge Philological Society from 1872 to 1879.—Transactions for 1879-1880.—Reviews—Appendix.

Vol. II. for 1881 and 1882. 8vo. pp. viii.-286, wrapper, 1883. 12*s*.

Cambridge Philological Society (Proceedings of the).—Parts I and II. 1882. 1*s*. 6*d*.; Parts III. 1*s*.; Parts IV.-VI., 2*s*. 6*d*.; Parts VII. and VIII. 2*s*.

China Review; or, Notes and Queries on the Far East. Published bi-monthly. 4to. Subscription £1 10*s*. per volume.

Chinese Recorder and Missionary Journal.—Shanghai. Subscription per volume (of 6 parts) 15*s*.

A complete set from the beginning. Vols. 1 to 10. 8vo. Foochow and Shanghai, 1861–1879. £9.

Containing important contributions on Chinese Philology, Mythology, and Geography, by Edkins, Giles, Bretschneider, Scarborough, etc. The earlier volumes are out of print.

Chrysanthemum (The).—A Monthly Magazine for Japan and the Far East. Vol. I. and II., complete. Bound £1 1*s*. Subscription £1 per volume

Geographical Society of Bombay.—JOURNAL AND TRANSACTIONS. A complete set. 19 vols. 8vo. Numerous Plates and Maps, some coloured. Bombay, 1844–70. £10 10*s*.

An important Periodical, containing grammatical sketches of several languages and dialects, as well as the most valuable contributions on the Natural Sciences of India. Since 1871 the above is amalgamated with the "Journal of the Bombay Branch of the Royal Asiatic Society."

Indian Antiquary (The).—A Journal of Oriental Research in Archæology, History, Literature, Languages, Philosophy, Religion, Folklore, etc. Edited by J. F. FLEET, C.I.E., M.R.A.S., etc., and CAPT. R. C. TEMPLE, F.R.G.S., M.R.A.S., etc. 4to. Published 12 numbers per anuum. Subscription £1 16*s*. A complete set. Vols. 1 to 11. £28 10*s*. (The earlier volumes are out of print.)

Indian Archipelago and Eastern Asia, Journal of the.—Edited by J. R. LOGAN, of Pinang. 9 vols. Singapore, 1847-55. New Series. Vols. I. to IV. Part 1, (all published), 1856-59. A complete set in 13 vols. 8vo. with many plates. £30.

Vol. I. of the New Series consists of 2 parts; Vol. II. of 4 parts; Vol. III. of No. 1 (never completed), and of Vol. IV. also only one number was published. A few copies remain of several volumes that may be had separately.

Japan, Transactions of the Seismological Society of, Vol. I. Parts i. and ii. April–June, 1880. 10*s.* 6*d.* Vol. II. July–December, 1880. 5*s.* Vol. III. January–December, 1881. 10*s.* 6*d.* Vol. IV. January–June. 1882. 9*s.*

Literature, Royal Society of.—See under "Royal."

Madras Journal of Literature and Science.—Published by the Committe of the Madras Literary Society and Auxiliary Royal Asiatic Society, and edited by MORRIS, COLE, and BROWN. A complete set of the Three Series (being Vols. I. to XVI., First Series; Vols. XVII. to XXII. Second Series; Vol. XXIII. Third Series, 2 Numbers, no more published). A fine copy, uniformly bound in 23 vols. With numerous plates, half calf. Madras, 1834–66. £42.

Equally scarce and important. On all South-Indian topics, especially those relating to Natural History and Science, Public Works and Industry, this Periodical is an unrivalled authority.

Madras Journal of Literature and Science. 1878. (I. Volume of the Fourth Series.) Edited by Gustav Oppert, Ph.D. 8vo. pp. vi. and 234, and xlvii. with 2 plates. 1879. 10*s.* 6*d.*

CONTENTS.- I. On the Classification of Languages. By Dr. G. Oppert.—II. On the Ganga Kings. By Lewis Rice.

Madras Journal of Literature and Science for the Year 1879. Edited by GUSTAV OPPERT, Ph.D., Professor of Sanskrit, Presidency College, Madras; Telugu Translator to Government, etc. 8vo. sewed, pp. 318. 10*s.* 6*d.*

Orientalia Antiqua.—See page 30.

Orientalist (The).—A Monthly Journal of Oriental Literature, Arts, and Science, Folk-lore, etc. Edited by W. GOONETELLIKE. Annual Subscription, 12*s.*

Pandit (The).—A Monthly Journal of the Benares College, devoted to Sanskrit Literature. Old Series. 10 vols. 1866–1876. New Series, vols. 1 to 5. 1876–1879. £1 4*s.* per volume.

Panjab Notes and Queries. A Monthly Periodical devoted to the Systematic Collection of Authentic Notes and Scraps of information regarding the Country and the People. Edited by Captain R. C. TEMPLE, etc. 4to. Subscription per annum. 10*s.*

Peking Gazette.—Translations of the Peking Gazette for 1872, 1873, 1874, 1875, 1876, 1877, and 1878. 8vo. cloth. 10*s.* 6*d.* each.

Philological Society (Transactions of The). A Complete Set, including the Proceedings of the Philological Society for the years 1842-1853. 6 vols. The Philological Society's Transactions, 1854 to 1876. 15 vols. The Philological Society's Extra Volumes. 9 vols. In all 30 vols. 8vo. £19 13*s.* 6*d.*

Proceedings (The) of the Philological Society 1842-1853. 6 vols. 8vo. £3.

Transactions of the Philological Society, 1854-1876. 15 vols. 8vo. £10 16*s.*

*** The Volumes for 1867, 1868-9, 1870-2, and 1873-4, are only to be had in complete sets, as above.

Separate Volumes.

For 1854: containing papers by Rev. J. W. Blakesley, Rev. T. O. Cockayne, Rev. J. Davies, Dr. J. W. Donaldson, Dr. Theod. Goldstücker, Prof. T. Hewitt Key, J. M. Kemble, Dr. R. G. Latham, J. M. Ludlow, Hensleigh Wedgwood, etc. 8vo. cl. £1 1*s.*

For 1855: with papers by Dr. Carl Abel, Dr. W. Bleek, Rev. Jno. Davies, Miss A. Gurney, Jas. Kennedy, Prof. T. H. Key, Dr. R. G. Latham, Henry Malden, W. Ridley, Thos. Watts, Hensleigh Wedgwood, etc. In 4 parts. 8vo. £1 1*s.*

*** Kamilaroi Language of Australia, by W. Ridley; and False Etymologies, by H. Wedgwood, separately. 1*s.*

For 1856-7: with papers by Prof. Aufrecht, Herbert Coleridge, Lewis Kr. Daa, M. de Haan, W. C. Jourdain, James Kennedy, Prof. Key, Dr. G. Latham, J. M. Ludlow, Rev. J. J. S. Perowne, Hensleigh Wedgwood, R. F. Weymouth, Jos. Yates, etc. 7 parts. 8vo. (The Papers relating to the Society's Dictionary are omitted.) £1 1*s.* each volume.

For 1858: including the volume of Early English Poems, Lives of the Saints, edited from MSS. by F. J. Furnivall; and papers by Ern. Adams, Prof. Aufrecht, Herbert Coleridge, Rev. Francis Crawford, M. de Haan Hettema, Dr. R. G. Latham, Dr. Lottner, etc. 8vo. cl. 12*s*.

For 1859: with papers by Dr. E. Adams, Prof. Aufrecht, Herb. Coleridge, F. J. Furnivall, Prof. T. H. Key, Dr. C. Lottner, Prof. De Morgan, F. Pulszky, Hensleigh Wedgwood, etc. 8vo. cl. 12*s*.

For 1860-1: including The Play of the Sacrament; and Pascon agau Arluth, the Passion of our Lord, in Cornish and English, both from MSS., edited by Dr. Whitley Stokes and papers by Dr. E. Adams, T. F. Barham, Rev. Derwent Coleridge, Herbert Coleridge, Sir John F. Davis, Danby P. Fry, Prof. T. H. Key, Dr. C. Lottner, Bishop Thirlwall, Hensleigh Wedgwood, R. F. Weymouth, etc. 8vo. cl. 12*s*.

For 1862-3: with papers by C. B. Cayley, D. P. Fry, Prof. Key, H. Malden, Rich. Morris, F. W. Newman, Robert Peacock, Hensleigh Wedgwood, R. F. Weymouth, etc. 8vo. cl. 12*s*.

For 1864: containing 1. Manning's (Jas.) Inquiry into the Character and Origin of the Possessive Augment in English, etc.; 2. Newman's (Francis W.) Text of the Iguvine Inscriptions, with Interlinear Latin Translation; 3. Barnes's (Dr. W.) Grammar and Glossary of the Dorset Dialect; 4. Gwreans An Bys—The Creation: a Cornish Mystery, Cornish and English, with Notes by Whitley Stokes, etc. 8vo. cl. 12*s*.

*** Separately: Manning's Inquiry, 3*s*.—Newman's Iguvine Inscription, 3*s*.—Stokes's Gwreans An Bys, 8*s*.

For 1865: including Wheatley's (H. B.) Dictionary of Reduplicated Words in the English Language; and papers by Prof. Aufrecht, Ed. Brock, C. B. Cayley, Rev. A. J. Church, Prof. T. H. Key, Rev. E. H. Knowles, Prof. H. Malden, Hon. G. P. Marsh, John Rhys, Guthbrand Vigfusson, Hensleigh Wedgwood, H. B. Wheatley, etc. 8vo. cl. 12*s*.

For 1866: including 1. Gregor's (Rev. Walter) Banffshire Dialect, with Glossary of Words omitted by Jamieson; 2. Edmondston's (T.) Glossary of the Shetland Dialect; and papers by Prof. Cassal, C. B. Cayley, Danby P. Fry, Prof. T. H Key, Guthbrand Vigfusson, Hensleigh Wedgwood, etc. 8vo. cl. 12*s*.

*** The Volumes for 1867, 1868–9, 1870–2, and 1873–4, are out of print. Besides contributions in the shape of valuable and interesting papers, the volume for 1867 also includes: 1. Peacock's (Rob. B.) Glossary of the Hundred of Lonsdale; and 2. Ellis (A. J.) On Palæotype representing Spoken Sounds; and on the Diphthong "Oy." The volume for 1868–9—1. Ellis's (A. J.) Only English Proclamation of Henry III. in Oct. 1258; to which are added "The Cuckoo's Song and "The Prisoner's Prayer," Lyrics of the XIII. Century, with Glossary; and 2. Stokes's (Whitley) Cornish Glossary. That for 1870-2—1. Murray's (Jas. A. H.) Dialect of the Southern Counties of Scotland, with a linguistical map. That for 1873–4—Sweet's (H.) History of English Sounds.

For 1875–6: containing the Rev. Richard Morris (President), Fourth and Fifth Annual Addresses. 1. Some Sources of Aryan Mythology by E. L. Brandreth; 2. C. B. Cayley on Certain Italian Diminutives; 3. Changes made by four young Children in Pronouncing English Words, by Jas. M. Menzies; 4. The Manx Language, by H. Jenner; 5. The Dialect of West Somerset, by F. T. Elworthy; 6. English Metre, by Prof. J. B. Mayor; 7. Words, Logic, and Grammar, by H. Sweet; 8. The Russian Language and its Dialects, by W. R. Morfill; 9. Relics of the Cornish Language in Mount's Bay, by H. Jenner. 10. Dialects and Prehistoric Forms of Old English. By Henry Sweet, Esq.; 11. On the Dialects of Monmouthshire, Herefordshire, Worcestershire, Gloucestershire, Berkshire, Oxfordshire, South Warwickshire, South Northamptonshire, Buckinghamshire, Hertfordshire, Middlesex, and Surrey, with a New Classification of the English Dialects. By Prince Louis Lucien Bonaparte (with Two Maps), Index, etc. Part I., 6*s*.; Part II., 6*s*.; Part III., 2*s*.

For 1877-8-9: containing the President's (Henry Sweet, Esq.) Sixth, Seventh, and (Dr. J. A. H. Murray) Eighth Annual Addresses. 1. Accadian Phonology, by Professor A. H. Sayce; 2. On *Here* and *There* in Chaucer, by Dr. R. Weymouth; 3. The Grammar of the Dialect of West Somerset, by F. T. Elworthy, Esq.; 4. English Metre, by Professor J. B. Mayor; 5. The Malagasy Language, by the Rev. W. E. Cousins; 6. The Anglo-Cymric Score, by A. J. Ellis, Esq., F.R.S. 7. Sounds and Forms of Spoken Swedish, by Henry Sweet, Esq.; 8. Russian Pronunciation, by Henry Sweet, Esq. Index, etc. Part I., 3*s.*; Part II., 7*s.* Part III. 8*s.*

For 1880-81: containing the President's (Dr. J. A. Murray) Ninth Annual Address. 1. Remarks on some Phonetic Laws in Persian, by Prof. Charles Rieu, Ph.D.; 2. On Portuguese Simple Sounds, compared with those of Spanish, Italian, French, English, etc., by H.I.H. Prince L. L. Bonaparte; 3. The Middle Voice in Virgil's Æneid, Book VI., by Benjamin Dawson, B.A.; 4. On a Difficulty in Russian Grammar, by C. B. Cayley; 5. The Polabes, by W. R. Morfill, M.A.; 6. Notes on the Makua Language, by Rev. Chauncy Maples, M.A.; 7. On the Distribution of English Place Names, by Walter R. Browne, M.A.; 8. *Dare*, "To Give"; and †-*Dere* "To Put," by Prof. Postgate, M.A.; 9. On som Differences between the Speech ov Edinboro' and London, by T. B. Sprague, M.A.; 10. Ninth Annual Address of the President (Dr. J. A. H. Murray) and Reports; 11. Sound-Notation, by H. Sweet, M.A.; 12 On Gender, by E. L. Brandreth; 13. Tenth Annual Address of the President, (A. J. Ellis, B.A.) and Reports; 14. Distribution of Place-Names in the Scottish Lowlands, by W. R. Browne, M.A.; 15. Some Latin and Greek Etymologies, and the change of *L* to *D* in Latin, by J. P. Postgate, M.A.; Supplement; Proceedings; Appendixes, etc.; 16. Notes on the N of AN, etc., in the Authorized and Revised Versions of the Bible. By B. Dawson, B.A.; 17. Notes on Translations of the New Testament. By B. Dawson, B.A.; 18. The Simple Sounds of all the Living Slavonic Languages compared with those of the Principal Neo-Latin and Germano-Scandinavian Tongues By H.I.H. Prince L.-L. Bonaparte; 19. On the Romonsch or Rhætian Languages in the Grisons and Tirol. By R. Martineau, M.A.—A Rough List of English Words found in Anglo-French, especially during the Thirteenth and Fourteenth Centuries; with numerous References. By the Rev. W. W. Skeat, M.A.; The Oxford MS. of the only English Proclamation of Henry III., 18 October, 1258. By the Rev. W. W. Skeat, MA.; and Errata in A. J. Ellis's copy of the only English Proclamation of Henry III., in Phil. Trans. 1869. Part I.; Postscript to Prince L.-L. Bonaparte's Paper on Neuter Neo-Latin Substantives; Index; Errata in Mr. Sweet's Paper on Sound Notation; List of Members. Part I. 12*s.* Part II. 8*s.* Part III. 7*s.*

For 1882-3-4: 1. Eleventh Annual Address of the President to the Philological Society, delivered at the Anniversary Meeting, Friday, 19th May, 1882. By A. J. Ellis, B.A., etc.; Obituary of Dr. J. Muir and Mr. H. Nicol. By the President; On the Work of the Philological Society. By the President; Reports; Conclusion. By the President. 2. Some Latin Etymologies. By Prof. Postgate, M.A. Initial Mutations in the Living Celtic, Basque, Sardinian, and Italian Dialects. By H. I. H. Prince Louis-Lucien Bonaparte. Spoken Portuguese. By H. Sweet, M.A. The Bosworth-Toller Anglo-Saxon Dictionary. By J. Platt, jun., Esq. The Etymology of "Surround." By the Rev. Prof. Skeat. Old English Verbs in *-egan* and their Subsequent History By Dr. J. A. H. Murray. Words connected with the Vine in Latin and the Neo-Latin Dialects. By H. I. H. Prince Louis-Lucien Bonaparte. Names of European Reptiles in the Living Neo-Latin Languages. By H. I. H. Prince Louis-Lucien Bonaparte. Appendices I. and II. Monthly Abstracts for the Session 1882-3. Part I. 10*s.* Part II. 10*s.*

The Society's Extra Volumes.

Early English Volume, 1862-64, containing: 1. Liber Cure Cocorum, A.D. *c.* 1440.—2. Hampole's (Richard Rolle) Pricke of Conscience, A.D. *c.* 1340.—3. The Castell off Love, A.D. *c.* 1320. 8vo. cloth. 1865. £1.

Or separately: Liber Cure Cocorum, Edited by Rich. Morris, 3s.; Hampole's (Rolle) Pricke of Conscience, edited by Rich. Morris, 12s.; and The Castell off Love, edited by Dr. R. F. Weymouth, 6s.

Dan Michel's Ayenbite of Inwyt, or Remorse of Conscience, in the Kentish Dialect, A.D. 1340. From the Autograph MS. in Brit. Mus. Edited with Introduction, Marginal Interpretations, and Glossarial Index, by Richard Morris. 8vo. cloth. 1866. 12s.

Levins's (Peter, A.D. 1570) Manipulus Vocabulorum: a Rhyming Dictionary of the English Language. With an Alphabetical Index by H. B. Wheatley. 8vo. cloth. 1867. 16s.

Skeat's (Rev. W. W.) Mœso-Gothic Glossary, with an Introduction, an Outline of Mœso-Gothic Grammar, and a List of Anglo-Saxon and old and modern English Words etymologically connected with Mœso-Gothic. 1868. 8vo. cl. 9s.

Ellis (A. J.) on Early English Pronunciation, with especial Reference to Shakspere and Chaucer: containing an Investigation of the Correspondence of Writing with Speech in England from the Anglo-Saxon Period to the Present Day, etc. 4 parts. 8vo. 1869–75. £2.

Mediæval Greek Texts: A Collection of the Earliest Compositions in Vulgar Greek, prior to A.D. 1500. With Prolegomena and Critical Notes by W. Wagner. Part I. Seven Poems, three of which appear for the first time. 1870. 8vo. 10s. 6d.

Poona Sarvajanik Sabha, Journal of the. Edited by S. H. Chiplonkar. Published quarterly. 3s. each number.

Royal Society of Literature of the United Kingdom (Transactions of The). First Series, 6 Parts in 3 Vols., 4to., Plates; 1827–39. Second Series, 11 Vols. or 33 Parts. 8vo., Plates; 1843–82. A complete set, as far as published, £10 10s. Very scarce. The first series of this important series of contributions of many of the most eminent men of the day has long been out of print and is very scarce. Of the Second Series, Vol. I.–IV., each containing three parts, are quite out of print, and can only be had in the complete series, noticed above. Three Numbers, price 4s. 6d. each, form a volume. The price of the volume complete, bound in cloth, is 13s. 6d.

Separate Publications.

I. Fasti Monastici Aevi Saxonici: or an Alphabetical List of the Heads of Religious Houses in England previous to the Norman Conquest, to which is prefixed a Chronological Catalogue of Contemporary Foundations. By Walter de Gray Birch. Royal 8vo. cloth. 1872. 7s. 6d.

II. Li Chantari di Lancellotto; a Troubadour's Poem of the XIV. Cent. Edited from a MS. in the possession of the Royal Society of Literature, by Walter de Gray Birch. Royal 8vo. cloth. 1874. 7s.

III. Inquisitio Comitatus Cantabrigiensis, nunc primum, è Manuscripto unico in Bibliothecâ Cottoniensi asservato, typis mandata: subjicitur Inquisitio Eliensis: curâ N. E. S. A. Hamilton. Royal 4to. With map and 3 facsimiles. 1876. £2 2s.

IV. A Commonplace-Book of John Milton. Reproduced by the autotype process from the original MS. in the possession of Sir Fred. U. Graham, Bart., of Netherby Hall. With an Introduction by A. J. Horwood. Sq. folio. Only one hundred copies printed. 1876. £2 2s.

V. Chronicon Adæ de Usk, A.D. 1377–1404. Edited, with a Translation and Notes, by Ed. Maunde Thompson. Royal 8vo. 1876. 10s. 6d.

Syro-Egyptian Society.—Original Papers read before the Syro-Egyptian Society of London. Volume I. Part 1. 8vo. sewed, 2 plates and a map, pp. 144. 3s. 6d.

Temple.—The Legends of the Panjab. By Captain R. C. Temple, Bengal Staff Corps, F.G.S., etc. Crown 8vo. Vols. I. Nos. 1 to 12, bound in cloth. £1 6s. Nos. 13 to 15, wrappers. 2s. each.

Theosophist (The). A Monthly Journal devoted to Oriental Philosophy, Art, Literature, and Occultism; embracing Mesmerism, Spiritualism, and other Secret Sciences. Conducted by H. P. Blavatsky. 4to. Subscription per annum £1.

Trübner's American, European and Oriental Literary Record.—A Register of the most important works published in North and South America, in India, China, Europe, and the British Colonies; with occasional Notes on German, Dutch, Danish, French, etc., books. 4to. In Monthly Numbers. Subscription 5*s.* per annum, or 6*d.* per number. A complete set, Nos. 1 to 142. London, 1865 to 1879. £12 12*s.*

Archæology, Ethnography, Geography, History, Law, Literature, Numismatics, and Travels.

Abel.—Slavic and Latin. Ilchester Lectures on Comparative Lexicography. Delivered at the Taylor Institution, Oxford. By Carl Abel, Ph.D. Post 8vo. pp. viii.-124, cloth. 1883. 5*s.*

Abel.—Linguistic Essays. See Trübner's Oriental Series, p. 5.

Ali.—The Proposed Political, Legal and Social Reforms in the Ottoman Empire and other Mohammedan States. By Moulaví Cherágh Ali, H.H. the Nizam's Civil Service. Demy 8vo. cloth, pp. liv.-184. 1883. 8*s.*

Arnold.—Indian Idylls. From the Sanskrit of the Mahâbhârata. By Edwin Arnold, C.S.I. Post 8vo. cloth, pp. xii.-282. 1883. 7*s.* 6*d.*

Arnold.—Indian Poetry. See "Trübner's Oriental Series," page 4.

Arnold.—Pearls of the Faith. See page 34.

Baden-Powell. — A Manual of the Jurisprudence for Forest Officers: being a Treatise on the Forest Law, and those branches of the general Civil and Criminal Law which are connected with Forest Administration; with a comparative Notice of the Chief Continental Laws. By B. H. Baden-Powell, B.C.S. 8vo. half-bound, pp. xxii-554. 1882. 12*s.*

Baden-Powell.—A Manual of the Land Revenue Systems and Land Tenures of British India. By B. H. Baden-Powell, B.C.S. Crown 8vo. half-bound, pp. xii.-788. 1882. 12*s.*

Badley.—Indian Missionary Record and Memorial Volume. By the Rev. B. H. Badley, of the American Methodist Mission. New Edition. 8vo. cloth. [*In Preparation*].

Balfour.—Waifs and Strays from the Far East. See p. 50.

Balfour.—The Divine Classic of Nan-Hua. See page 50.

Balfour.—Taoist Texts. See page 34.

Ballantyne.—Sankhya Aphorisms of Kapila. See "Trübner's Oriental Series," p. 6.

Beal.—See page 34.

Bellew.—From the Indus to the Tigris: a Narrative of a Journey through Balochistan, Afghanistan, Khorassan, and Iran, in 1872; with a Synoptical Grammar and Vocabulary of the Brahoe Language, and a Record of Meteorological Observations and Altitudes on the March from the Indus to the Tigris. By H. W. Bellew, C.S.I., Surgeon B.S.C., Author of "A Journal of a Mission to Afghanistan in 1857-58." Demy 8vo. cloth. pp. viii. and 496. 1874. 14*s.*

Bellew.—Kashmir and Kashgar. A Narrative of the Journey of the Embasy to Kashgar in 1873-74. By H. W. Bellew, C.S.I. Demy 8vo. cloth, pp. xxxii. and 420. 1875. 16*s.*

Bellew.—THE RACES OF AFGHANISTAN. Being a Brief Account of the Principal Nations inhabiting that Country. By Surgeon-Major H. W. BELLEW, C.S.I., late on Special Political Duty at Kabul. Crown 8vo. pp. 124, cloth. 1880. 7*s.* 6*d.*

Beveridge.—THE DISTRICT OF BAKARGANJ; its History and Statistics. By H. BEVERIDGE, B.C.S. 8vo. cloth, pp. xx. and 460. 1876. 21*s.*

Bibliotheca Orientalis: or, a Complete List of Books, Pamphlets, Essays, and Journals, published in France, Germany, England, and the Colonies, on the History and the Geography, the Religions, the Antiquities, Literature, and Languages of the East. Edited by CHARLES FRIEDERICI. Part I., 1876, sewed, pp. 86, 2*s.* 6*d.* Part II., 1877, pp. 100, 3*s.* 6*d.* Part III., 1878, 3*s.* 6*d.* Part IV., 1879, 3*s.* 6*d.* Part V., 1880. 3*s.*

Biddulph.—TRIBES OF THE HINDOO KOOSH. By Major J. BIDDULPH, B.S.C., Political Officer at Gilgit. 8vo. pp. 340, cloth. 1880. 15*s.*

Bleek.—REYNARD THE FOX IN SOUTH AFRICA; or, Hottentot Fables and Tales. See page 42.

Blochmann.—SCHOOL GEOGRAPHY OF INDIA AND BRITISH BURMAH. By H. BLOCHMANN, M.A. 12mo. wrapper, pp. vi. and 100. 2*s.* 6*d.*

Bombay Code, The.—Consisting of the Unrepealed Bombay Regulations, Acts of the Supreme Council, relating solely to Bombay, and Acts of the Governor of Bombay in Council. With Chronological Table. Royal 8vo. pp. xxiv.—774, cloth 1880. £1 1*s.*

Bombay Presidency.—GAZETTEER OF THE. Demy 8vo. half-bound. Vol. II., 14*s.* Vols. III.-VII., 8*s.* each; Vol. VIII., 9*s.*; X., XI., XII., XIV., XVI., 8*s.* each.

Bretschneider.—NOTES ON CHINESE MEDIÆVAL TRAVELLERS TO THE WEST. By E. BRETSCHNEIDER, M.D. Demy 8vo. sd., pp. 130. 5*s.*

Bretschneider. — ON THE KNOWLEDGE POSSESSED BY THE ANCIENT CHINESE OF THE ARABS AND ARABIAN COLONIES, and other Western Countries mentioned in Chinese Books. By E. BRETSCHNEIDER, M.D., Physician of the Russian Legation at Peking. 8vo. pp. 28, sewed. 1871. 1*s.*

Bretschneider.—NOTICES OF THE MEDIÆVAL GEOGRAPHY AND HISTORY OF CENTRAL AND WESTERN ASIA. Drawn from Chinese and Mongol Writings, and Compared with the Observations of Western Authors in the Middle Ages. By E. BRETSCHNEIDER, M.D. 8vo. sewed, pp. 233, with two Maps. 1876. 12*s.* 6*d.*

Bretschneider. — ARCHÆOLOGICAL AND HISTORICAL RESEARCHES ON PEKING AND ITS ENVIRONS. By E. BRETSCHNEIDER, M.D., Physician to the Russian Legation at Peking. Imp. 8vo. sewed, pp. 64, with 4 Maps. 1876. 5*s.*

Bretschneider.—BOTANICON SINICUM. Notes on Chinese Botany, from Native and Western Sources. By E. BRETSCHNEIDER, M.D. Crown 8vo. pp. 228, wrapper. 1882. 10*s.* 6*d.*

Budge.—ASSYRIAN TEXTS. See p. 47.

Budge.—HISTORY OF ESARHADDON. See Trübner's Oriental Series, p. 4.

Bühler.—ELEVEN LAND-GRANTS OF THE CHAULUKYAS OF AṆHILVÂḌ. A Contribution to the History of Gujarât. By G. BÜHLER. 16mo. sewed, pp. 126, with Facsimile. 3*s.* 6*d.*

Burgess.—ARCHÆOLOGICAL SURVEY OF WESTERN INDIA. By James Burgess, LL.D., etc., etc. Vol. 1. Report of the First Season's Operations in the Belgâm and Kaladgi Districts. Jan. to May, 1874. With 56 photographs and lith. plates. Royal 4to. pp. viii. and 45. 1875. £2 2*s.*

Vol. 2. Report of the Second Season's Operations. Report on the Antiquities of Kâthiâwâd and Kachh. 1874-5. With Map, Inscriptions, Photographs, etc. Roy. 4to. half bound, pp. x. and 242. 1876. £3 3*s.*

Vol. 3. Report of the Third Season's Operations. 1875–76. Report on the Antiquities in the Bidar and Aurangabad District. Royal 4to. half bound pp. viii. and 138, with 66 photographic and lithographic plates. 1878. £2 2*s.*

Vols. 4. and 5. Reports on the Buddhist Cave Temples and their Inscriptions; and the Elura Cave Temples and the Brahmanical and Jaina Caves in Western India: containing Views, Plans, Sections, and Elevations of Façades of Cave Temples; Drawings of Architectural and Mythological Sculptures; Facsimiles of Inscriptions, etc.; with Descriptive and Explanatory Text, and Translation, of Inscriptions, etc. Royal 4to. x.-140 and viii.-90, half morocco, gilt tops with 165 Plates and Woodcuts. 1883. £6 6*s*.

Burgess.—The Rock Temples of Elura or Verul. A Handbook for Visitors. By J. Burgess. 8vo. 3*s*. 6*d*., or with Twelve Photographs, 9*s*. 6*d*.

Burgess.—The Rock Temples of Elephanta Described and Illustrated with Plans and Drawings. By J. Burgess. 8vo. cloth, pp. 80, with drawings, price 6*s*.; or with Thirteen Photographs, price £1.

Burnell.—Elements of South Indian Palæography. From the Fourth to the Seventeenth Century A.D. By A. C. Burnell. Second Enlarged Edition, 35 Plates and Map. 4to. pp. xiv. and 148. 1878. £2 12*s*. 6*d*.

Carletti.—History of the Conquest of Tunis. Translated by J. T. Carletti. Crown 8vo. cloth, pp. 40. 1883. 2*s*. 6*d*.

Carpenter.—The Last Days in England of the Rajah Rammohun Roy. By Mary Carpenter, of Bristol. With Five Illustrations. 8vo. pp. 272, cloth. 7*s*. 6*d*.

Cesnola.—The History, Treasures, and Antiquities of Salamis, in the Island of Cyprus. By A. P. Di Cesnola, F.S.A. With an Introduction by S. Birch, D.C.L., Keeper of the Egyptian and Oriental Antiquities in the British Museum. With over 700 Illustrations and Map of Ancient Cyprus. Royal 8vo. pp. xlviii.-325, cloth, 1882. £1 11*s*. 6*d*.

Chamberlain.—Japanese Poetry. See "Trübner's Oriental Series," page 4.

Chattopadhyaya.—The Yatras; or the Popular Dramas of Bengal. Post 8vo. pp. 50, wrapper. 1882. 2*s*.

Clarke.—The English Stations in the Hill Regions of India: their Value and Importance, with some Statistics of their Produce and Trade. By Hyde Clarke, V.P.S.S. Post 8vo. paper, pp. 48. 1881. 1*s*.

Colebrooke.—The Life and Miscellaneous Essays of Henry Thomas Colebrooke. In 3 vols. Demy 8vo. cloth. 1873. Vol. I. The Biography by his Son, Sir T. E. Colebrooke, Bart., M.P. With Portrait and Map. pp. xii. and 492. 14*s*. Vols. II. and III. The Essays. A New Edition, with Notes by E. B. Cowell, Professor of Sanskrit in the University of Cambridge. pp. xvi.-544, and x.-520. 28*s*.

Crawford.—Recollections of Travels in New Zealand and Australia. By J. C. Crawford, F.G.S., Resident Magistrate, Wellington, etc., etc. With Maps and Illustrations. 8vo. cloth, pp. xvi. and 468. 1880. 18*s*.

Cunningham.—Corpus Inscriptionum Indicarum. Vol. I. Inscriptions of Asoka. Prepared by Alexander Cunningham, C.S.I., etc. 4to. cloth, pp. xiv. 142 and vi., with 31 plates. 1879. 32*s*.

Cunningham.—The Stupa of Bharhut. A Buddhist Monument, ornamented with numerous Sculptures illustrative of Buddhist Legend and History in the third century B.C. By Alexander Cunningham, C.S.I., C.I.E., Director-General Archæological Survey of India, etc. Royal 4to. cloth, gilt, pp. viii. and 144, with 51 Photographs and Lithographic Plates. 1879. £3 3*s*.

Cunningham.—The Ancient Geography of India. I. The Buddhist Period, including the Campaigns of Alexander, and the Travels of Hwen-Thsang. By Alexander Cunningham, Major-General, Royal Engineers (Bengal Retired). With thirteen Maps. 8vo. pp. xx. 590, cloth. 1870. 28*s*.

Cunningham.—Archæological Survey of India. Reports, made during the years 1862-1882. By A. Cunningham, C.S.I., Major-General, etc. With Maps and Plates. Vols. 1 to 18. 8vo. cloth. 10*s*. and 12*s*. each.

Cust.—Pictures of Indian Life. Sketched with the Pen from 1852 to 1881. By R. N. Cust, late of H.M. Indian Civil Service, and Hon. Sec. to the Royal Asiatic Society. Crown 8vo. cloth, pp. x. and 346. 1881. 7*s.* 6*d.*

Cust.—East Indian Languages. See "Trübner's Oriental Series," page 3.

Cust.—Languages of Africa. See "Trübner's Oriental Series," page 6.

Cust.—Linguistic and Oriental Essays. See "Trübner's Oriental Series," page 4.

Dalton.—Descriptive Ethnology of Bengal. By Edward Tuite Dalton, C.S.I., Colonel, Bengal Staff Corps, etc. Illustrated by Lithograph Portraits copied from Photographs. 38 Lithograph Plates. 4to. half-calf, pp. 340. £6 6*s.*

Da Cunha.—Notes on the History and Antiquities of Chaul and Bassein. By J. Gerson da Cunha, M.R.C.S. and L.M. Eng., etc. 8vo. cloth, pp. xvi. and 262. With 17 photographs, 9 plates and a map. £1 5*s.*

Da Cunha.—Contributions to the Study of Indo-Portuguese Numismatics. By J. G. Da Cunha, M.R.C.S., etc. Crown 8vo. stitched in wrapper. Fasc. I, pp. 18, with 1 plate; Fasc. II. pp. 16, with 1 plate, each 2*s.* 6*d.*

Das.—The Indian Ryot, Land Tax, Permanent Settlement, and the Famine. Chiefly compiled by Abhay Charan Das. Post 8vo. cloth, pp. iv.-662. 1881. 12*s.*

Davids.—Coins, etc., of Ceylon. See "Numismata Orientala," Vol. I. Part VI.

Dennys.—China and Japan. A complete Guide to the Open Ports of those countries, together with Pekin, Yeddo, Hong Kong, and Macao; forming a Guide Book and Vade Mecum for Travellers, Merchants, etc.; with 56 Maps and Plans. By W. F. Mayers, H.M.'s Consular Service; N. B. Dennys, late H.M.'s Consular Service; and C. King, Lieut. R.M.A. Edited by N. B. Dennys. 8vo. pp. 600, cloth. £2 2*s.*

Dowson.—Dictionary of Hindu Mythology, etc. See "Trübner's Oriental Series," page 3.

Egerton.—An Illustrated Handbook of Indian Arms; being a Classified and Descriptive Catalogue of the Arms exhibited at the India Museum; with an Introductory Sketch of the Military History of India. By the Hon. W. Egerton, M.A., M.P. 4to. sewed, pp. viii. and 162. 1880. 2*s.* 6*d.*

Elliot.—Memoirs on the History, Folklore, and Distribution of the Races of the North Western Provinces of India; being an amplified Edition of the original Supplementary Glossary of Indian Terms. By the late Sir H. M. Elliot, K.C.B. Edited, etc., by John Beames, B.C.S., etc. In 2 vols. demy 8vo., pp. xx., 370, and 396, cloth. With two Plates, and four coloured Maps. 1869. 36*s.*

Elliot.—Coins of Southern India. See "Numismata Orientalia." Vol. III. Part II. page 30.

Elliot.—The History of India, as told by its own Historians. The Muhammadan Period. Complete in Eight Vols. Edited from the Posthumous Papers of the late Sir H. M. Elliot, K.C.B., E. India Co.'s B.C.S., by Prof. J. Dowson, M.R.A.S., Staff College, Sandhurst. 8vo. cloth. 1867-1877.

Vol. I. pp xxxii. and 542. £4 4*s.*—Vol. II. pp. x. and 580. 18*s.*—Vol. III. pp. xii. and 627. 24*s.*—Vol. IV. pp. x. and 563. 21*s.*—Vol. V. pp. xii. and 576. 21*s.*—Vol. VI. pp. viii. and 574. 21*s.*—Vol. VII. pp. viii. and 574. 21*s.*—Vol. VIII. pp. xxxii., 444, and lxviii. 24*s.* Complete sets, £8 8*s.*

Farley.—EGYPT, CYPRUS, AND ASIATIC TURKEY. By J. L. FARLEY, Author of "The Resources of Turkey," etc. Demy 8vo. cl., pp. xvi.-270. 1878. 10*s.* 6*d.*

Featherman.—THE SOCIAL HISTORY OF THE RACES OF MANKIND. Vol. V. The Aramaeans. By A. FEATHERMAN. To be completed in about Ten Volumes. 8vo. cloth, pp. xvii. and 664. 1881. £1 1*s.*

Fenton.—EARLY HEBREW LIFE: a Study in Sociology. By JOHN FENTON. 8vo. cloth, pp. xxiv. and 102. 1880. 5*s.*

Fergusson and Burgess.—THE CAVE TEMPLES OF INDIA. By JAMES FERGUSSON, D.C.L., F.R.S., and JAMES BURGESS, F.R.G.S. Imp. 8vo. half bound, pp. xx. and 536, with 98 Plates. £2 2*s.*

Fergusson.—TREE AND SERPENT WORSHIP; or, Illustrations of Mythology and Art in India in the First and Fourth Centuries after Christ. From the Sculptures of Buddhist Topes at Sanchi and Amravati. Second revised Edition. By J. FERGUSSON, D.C.L. 4to. half bouud pp. xvi. and 276, with 101 plates. 1873. Out of print.

Fergusson.—ARCHÆOLOGY IN INDIA. With especial reference to the Works of Babu Rajendralala Mitra. By J. FERGUSSON, C.I.E. 8vo. pp. 116, with Illustrations, sewed. 1884. 5*s.*

Fornander.—AN ACCOUNT OF THE POLYNESIAN RACE: Its Origin and Migration, and the Ancient History of the Hawaiian People to the Times of Kamehameha I. By A. FORNANDER, Circuit Judge of the Island of Maui, H.I. Post 8vo. cloth. Vol. I., pp. xvi. and 248. 1877. 7*s.* 6*d.* Vol. II., pp. viii. and 400, cloth. 1880. 10*s.* 6*d.*

Forsyth.—REPORT OF A MISSION TO YARKUND IN 1873, under Command of SIR T. D. FORSYTH, K.C.S.I., C.B., Bengal Civil Service, with Historical and Geographical Information regarding the Possessions of the Ameer of Yarkund. With 45 Photographs, 4 Lithographic Plates, and a large Folding Map of Eastern Turkestan. 4to. cloth, pp. iv. and 573. £5 5*s.*

Gardner.—PARTHIAN COINAGE. See "Numismata Orientalia. Vol. I. Part V.

Garrett.—A CLASSICAL DICTIONARY OF INDIA, illustrative of the Mythology, Philosophy, Literature, Antiquities, Arts, Manners, Customs, etc., of the Hindus. By JOHN GARRETT. 8vo. pp. x. and 798. cloth. 28s.

Garrett.—SUPPLEMENT TO THE ABOVE CLASSICAL DICTIONARY OF INDIA. By J. GARRETT, Dir. of Public Instruction, Mysore. 8vo. cloth, pp. 160. 7*s.* 6*d.*

Gazetteer of the Central Provinces of India. Edited by CHARLES GRANT, Secretary to the Chief Commissioner of the Central Provinces. Second Edition. With a very large folding Map of the Central Provinces of India. Demy 8vo. pp. clvii. and 582, cloth. 1870. £1 4*s.*

Geiger.—CONTRIBUTIONS TO THE HISTORY OF THE DEVELOPMENT OF THE HUMAN RACE. Lectures and Dissertations by L. GEIGER. Translated from the German by D. Asher, Ph.D. Post 8vo. cloth, pp. x. and 156. 1880. 6*s.*

Goldstücker.—ON THE DEFICIENCIES IN THE PRESENT ADMINISTRATION OF HINDU LAW; being a paper read at the Meeting of the East India Association on the 8th June, 1870. By THEODOR GOLDSTÜCKER, Professor of Sanskrit in University College, London, &c. Demy 8vo. pp. 56, sewed. 1*s.* 6*d.*

Gover.—THE FOLK-SONGS OF SOUTHERN INDIA. By CHARLES E. GOVER. 8vo. pp. xxiii. and 299, cloth. 1872. 10*s.* 6*d.*

Griffin.—THE RAJAS OF THE PUNJAB. History of the Principal States in the Punjab, and their Political Relations with the British Government. By LEPEL H. GRIFFIN, B.C.S.; Under Sec. to Gov. of the Punjab, Author of "The Punjab Chiefs," etc. Second edition. Royal 8vo., pp. xiv. and 630. 1873. 21*s.*

Griffis.—THE MIKADO'S EMPIRE. Book I. History of Japan from 660 B.C. to 1872 A.D. Book II. Personal Experiences, Observations, and Studies in Japan, 1870–74. By W. E. GRIFFIS. Illustrated. Second Edition. 8vo. pp. 626, cloth. 1883. £1.

Growse.—MATHURA: A District Memoir. By F. S. GROWSE, B.C.S., C.I.E. Second Revised Edition. Illustrated. 4to. boards, pp. xxiv. and 520. 1880. 42*s.*

Hahn.—Tsuni||Goam. See Trübner's Oriental Series, page 5.

Head.—COINAGE OF LYDIA AND PERSIA. See "Numismata Orientalia." Vol. I, Part III.

Heaton.—AUSTRALIAN DICTIONARY OF DATES AND MEN OF THE TIME. Containing the History of Australasia, from 1542 to May, 1879. By I. H. HEATON. Royal 8vo. cloth pp. iv.—554. 1879. 15*s.*

Hebrew Literature Society. See page 71.

Hodgson.—ESSAYS ON THE LANGUAGES, LITERATURE, AND RELIGION OF NEPAL AND TIBET; together with further Papers on the Geography, Ethnology, and Commerce of those Countries. By B. H. HODGSON, late British Minister at Nepâl. Royal 8vo. cloth, pp. 288. 1874. 14*s.*

Hodgson.—ESSAYS ON INDIAN SUBJECTS. See "Trübner's Oriental Series," p. 4.

Hunter.—THE IMPERIAL GAZETTEER OF INDIA. By W. W. HUNTER, C.I.E., LL.D., Director-General of Statistics to the Government of India. Published by Command of the Secretary of State for India. 9 vols. 8vo. half morocco. 1881.

"A great work has been unostentatiously carried on for the last twelve years in India, the importance of which it is impossible to exaggerate. This is nothing less than a complete statistical survey of the entire British Empire in Hindostan. . . . We have said enough to show that the 'Imperial Gazetteer' is no mere dry collection of statistics; it is a treasury from which the politician and economist may draw countless stores of valuable information, and into which the general reader can dip with the certainty of always finding something both to interest and instruct him."—*Times.*

Hunter.—A STATISTICAL ACCOUNT OF BENGAL. By W. W. HUNTER, B.A., LL.D. Director-General of Statistics to the Government of India.

VOL.
I. 24 Parganás and Sundarbans.
II. Nadiyá and Jessor.
III. Midnapur, Húglí and Hourah.
IV. Bardwán, Birbhúm and Bánkurá.
V. Dacca, Bákarganj, Farídpur and Maimansinh.
VI. Chittagong Hill Tracts, Chittagong, Noákhálí, Tipperah, and Hill Tipperah State.
VII. Meldah, Rangpur and Dinájpur.
VIII. Rájsháhí and Bográ.
IX. Murshidábád and Pábná.
X. Dárjíling, Jalpáigurí and Kuch Behar State.
XI. Patná and Sáran.
XII. Gayá and Sháhábád.
XIII. Tirhut and Champáran.
XIV. Bhágalpur and Santál Parganás.
XV. Monghyr and Purniah.
XVI. Hazáribágh and Lohárdagá.
XVII. Singbhúm, Chutiá, Nágpur Tributary States and Mánbhúm.
XVIII. Cuttack and Balasor.
XIX. Purí, and Orissa Tributary States.
XX. Fisheries, Botany, and General Index

Published by command of the Government of India. In 20 Vols. 8vo. half-morocco. £5.

Hunter.—A STATISTICAL ACCOUNT OF ASSAM. By W. W. HUNTER, LL.D., C.I.E. 2 vols. 8vo. half morocco, pp. 420 and 490, with Two Maps. 1879. 10*s.*

Hunter.—FAMINE ASPECTS OF BENGAL DISTRICTS. A System of Famine Warnings. By W. W. HUNTER, LL.D. Crown 8vo. cloth, pp. 216. 1874. 7*s.* 6*d.*

Hunter.—THE INDIAN MUSALMANS. By W. W. HUNTER, LL.D., etc. Third Edition. 8vo. cloth, pp. 219. 1876. 10*s.* 6*d.*

Hunter.—AN ACCOUNT OF THE BRITISH SETTLEMENT OF ADEN in Arabia. Compiled by Captain F. M. HUNTER, Assistant Political Resident, Aden. Demy 8vo. half-morocco, pp. xii.-232. 1877. 7*s.* 6*d.*

Hunter.—A BRIEF HISTORY OF THE INDIAN PEOPLE. By W. W. Hunter, C.I.E., LL.D. Crown 8vo. pp. 222 with map, cloth. 1884. 3*s.* 6*d.*

Hunter.—Indian Empire. See Trübner's Oriental Series, page 5.

India.—Finance and Revenue Accounts of the Government of, for 1882-83. Fcp. 8vo. pp. viii.-220, boards. 1884. 2*s.* 6*d.*

Japan.—Map of Nippon (Japan): Compiled from Native Maps, and the Notes of recent Travellers. By R. H. Brunton, F.R.G.S., 1880. In 4 sheets, 21*s.*; roller, varnished, £1 11*s.* 6*d.*; Folded, in case, £1 5*s.* 6*d.*

Juvenalis Satiræ.—With a Literal English Prose Translation and and Notes. By J. D. Lewis, M.A. Second, Revised, and considerably Enlarged Edition. 2 Vols. post 8vo. pp. xii.-230, and 400, cloth. 1882. 12*s.*

Leitner.—Sinin-I-Islam. Being a Sketch of the History and Literature of Muhammadanism and their place in Universal History. *For the use of Maulvis.* By G. W. Leitner. Part I. The Early History of Arabia to the fall of the Abassides. 8vo. sewed. *Lahore.* 6*s.*

Leitner.—History of Indigenous Education in the Panjab since Annexation, and in 1882. By G. W. Leitner, LL.D., late on special duty with the Education Commission appointed by the Government of India. Fcap. folio, pp. 588, paper boards. 1883. £5.

Leland.—Fusang; or, the Discovery of America by Chinese Buddhist Priests in the Fifth Century. By Charles G. Leland. Crown 8vo. cloth, pp. xix. and 212. 1875. 7*s.* 6*d.*

Leland.—The Gypsies. See page 69.

Leonowens.—The Romance of Siamese Harem Life. By Mrs. Anna H. Leonowens, Author of "The English Governess at the Siamese Court." With 17 Illustrations, principally from Photographs, by the permission of J. Thomson, Esq. Crown 8vo. cloth, pp. viii. and 278. 1873. 14*s.*

Leonowens.—The English Governess at the Siamese Court: being Recollections of six years in the Royal Palace at Bangkok. By Anna Harriette Leonowens. With Illustrations from Photographs presented to the Author by the King of Siam. 8vo. cloth, pp. x. and 332. 1870 12*s.*

Long.—Eastern Proverbs and Emblems. See Trübner's Oriental Series, page 4.

Linde.—Tea in India. A Sketch, Index, and Register of the Tea Industry in India, published together with a Map of all the Tea Districts, etc. By F. Linde, Surveyor, Compiler of a Map of the Tea Localities of Assam, etc. Folio, wrapper, pp. xxii.-30, map mounted and in cloth boards. 1879. 63*s.*

McCrindle.—The Commerce and Navigation of the Erythræan Sea. Being a Translation of the Periplus Maris Erythraei, by an Anonymous Writer, and of Arrian's Account of the Voyage of Nearkhos, from the Mouth of the Indus to the Head of the Persian Gulf. With Introduction, Commentary, Notes, and Index. Post 8vo. cloth, pp. iv. and 238. 1879. 7*s.* 6*d.*

McCrindle.—Ancient India as Described by Megasthenês and Arrian. A Translation of Fragments of the Indika of Megasthenês collected by Dr. Schwanberk, and of the First Part of the Indika of Arrian. By J. W. McCrindle, M.A., Principal of Gov. College, Patna. With Introduction, Notes, and Map of Ancient India. Post 8vo. cloth, pp. xii.-224. 1877. 7*s.* 6*d.*

McCrindle.—Ancient India as described by Ktêsias, the Knidian, a translation of the abridgment of his "Indica," by Photios, and fragments of that work preserved in other writers. By J. W. McCrindle, M.A. With Introduction, Notes, and Index. 8vo. cloth, pp. viii.—104. 1882. 6*s.*

MacKenzie.—The History of the Relations of the Government with the Hill Tribes of the North-East Frontier of Bengal. By A. MacKenzie, B.C.S., Sec. to the Gov. Bengal. Royal. 8vo. pp. xviii.-586, cloth, with Map. 1884. 16*s.*

Madden.—COINS OF THE JEWS. See "Numismata Orientalia." Vol. II.

Malleson.—ESSAYS AND LECTURES ON INDIAN HISTORICAL SUBJECTS. By Col. G. B. MALLESON, C.S.I. Second Issue. Cr. 8vo. cloth, pp. 348. 1876. 5*s.*

Markham.—THE NARRATIVES OF THE MISSION OF GEORGE BOGLE, B.C.S., to the Teshu Lama, and of the Journey of T. Manning to Lhasa. Edited, with Notes, Introduction, and lives of Bogle and Manning, by C. R MARKHAM, C.B. Second Edition. 8vo. Maps and Illus., pp. clxi. 314, cl. 1879. 21*s.*

Marsden's Numismata Orientalia. New International Edition. *See* under NUMISMATA ORIENTALIA.

Marsden.—NUMISMATA ORIENTALIA ILLUSTRATA. The Plates of the Oriental Coins, Ancient and Modern, of the Collection of the late W. Marsden. Engraved from Drawings made under his Directions. 4to. 57 Plates, cl. 31*s.* 6*d.*

Mason.—BURMA: Its People and Productions; or, Notes on the Fauna, Flora, and Minerals of Tenasserim, Pegu and Burma. By the Rev. F. MASON, D.D. Vol. I. Geology, Mineralogy, and Zoology. Vol. II. Botany. Rewritten by W. THEOBALD, late Deputy-Sup. Geological Survey of India. 2 vols. Royal 8vo. pp. xxvi. and 560; xvi. and 781 and xxxvi. cloth. 1864. £3.

Matthews.—ETHNOLOGY AND PHILOLOGY OF THE HIDATSA INDIANS. By WASHINGTON MATTHEWS, Assistant Surgeon, U.S. Army. *Contents:*—Ethnography, Philology, Grammar, Dictionary, and English-Hidatsa Vocabulary. 8vo. cloth. £1 11*s.* 6*d.*

Mayers.—China and Japan. See DENNYS.

Mayers.—THE CHINESE GOVERNMENT. A Manual of Chinese Titles, categorically arranged and explained, with an Appendix. By W. F. MAYERS. Roy. 8vo. cloth, pp. viii.-160. 1878. £1 10*s.*

Metcalfe.—THE ENGLISHMAN AND THE SCANDINAVIAN; or, a Comparison of Anglo-Saxon and Old Norse Literature. By FREDERICK METCALFE, M.A., Author of "The Oxonian in Iceland, etc. Post 8vo. cloth, pp. 512. 1880. 18*s.*

Mitra.—THE ANTIQUITIES OF ORISSA. By RAJENDRALALA MITRA. Published under Orders of the Government of India. Folio, cloth. Vol. I. pp. 180. With a Map and 36 Plates. 1875. £6 6*s.* Vol. II. pp. vi. and 178. 1880. £4 4*s.*

Mitra.—BUDDHA GAYA; the Hermitage of Sákya Muni. By RAJENDRALALA MITRA, LL.D., C.I.E. 4to. cloth, pp. xvi. and 258, with 51 plates. 1878. £3.

Mitra.—THE SANSKRIT BUDDHIST LITERATURE OF NEPAL. By RAJENDRALALA MITRA, LL.D., C.I.E. 8vo. cloth, pp. xlviii.-340. 1882. 12*s.* 6*d.*

Moor.—THE HINDU PANTHEON. By EDWARD MOOR, F.R.S. A new edition, with additional Plates, Condensed and Annotated by the Rev. W. O. SIMPSON. 8vo. cloth, pp. xiii. and 401, with 62 Plates. 1864. £3.

Morris.—A DESCRIPTIVE AND HISTORICAL ACCOUNT OF THE GODAVERY DISTRICT in the Presidency of Madras. By H. MORRIS, formerly M.C.S. 8vo. cloth, with map, pp. xii. and 390. 1878. 12*s.*

Müller.—ANCIENT INSCRIPTIONS IN CEYLON. By Dr. EDWARD MÜLLER. 2 Vols. Text, crown 8vo., pp. 220, cloth and plates, oblong folio, cloth. 1883. 21*s.*

Notes, ROUGH, OF JOURNEYS made in the years 1868, 1869, 1870, 1871, 1872, 1873, in Syria, down the Tigris, India, Kashmir, Ceylon, Japan, Mongolia, Siberia, the United States, the Sandwich Islands, and Australasia. Demy 8vo. pp. 624, cloth. 1875. 14*s.*

Numismata Orientalia.—The International Numismata Orientalia. Edited by Edward Thomas, F.R.S., etc. Vol. I. Illustrated with 20 Plates and a Map. Royal 4to. cloth. 1878. £3 13*s*. 6*d*.

Also in 6 Parts sold separately, viz.:—

Part I.—Ancient Indian Weights. By E. Thomas, F.R.S., etc. Royal 4to. sewed, pp. 84, with a Plate and a Map of the India of Manu. 9*s*. 6*d*.

Part II.—Coins of the Urtuki Turkumans. By Stanley Lane Poole, Corpus Christi College Oxford. Royal 4to. sewed, pp. 44, with 6 Plates. 9*s*.

Part III. The Coinage of Lydia and Persia, from the Earliest Times to the Fall of the Dynasty of the Achæmenidæ. By Barclay V. Head, Assistant-Keeper of Coins, British Museum. Royal 4to. sewed, pp. viii. and 56, with three Autotype Plates. 10*s*. 6*d*.

Part IV. The Coins of the Tuluni Dynasty. By Edward Thomas Rogers. Royal 4to. sewed, pp. iv. and 22, and 1 Plate. 5*s*.

Part V. The Parthian Coinage. By Percy Gardner, M.A. Royal 4to. sewed, pp. iv. and 65, with 8 Autotype Plates. 18*s*.

Part VI. On the Ancient Coins and Measures of Ceylon. With a Discussion of the Ceylon Date of the Buddha's Death. By T. W. Rhys Davids, Barrister-at-Law, late of the Ceylon Civil Service. Royal 4to. sewed, pp. 60, with Plate. 10*s*.

Numismata Orientalia.—Vol. II. Coins of the Jews. Being a History of the Jewish Coinage and Money in the Old and New Testaments. By Frederick W. Madden, M.R.A.S., Member of the Numismatic Society of London, Secretary of the Brighton College, etc., etc. With 279 woodcuts and a plate of alphabets. Royal 4to. sewed, pp. xii. and 330. 1881. £2.
Or as a separate volume, cloth. £2 2*s*.

Numismata Orientalia.—Vol III. Part I. The Coins of Arakan, of Pegu, and of Burma. By Lieut.-General Sir Arthur Phayre, C.B., K.C.S.I., G.C.M.G., late Commissioner of British Burma. Royal 4to., pp. viii. and 48, with 5 Autotype Illustrations, sewed. 1882. 8*s*. 6*d*. Also contains the Indian Balhara and the Arabian Intercourse with India in the Ninth and following centuries. By Edward Thomas, F.R.S.

Numismata Orientalia.—Vol. III. Part II. The Coins of Southern India. By Sir W. Elliot. Royal 4to.

Olcott.—A Buddhist Catechism, according to the Canon of the Southern Church. By Colonel H. S. Olcott, President of the Theosophical Society. 24mo. pp. 32, wrapper. 1881. 1*s*.

Oppert.—On the Ancient Commerce of India: A Lecture. By Dr. G. Oppert. 8vo. paper, 50 pp. 1879. 1*s*.

Oppert.—Contributions to the History of Southern India. Part I. Inscriptions. By Dr. G. Oppert. 8vo. paper, pp. vi. and 74, with a Plate. 1882. 4*s*.

Orientalia Antiqua; or Documents and Researches relating to the History of the Writings, Languages, and Arts of the East. Edited by Terrien de La Couperie, M.R.A.S., etc., etc. Fcap. 4to. pp. 96, with 14 Plates, wrapper. Part I. pro Vol. I., complete in 6 parts, price 30*s*.

Osburn.—The Monumental History of Egypt, as recorded on the Ruins of her Temples, Palaces, and Tombs. By William Osburn. Illustrated with Maps, Plates, etc. 2 vols. 8vo. pp. xii. and 461; vii. and 643, cloth. £2 2*s*. Out of print.

Vol. I.—From the Colonization of the Valley to the Visit of the Patriarch Abram.
Vol. II.—From the Visit of Abram to the Exodus.

Oxley.—Egypt: and the Wonders of the Land of the Pharoahs. By W. Oxley. Illustrated by a New Version of the Bhagavat-Gita, an Episode of the Mahabharat, one of the Epic Poems of Ancient India. Crown 8vo. pp. viii-328, cloth. 1884. 7*s*. 6*d*.

Palestine.—Memoirs of the Survey of Western Palestine. Edited by W. BESANT, M.A., and E. H. PALMER, M.A., under the Direction of the Committee of the Palestine Exploration Fund. Complete in Seven Volumes. Demy 4to. cloth, with a Portfolio of Plans, and large scale Map. Second Issue. Price Twenty Guineas.

Palmer.—EGYPTIAN CHRONICLES, with a harmony of Sacred and Egyptian Chronology, and an Appendix on Babylonian and Assyrian Antiquities. By WILLIAM PALMER, M.A., and late Fellow of Magdalen College, Oxford. 2 vols., 8vo. cloth, pp. lxxiv. and 428, and viii. and 636. 1861. 12*s.*

Patell.—COWASJEE PATELL'S CHRONOLOGY, containing corresponding Dates of the different Eras used by Christians, Jews, Greeks, Hindûs, Mohamedans, Parsees, Chinese, Japanese, etc. By COWASJEE SORABJEE PATELL. 4to. pp. viii. and 184, cloth. 50*s.*

Pathya-Vakya, or Niti-Sastra. Moral Maxims extracted from the Writings of Oriental Philosophers. Corrected, Paraphrased, and Translated into English. By A. D. A. WIJAYASINHA. Foolscap 8vo. sewed, pp. viii. and 54. Colombo, 1881. 8*s.*

Paton.—A HISTORY OF THE EGYPTIAN REVOLUTION, from the Period of the Mamelukes to the Death of Mohammed Ali; from Arab and European Memoirs, Oral Tradition, and Local Research. By A. A. Paton. Second Edition. 2 vols. demy 8vo. cloth, pp. xii. and 395, viii. and 446. 1870. 7*s.* 6*d.*

Pfoundes.—Fu So Mimi Bukuro.—A BUDGET OF JAPANESE NOTES. By CAPT. PFOUNDES, of Yokohama. 8vo. sewed, pp. 184. 7*s.* 6*d.*

Phayre.—COINS OF ARAKAN, ETC. See "Numismata Orientalia." Vol. III. Part I.

Piry.—LE SAINT EDIT. LITTERATURE CHINOISE. See page 36.

Playfair.—THE CITIES AND TOWNS OF CHINA. A Geographical Dictionary by G. M. H. PLAYFAIR, of Her Majesty's Consular Service in China. 8vo. cloth, pp. 506. 1879. 25*s.*

Poole.—COINS OF THE URTUKÍ TURKUMÁNS. See "Numismata Orientalia." Vol. I. Part II.

Poole.—A SCHEME OF MOHAMMADAN DYNASTIES DURING THE KHALIFATE. By S. L. POOLE, B.A. Oxon., M.R.A.S., Author of "Selections from the Koran,' etc. 8vo. sewed, pp. 8, with a plate. 1880. 2*s.*

Poole.—AN INDEX TO PERIODICAL LITERATURE. By W. F. Poole, LL.D., Librarian of the Chicago Public Library. Third Edition, brought down to January, 1882. Royal 8vo. pp. xxviii. and 1442, cloth. 1883. £3 13*s.* 6*d.* Wrappers, £3 10*s.*

Ralston.—Tibetan Tales. See Trübner's Oriental Series, page 5.

Ram Raz.—ESSAY on the ARCHITECTURE of the HINDUS. By RAM RAZ, Native Judge and Magistrate of Bangalore. With 48 plates. 4to. pp. xiv. and 64, sewed. London, 1834. £2 2*s.*

Ravenstein.—THE RUSSIANS ON THE AMUR; its Discovery, Conquest, and Colonization, with a Description of the Country, its Inhabitants, Productions, and Commercial Capabilities, and Personal Accounts of Russian Travellers. By E. G. RAVENSTEIN, F.R.G.S. With 4 tinted Lithographs and 3 Maps. 8vo. cloth, pp. 500. 1861. 15.

Raverty.—NOTES ON AFGHANISTAN AND PART OF BALUCHISTAN, Geographical, Ethnographical, and Historical. By Major H. G. RAVERTY, Bombay Native Infantry (Retired). Fcap. folio, wrapper. Sections I. and II. pp. 98. 1880. 2*s.* Section III. pp. vi. and 218. 1881. 5*s.* Section IV. pp. x-136. 1883. 3*s.*

Rice.—MYSORE INSCRIPTIONS. Translated for the Government by LEWIS RICE. 8vo. pp. vii. 336, and xxx. With a Frontispiece and Map. Bangalore, 1879. £1 10*s.*

Rockhill.—LIFE OF THE BUDDHA. See "Trübner's Oriental Series, page 6.

Roe and Fryer.—TRAVELS IN INDIA IN THE SEVENTEENTH CENTURY. By Sir THOMAS ROE and Dr. JOHN FRYER. Reprinted from the "Calcutta Weekly Englishman." 8vo. cloth, pp. 474. 1873. 7*s.* 6*d.*

Rogers.—COINS OF THE TULUNI DYNASTY. See "Numismata Orientalia." Vol. I. Part. IV.

Routledge.—ENGLISH RULE AND NATIVE OPINION IN INDIA. From Notes taken in the years 1870-74. By JAMES ROUTLEDGE. Post 8vo. cloth, pp. 344. 1878. 10*s.* 6*d.*

Schiefner.—Tibetan Tales. See Trübner's Oriental Series, page 5.

Schlagintweit.—GLOSSARY OF GEOGRAPHICAL TERMS FROM INDIA AND TIBET, with Native Transcription and Transliteration. By HERMANN DE SCHLAGINTWEIT. Forming, with a "Route Book of the Western Himalaya, Tibet, and Turkistan," the Third Volume of H., A., and R. DE SCHLAGINTWEIT'S "Results of a Scientific Mission to India and High Asia." With an Atlas in imperial folio, of Maps, Panoramas, and Views. Royal 4to., pp. xxiv. and 293. 1863. £4.

Sewell.—REPORT ON THE AMARAVATI TOPE, and Excavations on its Site in 1877. By R. SEWELL, M.C.S. Royal 4to. 4 plates, pp. 70, boards. 1880. 3*s.*

Sewell.—ARCHÆOLOGICAL SURVEY OF SOUTHERN INDIA. Lists of the Antiquarian Remains in the Presidency of Madras. Compiled under the Orders of Government, by R. SEWELL, M.C.S. Vol. I., 4to. pp. xii-326, lxii., cloth. 1882. 20*s.*

Sherring.—Hindu Tribes and Castes as represented in Benares. By the Rev. M. A. SHERRING. With Illustrations. 4to. Cloth. Vol. I. pp. xxiv. and 408. 1872. *Now* £6 6*s.* Vol. II. pp. lxviii. and 376. 1879. £2 8*s.* Vol. III. pp. xii. and 336. 1881. £1 12*s.*

Sherring.—THE SACRED CITY OF THE HINDUS. An Account of Benares in Ancient and Modern Times. By the Rev. M. A. SHERRING, M.A., LL.D.; and Prefaced with an Introduction by FITZEDWARD HALL, Esq., D.C.L. 8vo. cloth, pp. xxxvi. and 388, with numerous full-page illustrations. 1868. 21*s.*

Sibree.—THE GREAT AFRICAN ISLAND. Chapters on Madagascar. A Popular Account of Recent Researches in the Physical Geography, Geology, and Exploration of the Country, and its Natural History and Botany, and in the Origin and Division, Customs and Language, Superstitions, Folk-Lore and Religious Belief, and Practices of the Different Tribes. Together with Illustrations of Scripture and Early Church History, from Native Statists and Missionary Experience. By the Rev. JAS. SIBREE, jun., F.R.G.S., of the London Missionary Society, etc. Demy 8vo. cloth, with Maps and Illustrations, pp. xii. and 372. 1880. 12*s.*

Smith.—CONTRIBUTIONS TOWARDS THE MATERIA MEDICA AND NATURAL HISTORY OF CHINA. For the use of Medical Missionaries and Native Medical Students. By F. PORTER SMITH, M.B. London, Medical Missionary in Central China. Imp. 4to. cloth, pp. viii. and 240. 1870. £1 1*s.*

Strangford.—ORIGINAL LETTERS AND PAPERS OF THE LATE VISCOUNT STRANGFORD, upon Philological and Kindred Subjects. Edited by VISCOUNTESS STRANGFORD. Post 8vo. cloth, pp. xxii. and 284. 1878. 12*s.* 6*d.*

Thomas.—ANCIENT INDIAN WEIGHTS. See Numismata Orientalia." Vol. I. Part I.

Thomas.—COMMENTS ON RECENT PEHLVI DECIPHERMENTS. With an Incidental Sketch of the Derivation of Aryan Alphabets, and contributions to the Early History and Geography of Tabaristân. Illustrated by Coins. By EDWARD THOMAS, F.R.S. 8vo. pp. 56, and 2 plates, cloth, sewed. 1872. 3*s.* 6*d.*

Thomas.—SASSANIAN COINS. Communicated to the Numismatic Society of London. By E. THOMAS, F.R.S. Two parts. With 3 Plates and a Woodcut. 12mo. sewed, pp. 43. 5*s.*

Thomas.—The Indian Balhará, and the Arabian intercourse with India in the ninth and following centuries. By EDWARD THOMAS. See Numismata Orientalia. Vol. III. Part I. page 30.

Thomas.—JAINISM; or, The Early Faith of Asoka. With Illustrations of the Ancient Religions of the East, from the Pantheon of the Indo-Scythians. With a Notice on Bactrian Coins and Indian Dates. By E THOMAS, F.R.S. 8vo. pp. viii., 24 and 82. With two Autotype Plates and Woodcuts. 7*s*. 6*d*.

Thomas.—RECORDS OF THE GUPTA DYNASTY. Illustrated by Inscriptions, Written History, Local Tradition and Coins. To which is added a Chapter on the Arabs in Sind. By EDWARD THOMAS, F.R.S. Folio, with a Plate, handsomely bound in cloth, pp. iv. and 64. 1876. Price 14*s*.

Thomas.—THE CHRONICLES OF THE PATHÁN KINGS OF DEHLI. Illustrated by Coins, Inscriptions, and other Antiquarian Remains. By EDWARD THOMAS, F.R.S. With numerous Copperplates and Woodcuts. Demy 8vo. cloth, pp. xxiv. and 467 1871. £1 8*s*.

Thomas.—THE REVENUE RESOURCES OF THE MUGHAL EMPIRE IN INDIA, from A.D. 1593 to A.D. 1707. A Supplement to "The Chronicles of the Pathán Kings of Delhi." By E. THOMAS, F.R.S. 8vo., pp. 60, cloth. 3*s*. 6*d*.

Thorburn.—BANNÚ; or, Our Afghán Frontier. By S. S. THORBURN, I.C.S., Settlement Officer of the Bannú District. 8vo. cloth, pp. x. and 480. 1876. 18*s*.

Vaughan. — THE MANNERS AND CUSTOMS OF THE CHINESE OF THE STRAITS SETTLEMENTS. By J. D. VAUGHAN, Advocate and Solicitor, Supreme Court, Straits Settlements. 8vo. pp. iv.-120, boards. 1879. 7*s*. 6*d*.

Watson.—INDEX TO THE NATIVE AND SCIENTIFIC NAMES OF INDIAN AND OTHER EASTERN ECONOMIC PLANTS AND PRODUCTS, By J. F. WATSON, M.A., M.D., etc. Imperial 8vo., cloth, pp. 650. 1868. £1 11*s*. 6*d*.

Wedgwood.—CONTESTED ETYMOLOGIES in the Dictionary of the Rev. W. W. Skeat. By HENSLEIGH WEDGWOOD. Crown 8vo. cloth, pp. viii.-194. 1882. 5*s*.

West and Buhler.—A DIGEST OF THE HINDU LAW of Inheritance, Partition, Adoption; Embodying the Replies of the Sastris in the Courts of the Bombay Presidency. With Introductions and Notes by the Hon. Justice RAYMOND WEST and J. G. BÜHLER, C.I.E. Third Edition. 8vo. pp. xc.-1450, wrapper. 1884. 36*s*.

Wheeler.—THE HISTORY OF INDIA FROM THE EARLIEST AGES. By J. TALBOYS WHEELER, Assistant Secretary to the Government of India in the Foreign Department, etc. etc. Demy 8vo. cl. 1867-1881.

Vol. I. The Vedic Period and the Maha Bharata. pp. lxxv. and 576. £3 10*s*. Vol. II., The Ramayana and the Brahmanic Period. pp. lxxxviii. and 680, with two Maps. 21*s*. Vol. III. Hindu, Buddhist, Brahmanical Revival. pp. 484, with two maps. 18*s*. Vol. IV. Part I. Mussulman Rule. pp. xxxii. and 320. 14*s*. Vol. IV. Part II. Moghul Empire—Aurangzeb. pp. xxviii. and 280. 12*s*.

Wheeler.—EARLY RECORDS OF BRITISH INDIA. A History of the English Settlement in India, as told in the Government Records, the works of old travellers and other contemporary Documents, from the earliest period down to the rise of British Power in India. By J. TALBOYS WHEELER. Royal 8vo. cloth, pp. xxxii. and 392. 1878. 15*s*.

Williams.—MODERN INDIA AND THE INDIANS. See Trübner's Oriental Series, p. 4.

Wise.—COMMENTARY ON THE HINDU SYSTEM OF MEDICINE. By T. A. WISE, M.D., Bengal Medical Service. 8vo., pp. xx. and 432, cloth. 7*s*. 6*d*.

Wise.—REVIEW OF THE HISTORY OF MEDICINE. By THOMAS A. WISE, M.D. 2 vols. 8vo. cloth. Vol. I., pp. xcviii. and 397; Vol. II., pp. 574. 10*s*.

THE RELIGIONS OF THE EAST.

Adi Granth (The); OR, THE HOLY SCRIPTURES OF THE SIKHS, translated from the original Gurmukhī, with Introductory Essays, by Dr. ERNEST TRUMPP, Prof. Oriental Languages Munich, Roy. 8vo. cl. pp. 866. £2 12*s.* 6*d.*

Alabaster.—THE WHEEL OF THE LAW: Buddhism illustrated from Siamese Sources by the Modern Buddhist, a Life of Buddha, and an account of the Phrabat. By HENRY ALABASTER, Interpreter of H.M. Consulate-General in Siam. Demy 8vo. pp. lviii. and 324, cloth. 1871. 14*s.*

Amberley.—AN ANALYSIS OF RELIGIOUS BELIEF. By VISCOUNT AMBERLEY. 2 vols. 8vo. cl., pp. xvi. 496 and 512. 1876. 30*s.*

Apastambíya Dharma Sutram.—APHORISMS OF THE SACRED LAWS OF THE HINDUS, by Apastamba. Edited, with a Translation and Notes, by G. Bühler. 2 parts. 8vo. cloth, 1868–71. £1 4*s.* 6*d.*

Arnold.—THE LIGHT OF ASIA; or, The Great Renunciation (Mahabhinishkramana). Being the Life and Teaching of Gautama, Prince of India, and Founder of Buddhism (as told by an Indian Buddhist). By EDWIN ARNOLD, C.S.I., etc. Cheap Edition. Crown 8vo. parchment, pp. xvi. and 238. 1882. 2*s.* 6*d.* Library Edition, post 8vo. cloth. 7*s.* 6*d.* Illustrated Edition. 4to. pp. xx.-196, cloth. 1884. 21*s.*

Arnold.—INDIAN POETRY. See "Trübner's Oriental Series," page 4.

Arnold.—PEARLS OF THE FAITH; or, Islam's Rosary. Being the Ninety-nine Beautiful Names of Allah (Asmâ-el-'Husnâ), with Comments in Verse from various Oriental sources as made by an Indian Mussulman. By E. ARNOLD, C.S.I., etc. Third Ed. Cr. 8vo. cl., pp. xvi.-320. 1884. 7*s.* 6*d.*

Balfour.—TAOIST TEXTS; Ethical, Political, and Speculative. By FREDERICK HENRY BALFOUR, Editor of the North-China Herald. Imp. 8vo. pp. vi.-118, cloth [1884], price 10*s.* 6*d.*

Ballantyne.—The Sanlhya Aphorisms of Kapila. See "Trübner's Oriental Series," p. 6.

Banerjea.—THE ARIAN WITNESS, or the Testimony of Arian Scriptures in corroboration of Biblical History and the Rudiments of Christian Doctrine. Including Dissertations on the Original Home and Early Adventures of Indo-Arians. By the Rev. K. M. BANERJEA. 8vo. sewed, pp. xviii. and 236. 8*s.* 6*d.*

Barth.—RELIGIONS OF INDIA. See "Trübner's Oriental Series," page 4.

Beal.—TRAVELS OF FAH HIAN AND SUNG-YUN, Buddhist Pilgrims from China to India (400 A.D. and 518 A.D.) Translated from the Chinese, by S. BEAL, B.A. Crown 8vo. pp. lxxiii. and 210, cloth, with a coloured map. Out of print.

Beal.—A CATENA OF BUDDHIST SCRIPTURES FROM THE CHINESE. By S. BEAL, B.A. 8vo. cloth, pp. xiv. and 436. 1871. 15*s.*

Beal.—THE ROMANTIC LEGEND OF SÂKHYA BUDDHA. From the Chinese-Sanscrit by the Rev. S. BEAL. Crown 8vo. cloth, pp. 400. 1875. 12*s.*

Beal.—THE DHAMMAPADA. See "Trübner's Oriental Series," page 3.

Beal.—ABSTRACT OF FOUR LECTURES ON BUDDHIST LITERATURE IN CHINA, Delivered at University College, London. By SAMUEL BEAL. Demy 8vo. cloth, pp. 208. 1882. 10*s.* 6*d.*

Beal.—Buddhist Records of the Western World. See "Trübner's Oriental Series," p. 6.

Bigandet.—GAUDAMA, the Buddha of the Burmese. See "Trübner's Oriental Series," page 4.

Brockie.—INDIAN PHILOSOPHY. Introductory Paper. By WILLIAM BROCKIE. 8vo. pp. 26, sewed. 1872. 6*d.*

Brown.—THE DERVISHES; or, ORIENTAL SPIRITUALISM. By JOHN P. BROWN, Sec. and Dragoman of Legation of U.S.A. Constantinople. With twenty-four Illustrations. 8vo. cloth, pp. viii. and 415. 14*s.*

Burnell.—THE ORDINANCES OF MANU. See "Trübner's Oriental Series," page 6.

Callaway.—THE RELIGIOUS SYSTEM OF THE AMAZULU.

Part I.—Unkulunkulu; or, the Tradition of Creation as existing among the Amazulu and other Tribes of South Africa, in their own words, with a translation into English, and Notes. By the Rev. Canon CALLAWAY, M.D. 8vo. pp. 128, sewed. 1868. 4*s.*

Part II.—Amatongo; or, Ancestor Worship, as existing among the Amazulu, in their own words, with a translation into English, and Notes. By the Rev. CANON CALLAWAY, M.D. 1869. 8vo. pp. 197, sewed. 1869. 4*s.*

Part III.—Izinyanga Zokubula; or, Divination, as existing among the Amazulu, in their own words. With a translation into English, and Notes. By the Rev. CANON CALLAWAY, M.D. 8vo. pp. 150, sewed. 1870. 4*s.*

Part IV.—Abatakati, or Medical Magic and Witchcraft, 8vo. pp. 40, sewed. 1*s.* 6*d.*

Chalmers.—THE ORIGIN OF THE CHINESE; an Attempt to Trace the connection of the Chinese with Western Nations in their Religion, Superstitions Arts, Language, and Traditions. By JOHN CHALMERS, A.M. Foolscap 8vo. cloth, pp. 78. 5*s.*

Clarke.—TEN GREAT RELIGIONS: an Essay in Comparative Theology. By JAMES FREEMAN CLARKE. 8vo. cloth, pp. x. and 528. 1871. 15*s.*

Clarke.—TEN GREAT RELIGIONS. Part II. A Comparison of All Religions. By J. F. CLARKE. Demy 8vo., pp. xxviii.-414, cloth. 1883. 10*s.* 6*d.*

Clarke.—SERPENT AND SIVA WORSHIP, and Mythology in Central America, Africa and Asia. By HYDE CLARKE, Esq. 8vo. sewed. 1*s.*

Conway.—THE SACRED ANTHOLOGY. A Book of Ethnical Scriptures. Collected and edited by M. D. CONWAY. 5th edition. Demy 8vo. cloth, pp. xvi. and 480. 1876. 12*s.*

Coomára Swamy.—THE DATHÁVANSA; or, the History of the Tooth-Relic of Gotama Buddha. The Pali Text and its Translation into English, with Notes. By Sir M. COOMÁRA SWÁMY, Mudeliár. Demy 8vo. cloth, pp. 174. 1874. 10*s.* 6*d.*

Coomára Swamy.—THE DATHÁVANSA; or, the History of the Tooth-Relic of Gotama Buddha. English Translation only. With Notes. Demy 8vo. cloth, pp. 100. 1874. 6*s.*

Coomára Swamy.—SUTTA NÍPÁTA; or, the Dialogues and Discourses of Gotama Buddha. Translated from the Pali, with Introduction and Notes. By Sir M. COOMÁRA SWAMY. Cr. 8vo. cloth, pp. xxxvi. and 160. 1874. 6*s.*

Coran.—EXTRACTS FROM THE CORAN IN THE ORIGINAL, WITH ENGLISH RENDERING. Compiled by Sir WILLIAM MUIR, K.C.S.I., LL.D., Author of the "Life of Mahomet." Crown 8vo. cloth, pp. 58. 1880. 3*s.* 6*d.*

Cowell.—THE SARVA DARSANA SAMGRAHA. See "Trübner's Oriental Series," p. 5.

Cunningham.—THE BHILSA TOPES; or, Buddhist Monuments of Central India: comprising a brief Historical Sketch of the Rise, Progress, and Decline of Buddhism; with an Account of the Opening and Examination of the various Groups of Topes around Bhilsa. By Brev.-Major A. Cunningham. Illustrated. 8vo. cloth, 33 Plates, pp. xxxvi. 370. 1854. £2 2*s.*

Da Cunha.—MEMOIR ON THE HISTORY OF THE TOOTH-RELIC OF CEYLON; with an Essay on the Life and System of Gautama Buddha. By J. GERSON DA CUNHA. 8vo. cloth, pp. xiv. and 70. With 4 photographs and cuts. 7*s.* 6*d.*

Davids.—BUDDHIST BIRTH STORIES. See Trübner's Oriental Series," page 4.

Davies.—HINDU PHILOSOPHY. See Trübner's Oriental Series," page 5.

Dowson.—DICTIONARY OF HINDU MYTHOLOGY, ETC. See Trübner's Oriental Series," page 4.

Dickson.—THE PÂTIMOKKHA, being the Buddhist Office of the Confession of Priests. The Pali Text, with a Translation, and Notes, by J. F. DICKSON, M.A. 8vo. sd., pp. 69. 2*s*.

Edkins.—CHINESE BUDDHISM. See "Trübner's Oriental Series," page 4.

Edkins.—RELIGION IN CHINA. See "Trübner's Oriental Series," p. 6.

Eitel.—HANDBOOK FOR THE STUDENT OF CHINESE BUDDHISM. By the Rev. E. J. EITEL, L. M. S. Crown 8vo. cloth, pp. viii. and 224. 1870. 18*s*.

Eitel.—BUDDHISM: its Historical, Theoretical, and Popular Aspects. In Three Lectures. By Rev. E. J. EITEL, M.A. Ph.D. Second Edition. Demy 8vo. sewed, pp. 130. 1873. 5*s*.

Examination (Candid) of Theism.—By Physicus. Post 8vo. cloth, pp. xviii. and 198. 1878. 7*s*. 6*d*.

Faber.—A SYSTEMATICAL DIGEST OF THE DOCTRINES OF CONFUCIUS, according to the ANALECTS, GREAT LEARNING, and DOCTRINE of the MEAN. with an Introduction on the Authorities upon CONFUCIUS and Confucianism. By ERNST FABER, Rhenish Missionary. Translated from the German by P. G. von Möllendorff. 8vo. sewed, pp. viii. and 131. 1875. 12*s*. 6*d*.

Faber.—INTRODUCTION TO THE SCIENCE OF CHINESE RELIGION. A Critique of Max Müller and other Authors. By the Rev. ERNST FABER, Rhenish Missionary in Canton. Crown 8vo. stitched in wrapper, pp. xii. and 154. 1880. 7*s*. 6*d*.

Faber.—THE MIND OF MENCIUS. See "Trübner's Oriental Series," p. 4.

Giles.—RECORD OF THE BUDDHIST KINGDOMS. Translated from the Chinese by H. A. GILES, of H.M. Consular Service. 8vo. sewed, pp. x.–129. 5*s*.

Gough.—THE PHILOSOPHY OF THE UPANISHADS. See "Trübner's Oriental Series," p. 6.

Gubernatis.—ZOOLOGICAL MYTHOLOGY; or, the Legends of Animals. By ANGELO DE GUBERNATIS, Professor of Sanskrit and Comparative Literature in the Instituto di Studii Superiori e di Perfezionamento at Florence, etc. In 2 vols. 8vo. pp. xxvi. and 432, vii. and 442. 28*s*.

Gulshan I. Raz: THE MYSTIC ROSE GARDEN OF SA'D UD DIN MAHMUD SHABISTARI. The Persian Text, with an English Translation and Notes, chiefly from the Commentary of Muhammed Bin Yahya Lahiji. By E. H. WHINFIELD, M.A., late of H.M.B.C.S. 4to. cloth, pp. xvi. 94 and 60. 1880. 10*s*. 6*d*.

Hardy.—CHRISTIANITY AND BUDDHISM COMPARED. By the late REV. R. SPENCE HARDY, Hon. Member Royal Asiatic Society. 8vo. sd. pp. 138. 6*s*.

Haug.—THE PARSIS. See "Trübner's Oriental Series," p. 3.

Haug.—THE AITAREYA BRAHMANAM OF THE RIG VEDA: containing the Earliest Speculations of the Brahmans on the meaning of the Sacrificial Prayers and on the Origin, Performance, and Sense of the Rites of the Vedic Religion. Edited, Translated, and Explained by MARTIN HAUG, Ph.D., Superintendent of Sanskrit Studies in the Poona College, etc., etc. In 2 Vols. Crown 8vo. Vol. I. Contents, Sanskrit Text, with Preface, Introductory Essay, and a Map of the Sacrificial Compound at the Soma Sacrifice, pp. 312. Vol. II. Translation with Notes, pp. 544. £2 2*s*.

Hawken.—UPA-SASTRA: Comments, Linguistic and Doctrinal, on Sacred and Mythic Literature. By J. D. HAWKEN. 8vo. cloth, pp. viii. –288. 7*s*. 6*d*.

Hershon.—A TALMUDIC MISCELLANY. See "Trübner's Oriental Series," p. 4.

Hodgson.—ESSAYS RELATING TO INDIAN SUBJECTS. See "Trübner's Oriental Series," p. 4.

Inman.—ANCIENT PAGAN AND MODERN CHRISTIAN SYMBOLISM EXPOSED AND EXPLAINED. By THOMAS INMAN, M.D. Second Edition. With Illustrations. Demy 8vo. cloth, pp. xl. and 148. 1874. *7s. 6d.*

Johnson.—ORIENTAL RELIGIONS and their Relation to Universal Religion. By SAMUEL JOHNSON. First Section—India. In 2 Volumes, post 8vo. cloth. pp. 408 and 402. *21s*

Journal of the Ceylon Branch of the Royal Asiatic Society.—For Papers on Buddhism contained in it, see page 11.

Kistner.—BUDDHA AND HIS DOCTRINES. A Bibliographical Essay. By OTTO KISTNER. Imperial 8vo., pp. iv. and 32, sewed. *2s. 6d.*

Koran (The); commonly called THE ALCORAN OF MOHAMMED. Translated into English immediately from the original Arabic. By GEORGE SALE, Gent. To which is prefixed the Life of Mohammed. Crown 8vo. cloth, pp. 472. *7s.*

Koran.—Arabic text. Lithographed in Oudh. Foolscap 8vo. pp. 502. sewed. Lucknow, A.H. 1295 (1877). *9s.*

Lane.—THE KORAN. See "Trübner's Oriental Series," p. 3.

Legge.—CONFUCIANISM IN RELATION TO CHRISTIANITY. A Paper read before the Missionary Conference in Shanghai, on May 11, 1877. By Rev. JAMES LEGGE, D.D. 8vo. sewed, pp. 12. 1877. *1s. 6d.*

Legge.—THE LIFE AND TEACHINGS OF CONFUCIUS. With Explanatory Notes. By JAMES LEGGE, D.D. Fifth Edition. Crown 8vo. cloth, pp. vi. and 338. 1877. *10s. 6d.*

Legge.—THE LIFE AND WORKS OF MENCIUS. With Essays and Notes. By JAMES LEGGE. Crown 8vo. cloth, pp. 402. 1875. *12s.*

Legge.—CHINESE CLASSICS. *v.* under "Chinese," p. 51.

Leigh.—THE RELIGION OF THE WORLD. By H. STONE LEIGH. 12mo. pp. xii. 66, cloth. 1869. *2s. 6d.*

M'Clatchie.—CONFUCIAN COSMOGONY. A Translation (with the Chinese Text opposite) of Section 49 (Treatise on Cosmogony) of the "Complete Works" of the Philosopher Choo-Foo-Tze. With Explanatory Notes by the Rev. TH. M'CLATCHIE, M.A. Small 4to. pp. xviii. and 162. 1874. *12s. 6d.*

Mills.—THE INDIAN SAINT; or, Buddha and Buddhism.—A Sketch Historical and Critical. By C. D. B. MILLS. 8vo. cl., pp. 192. *7s. 6d.*

Mitra.—BUDDHA GAYA, the Hermitage of Sákya Muni. By RAJENDRALALA MITRA, LL.D., C.I.E. 4to. cloth, pp. xvi. and 258, with 51 Plates. 1878. *£3.*

Muhammed.—THE LIFE OF MUHAMMED. Based on Muhammed Ibn Ishak. By Abd El Malik Ibn Hisham. Edited by Dr. FERDINAND WÜSTENFELD. The Arabic Text. 8vo. pp. 1026, sewed. Price *21s.* Introduction, Notes, and Index in German. 8vo. pp. lxxii. and 266, sewed. *7s. 6d.* Each part sold separately.

The text based on the Manuscripts of the Berlin, Leipsic, Gotha and Leyden Libraries, has been carefully revised by the learned editor, and printed with the utmost exactness.

Müller.—THE HYMNS OF THE RIG VEDA IN THE SAMHITA AND PADA TEXTS. Reprinted from the Editio Princept by F. MAX MULLER, M.A. Second Edition. With the two texts on parallel pages. 2 vols., 8vo, pp. 800-828, stitched in wrapper. 1877. £1 12*s*.

Muir.—TRANSLATIONS FROM THE SANSKRIT. See "Trübner's Oriental Series," p. 3.

Muir.—ORIGINAL SANSKRIT TEXTS—*v.* under Sanskrit.

Muir.—EXTRACTS FROM THE CORAN. In the Original, with English rendering. Compiled by Sir WILLIAM MUIR, K.C.S.I., LL.D., Author of "The Life of Mahomet." Crown 8vo, pp. viii. and 64, cloth. 1880. 3*s*. 6*d*.

Müller.—THE SACRED HYMNS OF THE BRAHMINS, as preserved to us in the oldest collection of religious poetry, the Rig-Veda-Sanhita, translated and explained. By F. MAX MÜLLER, M.A., Oxford. Volume I. Hymns to the Maruts or the Storm Gods. 8vo. pp. clii. and 264. 12*s*. 6*d*.

Müller.—LECTURE ON BUDDHIST NIHILISM. By F. MAX MÜLLER, M.A. Delivered before the Association of German Philologists, at Kiel, 28th September, 1869. (Translated from the German.) Sewed. 1869. 1*s*.

Müller.—RIG VEDA SAMHITA AND PADA TEXTS. See page 89.

Newman.—HEBREW THEISM. By F. W. NEWMAN. Royal 8vo. stiff wrappers, pp. viii. and 172. 1874. 4*s*. 6*d*.

Piry.—LE SAINT EDIT, ÉTUDE DE LITTERATURE CHINOISE. Préparée par A. THÉOPHILE PIRY, du Service des Douanes Maritimes de Chine. 4to. pp. xx. and 320, cloth. 1879. 21*s*.

Priaulx.—QUÆSTIONES MOSAICÆ; or, the first part of the Book of Genesis compared with the remains of ancient religions. By OSMOND DE BEAUVOIR PRIAULX. 8vo. pp. viii. and 548, cloth. 12*s*.

Redhouse.—THE MESNUVI. See "Trübner's Oriental Series," p. 4.

Rig-Veda Sanhita.—A COLLECTION OF ANCIENT HINDU HYMNS. Constituting the First Ashtaka, or Book of the Rig-veda; the oldest authority for the religious and social institutions of the Hindus. Translated from the Original Sanskrit by the late H. H. WILSON, M.A. 2nd Ed., with a Postscript by Dr. FITZEDWARD HALL. Vol. I. 8vo. cloth, pp. lii. and 348, price 21*s*.

Rig-Veda Sanhita.—A Collection of Ancient Hindu Hymns, constituting the Fifth to Eighth Ashtakas, or books of the Rig-Veda, the oldest Authority for the Religious and Social Institutions of the Hindus. Translated from the Original Sanskrit by the late HORACE HAYMAN WILSON, M.A., F.R.S., etc. Edited by E. B. COWELL, M.A., Principal of the Calcutta Sanskrit College. Vol. IV., 8vo., pp. 214, cloth. 14*s*.
A few copies of Vols. II. and III. still left. [*Vols. V. and VI. in the Press.*

Rockhill.—THE LIFE OF THE BUDDHA. See "Trübner's Oriental Series," page 6.

Sacred Books (The) OF THE EAST. Translated by various Oriental Scholars, and Edited by F. Max Müller. All 8vo. cloth.

Vol. I. The Upanishads. Translated by F. Max Müller. Part I. The Khândogya-Upanishad. The Talavakâra-Upanishad. The Aitareya-Aranyaka. The Kaushîtaki-Brâhmana-Upanishad and the Vâgasaneyi-Samhitâ-Upanishad. pp. xii. and 320. 10*s*. 6*d*.

Vol. II. The Sacred Laws of the Âryas, as taught in the Schools of Âpastamba, Gautama, Vâsishtha, and Baudhâyana. Translated by Georg Bühler. Part I. Apastamba and Gautama. pp. lx. and 312. 1879. 10*s*. 6*d*.

Vol. III. The Sacred Books of China. The Texts of Confucianism. Translated by James Legge. Part I. The Shû King. The Religious Portions of the Shih King The Hsiâo King. pp. xxxii. and 492. 1879. 12*s*. 6*d*.

Vol. IV. The Zend-Avesta. Part I. The Vendîdâd. Translated by James Darmesteter. pp. civ. and 240. 10*s*. 6*d*.

Vol. V. Pahlavi Texts. Part I. The Bundahis, Bahman Yast, and Shâyast-la-Shâyast. Translated by E. W. West. pp. lxxiv. and 438. 12*s*. 6*d*.

Vol. VI. The Qur'ân. Part I. Translated by E. H. Palmer. pp. cxx. and 268, cloth. 10*s*. 6*d*.

Vol. VII. The Institutes of Vishnu. Translated by Julius Jolly. pp. xl. and 316. 10*s*. 6*d*.

Vol. VIII. The Bhagavadgitâ with other extracts from the Mahâbhârata. Translated by Kashinath Trunbak Telang. pp. 446. 10*s*. 6*d*.

Vol. IX. The Qur'ân. Part II. Translated by E. H. Palmer. pp. x. and 362. 10*s*. 6*d*.

Vol. X. The Suttanipâta, etc. Translated by V. Fausböll. pp. lvi. and 224, 10*s*. 6*d*.

Vol. XI. The Mahâparinibbâna Sutta. The Tevigga Sutta. The Mahâsudassana Sutta. The Dhamma-Kakkappavattana Sutta. Translated by T. W. Rhys Davids. pp. xlviii.-320. 10*s*. 6*d*.

Vol. XII. The Satapatha-Brâhmana. Translated by Prof. Eggeling. Vol. I. pp. xlviii. and 456. 12*s*. 6*d*.

Vol. XIII. The Pâtimokkha. Translated by T. W. Rhys Davids. The Mahavagga. Part I. Translated by Dr. H. Oldenberg. pp. xxxviii. and 360. 10*s*. 6*d*.

Vol. XIV. The Sacred Laws of the Aryas, as taught in the Schools of Vâsishtha and Baudhâyana. Translated by Prof. Georg Buhler.

Vol. XV. The Upanishads. Part II. Translated by F. Max Müller. [*In preparation*.

Vol. XVI. The Yî King. Translated by James Legge. pp. xxii. and 448. 10*s*. 6*d*.

Vol. XVII. The Mahâvagga. Part II. Translated by T. W. Rhys Davids, and Dr. H. Oldenberg.

Vol. XVIII. The Dâdistân-î Dînîk and Mainyô-i Khard. Pahlavi Texts. Part II. Translated by E. W. West.

Vol. XIX. The Fo-sho-hing-tsan-king. Translated by Samuel Beal.

Vol. XX. The Yâyu-Purâna. Translated by Prof. Bhandarkar, of Elphinstone College, Bombay.

Vol. XXI. The Saddharma-pundarika. Translated by Prof. Kern.

Vol. XXII. The Akârânga-Sûtra. Translated by Prof. Jacobi.

Schlagintweit.—Buddhism in Tibet. Illustrated by Literary Documents and Objects of Religious Worship. With an Account of the Buddhist Systems preceding it in India. By Emil Schlagintweit, LL.D. With a Folio Atlas of 20 Plates, and 20 Tables of Native Prints in the Text. Royal 8vo., pp. xxiv. and 404. £2 2*s*.

Sell.—The Faith of Islam. By the Rev. E. Sell, Fellow of the University of Madras. Demy 8vo. cloth, pp. xiv. and 270. 1880. 6*s*. 6*d*.

Sell.—Ihn-i-Tajwid; or, Art of Reading the Quran. By the Rev. E. Sell, B.D. 8vo., pp. 48, wrappers. 1882. 2*s*. 6*d*.

Sherring.—The Hindoo Pilgrims. By the Rev. M. A. Sherring, Fcap. 8vo. cloth, pp. vi. and 125. 5*s*.

Singh.—Sakhee Book; or, the Description of Gooroo Gobind Singh's Religion and Doctrines, translated from Gooroo Mukhi into Hindi, and afterwards into English. By Sirdar Attar Singh, Chief of Bhadour. With the Author's photograph. 8vo. pp. xviii. and 205. Benares, 1873. 15*s*.

Sinnett.—The Occult World. By A. P. Sinnett, President of the Simla Eclectic Theosophical Society. Fourth Edition. Fcap. 8vo., pp. xiv. and 140, cloth. 1884. 3*s*. 6*d*.

Sinnett.—Esoteric Buddhism. By A. P. Sinnett, Author of the "Occult World," President of the Simla Eclectic Theosophical Society. Fourth Edition. Crown 8vo., pp. xx. and 216, cloth. 1884. 7*s*. 6*d*.

Syed Ahmad.—A SERIES OF ESSAYS ON THE LIFE OF MOHAMMED, and Subjects subsidiary thereto. By SYED AHMAD KHAN BAHADOR, C.S.I., Author of the "Mohammedan Commentary on the Holy Bible," Honorary Member of the Royal Asiatic Society, and Life Honorary Secretary to the Allygurh Scientific Society. 8vo. pp. 532, with 4 Genealogical Tables, 2 Maps, and a Coloured Plate, handsomely bound in cloth. £1 10*s.*

Thomas.—JAINISM. See page 28.

Tiele.—OUTLINES OF THE HISTORY OF RELIGION. See "Trübner's Oriental Series," page 6.

Tiele.—History of Egyptian Religion. See Trübner's Oriental Series, page 5.

Vishnu-Purana (The); a System of Hindu Mythology and Tradition. Translated from the original Sanskrit, and Illustrated by Notes derived chiefly from other Purâṇas. By the late H. H. WILSON, M.A., F.R.S., Boden Professor of Sanskrit in the University of Oxford, etc.,etc. Edited by FITZEDWARD HALL. In 6 vols. 8vo. Vol. I. pp. cxl. and 200; Vol. II. pp. 343; Vol. III., pp. 348; Vol. IV. pp. 346, cloth; Vol. V. Part I. pp. 392, cloth. 10*s.* 6*d.* each. Vol. V., Part 2, containing the Index, compiled by F. Hall. 8vo. cloth, pp. 268. 12*s.*

Wake.—THE EVOLUTION OF MORALITY. Being a History of the Development of Moral Culture. By C. STANILAND WAKE, author of "Chapters on Man," etc. Two vols. 8vo. cloth, pp. xvi. and 506, xii. and 474. 21*s.*

Wherry.—Commentary on the Quran. See Trübner's Oriental Series, page 5.

Wilson.—Works of the late HORACE HAYMAN WILSON, M.A., F.R.S., Member of the Royal Asiatic Societies of Calcutta and Paris, and of the Oriental Soc. of Germany, etc., and Boden Prof. of Sanskrit in the University of Oxford, Vols I. and II. ESSAYS AND LECTURES chiefly on the Religion of the Hindus. by the late H. H. WILSON, M.A., F.R.S., etc. Collected and edited by Dr. REINHOLD ROST. 2 vols. cloth, pp. xiii. and 399, vi. and 416. 21*s.*

COMPARATIVE PHILOLOGY.

POLYGLOTS.

Beames.—OUTLINES OF INDIAN PHILOLOGY. With a Map, showing the Distribution of the Indian Languages. By JOHN BEAMES. Second enlarged and revised edition. Crown 8vo. cloth, pp. viii. and 96. 1868. 5*s.*

Beames.—A COMPARATIVE GRAMMAR OF THE MODERN ARYAN LANGUAGES OF INDIA (to wit), Hindi, Panjabi, Sindhi, Gujarati, Marathi, Uriya, and Bengali. By JOHN BEAMES, Bengal C.S., M.R.A.S., &c. 8vo. cloth. Vol. I. On Sounds. pp. xvi. and 360. 1872. 16*s.* Vol. II. The Noun and the Pronoun. pp. xii. and 348. 1875. 16*s.* Vol III. The Verb. pp. xii. and 316. 1879. 16*s.*

Bellows.—ENGLISH OUTLINE VOCABULARY, for the use of Students of the Chinese, Japanese, and other Languages. Arranged by JOHN BELLOWS. With Notes on the writing of Chinese with Roman Letters. By Professor SUMMERS, King's College, London. Crown 8vo., pp. 6 and 368, cloth. 6*s.*

Bellows.—OUTLINE DICTIONARY, FOR THE USE OF MISSIONARIES, Explorers, and Students of Language. By MAX MÜLLER, M.A., Taylorian Professor in the University of Oxford. With an Introduction on the proper use of the ordinary English Alphabet in transcribing Foreign Languages. The Vocabulary compiled by JOHN BELLOWS. Crown 8vo. Limp morocco, pp. xxxi. and 368. 7*s.* 6*d.*

Caldwell.—A COMPARATIVE GRAMMAR OF THE DRAVIDIAN, OR SOUTH-INDIAN FAMILY OF LANGUAGES. By the Rev. R. CALDWELL, LL.D. A Second, corrected, and enlarged Edition. Demy 8vo. pp. 805. 1875. 28*s.*

Calligaris.—LE COMPAGNON DE TOUS, OU DICTIONNAIRE POLYGLOTTE. Par le Colonel LOUIS CALLIGARIS, Grand Officier, etc. (French—Latin—Italian—Spanish—Portuguese—German—English—Modern Greek—Arabic—Turkish.) 2 vols. 4to., pp. 1157 and 746. Turin. £4 4*s.*

Campbell.—SPECIMENS OF THE LANGUAGES OF INDIA, including Tribes of Bengal, the Central Provinces, and the Eastern Frontier. By Sir G. CAMPBELL, M.P. Folio, paper, pp. 308. 1874. £1 11*s.* 6*d.*

Clarke.—RESEARCHES IN PRE-HISTORIC AND PROTO-HISTORIC COMPARATIVE PHILOLOGY, MYTHOLOGY, AND ARCHÆOLOGY, in connexion with the Origin of Culture in America and the Accad or Sumerian Families. By HYDE CLARKE. Demy 8vo. sewed, pp. xi. and 74. 1875. 2*s.* 6*d.*

Cust.—LANGUAGES OF THE EAST INDIES. See Trübner's Oriental Series," page 3.

Douse.—GRIMM'S LAW; A STUDY: or, Hints towards an Explanation of the so-called "Lautverschiebung." To which are added some Remarks on the Primitive Indo-European *K*, and several Appendices. By T. LE MARCHANT DOUSE. 8vo. cloth, pp. xvi. and 230. 10*s.* 6*d.*

Dwight.—MODERN PHILOLOGY: Its Discovery, History, and Influence. New edition, with Maps, Tabular Views, and an Index. By BENJAMIN W DWIGHT. In two vols. cr. 8vo. cloth. First series, pp. 360; second series, pp. xi. and 554. £1.

Edkins.—CHINA'S PLACE IN PHILOLOGY. An Attempt to show that the Languages of Europe and Asia have a Common Origin. By the Rev. JOSEPH EDKINS. Crown 8vo. cloth, pp. xxiii. and 403. 10*s.* 6*d.*

Ellis.—ETRUSCAN NUMERALS. By ROBERT ELLIS, B.D. 8vo. sewed, pp. 52. 2*s.* 6*d.*

Ellis.—THE ASIATIC AFFINITIES OF THE OLD ITALIANS. By ROBERT ELLIS, B.D., Fellow of St. John's College, Cambridge, and author of "Ancient Routes between Italy and Gaul." Crown 8vo. pp. iv. 156, cloth. 1870. 5*s.*

Ellis.—ON NUMERALS, as Signs of Primeval Unity among Mankind. By ROBERT ELLIS, B.D., Late Fellow of St. John's College, Cambridge. Demy 8vo. cloth, pp. viii. and 94. 3*s.* 6*d.*

Ellis.—PERUVIA SCYTHICA. The Quichua Language of Peru: its derivation from Central Asia with the American languages in general, and with the Turanian and Iberian languages of the Old World, including the Basque, the Lycian, and the Pre-Aryan language of Etruria. By ROBERT ELLIS, B.D. 8vo. cloth, pp. xii. and 219. 1875. 6*s.*

English and Welsh Languages.—THE INFLUENCE OF THE ENGLISH AND Welsh Languages upon each other, exhibited in the Vocabularies of the two Tongues. Intended to suggest the importance to Philologers, Antiquaries, Ethnographers, and others, of giving due attention to the Celtic Branch of the Indo-Germanic Family of Languages. Square, pp. 30, sewed. 1869. 1*s.*

Geiger.—CONTRIBUTIONS TO THE HISTORY OF THE DEVELOPMENT OF THE HUMAN RACE. Lectures and Dissertations. By LAZARUS GEIGER. Translated from the Second German Edition by DAVID ASHER, Ph.D. Post 8vo. cloth, pp. x. and 156. 1880. 6*s.*

Grey.—HANDBOOK OF AFRICAN, AUSTRALIAN, AND POLYNESIAN PHILOLOGY, as represented in the Library of His Excellency Sir George Grey, K.C.B., Her Majesty's High Commissioner of the Cape Colony. Classed, Annotated, and Edited by Sir GEORGE GREY and Dr. H. I. BLEEK.

Vol. I. Part 1.—South Africa. 8vo. pp. 186. 20*s.*
Vol. I. Part 2.—Africa (North of the Tropic of Capricorn). 8vo. pp. 70. 4*s.*
Vol. I. Part 3.—Madagascar. 8vo. pp. 24. 2*s.*
Vol. II. Part 1.—Australia. 8vo. pp. iv. and 44. 3*s.*
Vol. II. Part 2.—Papuan Languages of the Loyalty Islands and New Hebrides, comprising those of the Islands of Nengone, Lifu, Aneitum, Tana, and others. 8vo. p. 12. 1*s.*
Vol. II. Part 3.—Fiji Islands and Rotuma (with Supplement to Part II., Papuan Languages, and Part I., Australia). 8vo. pp. 34. 2*s.*
Vol. II. Part 4.—New Zealand, the Chatham Islands, and Auckland Islands. 8vo. pp. 76. 7*s.*
Vol. II. Part 4 (*continuation*).—Polynesia and Borneo. 8vo. pp. 77-154. 7*s.*
Vol. III. Part 1.—Manuscripts and Incunables. 8vo. pp. viii. and 24. 2*s.*
Vol. IV. Part 1.—Early Printed Books. England. 8vo. pp. vi. and 266. 12*s.*

Gubernatis.—ZOOLOGICAL MYTHOLOGY; or, the Legends of Animals. By ANGELO DE GUBERNATIS, Professor of Sanskrit and Comparative Literature in the Instituto di Studii Superiori e di Perfezionamento at Florence, etc. In 2 vols. 8vo. pp. xxxvi. and 432, vii. and 442. 28*s.*

Hoernle.—A COMPARATIVE GRAMMAR OF THE GAUDIAN LANGUAGE, with Special Reference to the Eastern Hindi. Accompanied by a Language Map and a Table of Alphabets. By A. F. R. HOERNLE. Demy 8vo. pp. 474 1880. 18*s.*

Hunter.—A Comparative Dictionary of the Non-Aryan Languages of India and High Asia. With a Dissertation, Political and Linguistic, on the Aboriginal Races. By W. W. HUNTER, B.A., of H.M.'s Civil Service. Being a Lexicon of 144 Languages, illustrating Turanian Speech. Compiled from the Hodgson Lists, Government Archives, and Original MSS., arranged with Prefaces and Indices in English, French, German, Russian, and Latin. Large 4to. cloth, toned paper, pp. 230. 1869. 42*s.*

Kilgour.—THE HEBREW OR IBERIAN RACE, including the Pelasgians, the Phenicians, the Jews, the British, and others. By HENRY KILGOUR. 8vo. sewed, pp. 76. 1872. 2*s.* 6*d.*

March.—A COMPARATIVE GRAMMAR OF THE ANGLO-SAXON LANGUAGE; in which its forms are illustrated by those of the Sanskrit, Greek, Latin, Gothic, Old Saxon, Old Friesic, Old Norse, and Old High-German. By FRANCIS A. MARCH, LL.D. Demy 8vo. cloth, pp. xi. and 253. 1877. 10*s.*

Notley.—A COMPARATIVE GRAMMAR OF THE FRENCH, ITALIAN, SPANISH, AND PORTUGUESE LANGUAGES. By EDWIN A. NOTLEY. Crown oblong 8vo. cloth, pp. xv. and 396. 7*s.* 6*d.*

Oppert.—On the Classification of Languages. A Contribution to Comparative Philology. By Dr. G. OPPERT. 8vo. paper, pp. vi. and 146. 1879. 7*s.* 6*d.*

Oriental Congress.—Report of the Proceedings of the Second International Congress of Orientalists held in London, 1874. Roy. 8vo. paper, pp. 76. 5*s.*

Oriental Congress.—TRANSACTIONS OF THE SECOND SESSION OF THE INTERNATIONAL CONGRESS OF ORIENTALISTS, held in London in September, 1874. Edited by ROBERT K. DOUGLAS, Honorary Secretary. Demy 8vo. cloth, pp. viii. and 456. 21*s.*

Pezzi.—ARYAN PHILOLOGY, according to the most recent Researches (Glottologia Aria Recentissima), Remarks Historical and Critical. By DOMENICO PEZZI, Membro della Facolta de Filosofia e lettere della R. Universit. di Torino. Translated by E. S. ROBERTS, M.A., Fellow and Tutor of Gonville and Caius College. Crown 8vo. cloth, pp. xvi. and 199. 6*s.*

Sayce.—An Assyrian Grammar for Comparative Purposes. By A. H. Sayce, M.A. 12mo. cloth, pp. xvi. and 188. 1872. 7*s*. 6*d*.

Sayce.—The Principles of Comparative Philology. By A. H. Sayce, Fellow and Tutor of Queen's College, Oxford. Second Edition. Cr. 8vo. cl., pp. xxxii. and 416. 10*s*. 6*d*.

Schleicher.—Compendium of the Comparative Grammar of the Indo-European, Sanskrit, Greek, and Latin Languages. By August Schleicher. Translated from the German by H. Bendall, B.A., Chr. Coll. Camb. 8vo. cloth, Part I. Grammar. pp. 184. 1874. 7*s*. 6*d*. Part II. Morphology. pp. viii. and 104. 1877. 6*s*.

Singer.—Grammar of the Hungarian Language simplified. By Ignatius Singer. Crown 8vo. cloth, pp. vi.-88. 1882.

Trübner's Collection of Simplified Grammars of the principal Asiatic and European Languages. Edited by Reinhold Rost, LL.D., Ph.D. Crown 8vo. cloth, uniformly bound.

I.—Hindustani, Persian, and Arabic. By the late E. H. Palmer, M.A. Pp. 112. 5*s*.
II.—Hungarian. By I. Singer, of Buda-Pesth. Pp. vi. and 88. 4*s*. 6*d*.
III.—Basque. By W. Van Eys. Pp. xii. and 52. 3*s*. 6*d*.
IV.—Malagasy. By G. W. Parker. Pp. 66. 5*s*.
V.—Modern Greek. By E. M. Geldart, M.A. Pp. 68. 2*s*. 6*d*.
VI.—Roumanian. By M. Torceanu. Pp. viii. and 72. 5*s*.
VII.—Tibetan. By H. A. Jäschke. Pp. viii. and 104. 5*s*.
VIII.—Danish. By E. C. Otté. Pp. viii. and 66. 2*s*. 6*d*.
IX.—Turkish. By J. W. Redhouse. Pp. xii. and 204. 10*s*. 6*d*.
X.—Swedish. By E. C. Otté. Pp. xii. and 70. 2*s*. 6*d*.
XI.—Polish. By W. R. Morfill, M.A. Pp. viii. and 64. 3*s*. 6*d*.
XII.—Pali. By E. Müller, Ph.D. Pp. xvi. and 144. 7*s*. 6*d*.

Trübner's Catalogue of Dictionaries and Grammars of the Principal Languages and Dialects of the World. Considerably enlarged and revised, with an Alphabetical Index. A Guide for Students and Booksellers. Second Edition, 8vo. pp. viii. and 170, cloth. 1882. 5*s*.

*** The first edition, consisting of 64 pp., contained 1,100 titles; the new edition consists of 170 pp., and contains 3,000 titles.

Trumpp.—Grammar of the Paṣ̊to, or Language of the Afghans, compared with the Irānian and North-Indian Idioms. By Dr. Ernest Trumpp. 8vo. sewed, pp. xvi. and 412. 21*s*.

Weber.—Indian Literature. See "Trübner's Oriental Series," p. 3.

Wedgwood.—On the Origin of Language. By Hensleigh Wedgwood, late Fellow of Christ's College, Cambridge. Fcap. 8vo. pp. 172, cloth. 3*s*. 6*d*.

Whitney.—Language and its Study, with especial reference to the Indo-European Family of Languages. Seven Lectures by W. D. Whitney, Professor of Sanskrit, Yale College. Edited with Introduction, Notes, Grimm's Law with Illustration, Index, etc., by the Rev. R. Morris, M.A., LL.D. Second Edition. Cr. 8vo. cl., pp. xxii. and 318. 1881. 5*s*.

Whitney.—Language and the Study of Language: Twelve Lectures on the Principles of Linguistic Science. By W. D. Whitney. Fourth Edition, augmented by an Analysis. Crown 8vo. cloth, pp. xii. and 504. 1884. 10*s*. 6*d*.

Whitney.—Oriental and Linguistic Studies. By W. D. Whitney, Cr. 8vo. cl. 1874. Pp. x. and 418. 12*s*.

First Series. The Veda; the Avesta; the Science of Language.

Second Series.—The East and West—Religion and Mythology—Orthography and Phonology—Hindú Astronomy. Pp. 446. 12*s*.

GRAMMARS, DICTIONARIES, TEXTS, AND TRANSLATIONS.

AFRICAN LANGUAGES.

Bleek.—A Comparative Grammar of South African Languages. By W. H. I. Bleek, Ph.D. Volume I. I. Phonology. II. The Concord. Section 1. The Noun. 8vo. pp. xxxvi. and 322, cloth. 1869. £4 4*s.*

Bleek.—A Brief Account of Bushman Folk Lore and other Texts. By W. H. I. Bleek, Ph.D., etc., etc. Folio sd., pp. 21. 1875. 2*s.* 6*d.*

Bleek.—Reynard the Fox in South Africa; or, Hottentot Fables. Translated from the Original Manuscript in Sir George Grey's Library. By Dr. W. H. I. Bleek, Librarian to the Grey Library, Cape Town, Cape of Good Hope. Post. 8vo., pp. xxxi. and 94, cloth. 1864. 3*s.* 6*d.*

Callaway.—Izinganekwane, Nensumansumane, Nezindaba, Zabantu (Nursery Tales, Traditions, and Histories of the Zulus). In their own words, with a Translation into English, and Notes. By the Rev. Henry Callaway, M.D. Volume I., 8vo. pp. xiv. and 378, cloth. Natal, 1866 and 1867. 16*s.*

Callaway.—The Religious System of the Amazulu.

Part I.—Unkulunkulu; or, the Tradition of Creation as existing among the Amazulu and other Tribes of South Africa, in their own words, with a translation into English, and Notes. By the Rev. Canon Callaway, M.D. 8vo. pp. 128, sewed. 1868. 4s.

Part II.—Amatongo; or, Ancestor Worship, as existing among the Amazulu, in their own words, with a translation into English, and Notes. By the Rev. Canon Callaway, M.D. 1869. 8vo. pp. 127, sewed. 1869. 4*s.*

Part III.—Izinyanga Zokubula; or, Divination, as existing among the Amazulu, in their own words. With a Translation into English, and Notes. By the Rev. Canon Callaway, M.D. 8vo. pp. 150, sewed. 1870. 4*s.*

Part IV.—Abatakati, or Medical Magic and Witchcraft. 8vo. pp. 40, sewed. 1*s.* 6*d.*

Christaller.—A Dictionary, English, Tshi, (Asante), Akra; Tshi (Chwee), comprising as dialects Akán (Asànté, Akém, Akuapém, etc.) and Fànté; Akra (Accra), connected with Adangme; Gold Coast, West Africa.

Enyiresi, Twi né Ňkrań | Eńliši, Otšŭi ke̱ Gā
nse̱m - asekyere̱ - ṅhōma. | wiemo̱i - ašišitšōmo̱ - wolo.

By the Rev. J. G. Christaller, Rev. C. W. Locher, Rev. J. Zimmermann. 16mo. 7*s.* 6*d.*

Christaller.—A Grammar of the Asante and Fante Language, called Tshi (Chwee, Twi): based on the Akuapem Dialect, with reference to the other (Akan and Fante) Dialects. By Rev. J. G. Christaller. 8vo. pp. xxiv. and 203. 1875. 10*s.* 6*d.*

Christaller.—Dictionary of the Asante and Fante Language, called Tshi (Chwee Twi). With a Grammatical Introduction and Appendices on the Geography of the Gold Coast, and other Subjects. By Rev. J. G. Christaller. Demy 8vo. pp. xxviii. and 672, cloth. 1882. £1 5*s.*

Cust.—Sketch of the Modern Languages of Africa. See "Trübner's Oriental Series," page 6.

Döhne.—The Four Gospels in Zulu. By the Rev. J. L. Döhne, Missionary to the American Board, C.F.M. 8vo. pp. 208, cloth. Pietermaritzburg, 1866. 5*s.*

Döhne.—A Zulu-Kafir Dictionary, etymologically explained, with copious Illustrations and examples, preceded by an introduction on the Zulu-Kafir Language. By the Rev. J. L. Döhne. Royal 8vo. pp. xlii. and 418, sewed. Cape Town, 1857. 21*s.*

Grey.—Handbook of African, Australian, and Polynesian Philology, as represented in the Library of His Excellency Sir George Grey, K.C.B., Her Majesty's High Commissioner of the Cape Colony. Classed, Annotated, and Edited by Sir George Grey and Dr. H. I. Bleek.

Vol. I. Part 1.—South Africa. 8vo. pp. 186. 20*s.*
Vol. I. Part 2.—Africa (North of the Tropic of Capricorn). 8vo. pp. 70. 4*s.*
Vol. I. Part 3.—Madagascar. 8vo. pp. 24. 5*s.*
Vol. II. Part 1.—Australia. 8vo. pp. iv. and 44.
Vol. II. Part 2.—Papuan Languages of the Loyalty Islands and New Hebrides, comprising those of the Islands of Nengone, Lifu, Aneitum, Tana, and others. 8vo. pp. 12. 1*s.*
Vol. II. Part 3.—Fiji Islands and Rotuma (with Supplement to Part II., Papuan Languages, and Part I., Australia). 8vo. pp. 34. 2*s.*
Vol. II. Part 4.—New Zealand, the Chatham Islands, and Auckland Islands. 8vo. pp. 76. 7*s.*
Vol. II. Part 4 (*continuation*).—Polynesia and Borneo. 8vo. pp. 77-154. 7*s.*
Vol. III. Part 1.—Manuscripts and Incunables. 8vo. pp. viii. and 24. 2*s.*
Vol. IV. Part 1.—Early Printed Books. England. 8vo. pp. vi. and 266. 12*s.*

Grout.—The Isizulu: a Grammar of the Zulu Language; accompanied with an Historical Introduction, also with an Appendix. By Rev. Lewis Grout. 8vo. pp. lii. and 432, cloth. 21*s.*

Hahn.—Tsuni-||Goam. See Trübner's Oriental Series, page 5.

Krapf.—Dictionary of the Suahili Language. Compiled by the Rev. Dr. L. Krapf, Missionary of the Church Missionary Society in East Africa. With an Appendix, containing an Outline of a Suahili Grammar. Royal 8vo. cloth, pp. xl.-434. 1882. 30*s.*

Steere.—Short Specimens of the Vocabularies of Three Unpublished African Languages (Gindo, Zaramo, and Angazidja). Collected by Edward Steere, LL.D. 12mo. pp. 20. 6*d.*

Steere.—Collections for a Handbook of the Nyamwezi Language, as spoken at Unyanyembe. By Edward Steere, LL.D. Fcap. cloth, pp. 100. 1*s.* 6*d.*

Tindall.—A Grammar and Vocabulary of the Namaqua-Hottentot Language. By Henry Tindall, Wesleyan Missionary. 8vo. pp. 124, sewed. 6*s.*

Zulu Izaga; That is, Proverbs, or Out-of-the-Way Sayings of the Zulus. Collected, Translated, and interpreted by a Zulu Missionary. Crown 8vo. pp. iv. and 32, sewed. 2*s.* 6*d.*

AMERICAN LANGUAGES.

Byington.—Grammar of the Choctaw Language. By the Rev. Cyrus Byington. Edited from the Original MSS. in Library of the American Philosophical Society, by D. G. Brinton, M.D. Cr. 8vo. sewed, pp. 56. 7*s.* 6*d.*

Ellis.—Peruvia Scythica. The Quichua Language of Peru: its derivation from Central Asia with the American languages in general, and with the Turanian and Iberian languages of the Old World, including the Basque, the Lycian, and the Pre-Aryan language of Etruria. By Robert Ellis, B.D. 8vo. cloth, pp. xii. and 219. 1875. 6*s.*

Howse.—A GRAMMAR OF THE CREE LANGUAGE. With which is combined an analysis of the Chippeway Dialect. By JOSEPH HOWSE, Esq., F.R.G.S. 8vo. pp. xx. and 324, cloth. 7*s*. 6*d*.

Markham.—OLLANTA: A DRAMA IN THE QUICHUA LANGUAGE. Text, Translation, and Introduction, By CLEMENTS R. MARKHAM, F.R.G.S. Crown 8vo., pp. 128, cloth. 7*s*. 6*d*.

Matthews.—ETHNOLOGY AND PHILOLOGY OF THE HIDATSA INDIANS. By WASHINGTON MATTHEWS, Assistant Surgeon, U.S. Army. 8vo. cloth. £1 11*s*. 6*d*.

CONTENTS:—Ethnography, Philology, Grammar, Dictionary, and English-Hidatsa Vocabulary.

Nodal.—LOS VINCULOS DE OLLANTA Y CUSI-KCUYLLOR. DRAMA EN QUICHUA. Obra Compilada y Espurgada con la Version Castellana al Frente de su Testo por el Dr. JOSÉ FERNANDEZ NODAL, Abogado de los Tribunales de Justicia de la República del Perú. Bajo los Auspicios de la Redentora Sociedad de Filántropos para Mejoror la Suerte de los Aboríjenes Peruanos. Roy. 8vo. bds. pp. 70. 1874. 7*s*. 6*d*.

Nodal.—ELEMENTOS DE GRAMÁTICA QUICHUA Ó IDIOMA DE LOS YNCAS. Bajo los Auspicios de la Redentora, Sociedad de Filántropos para mejorar la suerte de los Aboríjenes Peruanos. Por el Dr. JOSE FERNANDEZ NODAL, Abogado de los Tribunales de Justicia de la República del Perú. Royal 8vo. cloth, pp. xvi. and 441. Appendix, pp. 9. £1 1*s*.

Ollanta: A DRAMA IN THE QUICHUA LANGUAGE. See under MARKHAM and under NODAL.

Pimentel. — CUADRO DESCRIPTIVO Y COMPARATIVO DE LAS LENGUAS INDÍGENAS DE MÉXICO, o Tratado de Filologia Mexicana. Par FRANCISCO PIMENTEL. 2 Edicion unica completa. 3 Volsume 8vo. *Mexico*, 1875. £2 2*s*.

Thomas.—THE THEORY AND PRACTICE OF CREOLE GRAMMAR. By J. J. THOMAS. Port of Spain (Trinidad), 1869. 1 vol. 8vo. bds. pp. viii. and 135. 12*s*.

ANGLO-SAXON.

March.—A COMPARATIVE GRAMMAR OF THE ANGLO-SAXON LANGUAGE; in which its forms are illustrated by those of the Sanskrit, Greek, Latin, Gothic, Old Saxon, Old Friesic, Old Norse, and Old High-German. By FRANCIS A. MARCH, LL.D. Demy 8vo. cloth, pp. xi. and 253. 1877. 10*s*.

Rask.—A GRAMMAR OF THE ANGLO-SAXON TONGUE. From the Danish of Erasmus Rask, Professor of Literary History in, and Librarian to, the University of Copenhagen, etc. By BENJAMIN THORPE. Third edition, corrected and improved, with Plate. Post 8vo. cloth, pp. vi. and 192. 1879. 5*s*. 6*d*.

Wright.—ANGLO-SAXON AND OLD-ENGLISH VOCABULARIES, Illustrating the Condition and Manners of our Forefathers, as well as the History of the Forms of Elementary Education, and of the Languages spoken in this Island from the Tenth Century to the Fifteenth. Edited by THOMAS WRIGHT, Esq., M.A., F.S.A., etc. Second Edition, edited, and collated, by RICHARD WULCKER. 8vo. pp. xii.–420 and iv.–486, cloth. 1884. 28*s*.

ARABIC.

Ahlwardt.—THE DIVÂNS OF THE SIX ANCIENT ARABIC POETS, Ennâbiga, 'Antara, Tarafa, Zuhair, 'Algama, and Imruolgais; chiefly according to the MSS. of Paris, Gotha, and Leyden, and the collection of their Fragments: with a complete list of the various readings of the Text. Edited by W. AHLWARDT, 8vo. pp. xxx. 340, sewed. 1870. 12*s.*

Alif Laîlat wa Laîlat.—THE ARABIAN NIGHTS. 4 vols. 4to. pp. 495, 493, 442, 434. Cairo, A.H. 1279 (1862). £3 3*s.*

This celebrated Edition of the Arabian Nights is now, for the first time, offered at a price which makes it accessible to Scholars of limited means.

Athar-ul-Adhâr—TRACES OF CENTURIES; or, Geographical and Historical Arabic Dictionary, by SELIM KHURI and SELIM SH-HADE. Geographical Parts I. to IV., Historical Parts I. and II. 4to. pp. 788 and 384. Price 7*s.* 6*d.* each part. [*In course of publication.*

Badger.—AN ENGLISH-ARABIC LEXICON, in which the equivalents for English words and Idiomatic Sentences are rendered into literary and colloquial Arabic. By GEORGE PERCY BADGER, D.C.L. 4to. cloth, pp. xii. and 1248. 1880. £4.

Butrus-al-Bustâny.—كتاب دائِرَةِ المَعارِفِ An Arabic Encylopædia of Universal Knowledge, by BUTRUS-AL-BUSTÂNY, the celebrated compiler of Mohît ul Mohît (محيط المحيط), and Katr el Mohît (قطر المحيط). This work will be completed in from 12 to 15 Vols., of which Vols. I. to VII. are ready, Vol. I. contains letter ا to اب; Vol. II. اب to ار; Vol. III. ار to اغ Vol. IV. اغ to اي Vol. V. با to بي Vol. VI. با to حر. Vol. VII. حر to دم. Small folio, cloth, pp. 800 each. £1 11*s.* 6*d.* per Vol.

Cotton.—ARABIC PRIMER. Consisting of 180 Short Sentences containing 30 Primary Words prepared according to the Vocal System of Studying Language. By General SIR ARTHUR COTTON, K.C.S.I. Cr. 8vo. cloth, pp. 38. 2*s.*

Hassoun.—THE DIWAN OF HATIM TAI. An Old Arabic Poet of the Sixth Century of the Christian Era. Edited by R. HASSOUN. With Illustrations. 4to. pp. 43. 3*s.* 6*d.*

Jami, Mulla.—SALAMAN U ABSAL. An Allegorical Romance; being one of the Seven Poems entitled the Haft Aurang of Mullā Jāmī, now first edited from the Collation of Eight Manuscripts in the Library of the India House, and in private collections, with various readings, by FORBES FALCONER, M.A., M.R.A.S. 4to. cloth, pp. 92. 1850. 7*s.* 6*d.*

Koran (The). Arabic text, lithographed in Oudh, A.H. 1284 (1867). 16mo. pp. 942. 9*s.*

Koran (The); commonly called The Alcoran of Mohammed. Translated into English immediately from the original Arabic. By GEORGE SALE, Gent. To which is prefixed the Life of Mohammed. Crown 8vo. cloth, pp. 472. 7*s.*

Koran.—EXTRACTS FROM THE CORAN IN THE ORIGINAL, WITH ENGLISH RENDERING. Compiled by Sir WILLIAM MUIR, K.C.S.I., LL.D., Author of the "Life of Mahomet." Crown 8vo. pp. 58, cloth. 1880. 3*s.* 6*d.*

Ko-ran (Selections from the).—See "Trübner's Oriental Series." p. 3.

Leitner.—INTRODUCTION TO A PHILOSOPHICAL GRAMMAR OF ARABIC. Being an Attempt to Discover a Few Simple Principles in Arabic Grammar. By G. W. LEITNER. 8vo. sewed, pp. 52. *Lahore.* 4*s.*

Morley.—A DESCRIPTIVE CATALOGUE of the HISTORICAL MANUSCRIPTS in the ARABIC and PERSIAN LANGUAGES preserved in the Library of the Royal Asiatic Society of Great Britain and Ireland. By WILLIAM H. MORLEY, M.R.A.S. 8vo. pp. viii. and 160, sewed. London, 1854. 2*s*. 6*d*.

Muhammed.—THE LIFE OF MUHAMMED. Based on Muhammed Ibn Ishak. By Abd El Malik Ibn Hisham. Edited by Dr. FERDINAND WÜSTENFELD. The Arabic Text. 8vo. pp. 1026, sewed. Price 21*s*. Introduction, Notes, and Index in German. 8vo. pp. lxxii. and 266, sewed. 7*s*. 6*d*. Each part sold separately.

The text based on the Manuscripts of the Berlin, Leipsic, Gotha and Leyden Libraries, has been carefully revised by the learned editor, and printed with the utmost exactness.

Newman.—A HANDBOOK OF MODERN ARABIC, consisting of a Practical Grammar, with numerous Examples, Dialogues, and Newspaper Extracts, in a European Type. By F. W. NEWMAN, Emeritus Professor of University College, London; formerly Fellow of Balliol College, Oxford. Post 8vo. pp. xx. and 192, cloth. 1866. 6*s*.

Newman.—A DICTIONARY OF MODERN ARABIC.—1. Anglo-Arabic Dictionary. 2. Anglo-Arabic Vocabulary. 3. Arabo-English Dictionary. By F. W. NEWMAN, Emeritus Professor of University College, London. In 2 vols. crown 8vo., pp. xvi. and 376—464, cloth. £1 1*s*.

Palmer.—THE SONG OF THE REED; and other Pieces. By E. H. PALMER, M.A., Cambridge. Crown 8vo. cloth, pp. 208. 1876. 5*s*.

Among the Contents will be found translations from Hafiz, from Omer el Kheiyâm, and from other Persian as well as Arabic poets.

Palmer.—HINDUSTANI, PERSIAN, AND ARABIC GRAMMAR SIMPLIFIED. B. E. H. PALMER, M.A., Professor of Arabic at the University of Cambridge, and Examiner in Hindustani for H.M. Civil Service Commissioners. Crown 8vo. pp. viii.-104, cloth. 1882. 5*s*.

Rogers.—NOTICE ON THE DINARS OF THE ABBASSIDE DYNASTY. By EDWARD THOMAS ROGERS, late H.M. Consul, Cairo. 8vo. pp. 44, with a Map and four Autotype Plates. 5*s*.

Schemeil.—EL MUBTAKER; or, First Born. (In Arabic, printed at Beyrout). Containing Five Comedies, called Comedies of Fiction, on Hopes and Judgments, in Twenty-six Poems of 1092 Verses, showing the Seven Stages of Life, from man's conception unto his death and burial. By EMIN IBRAHIM SCHEMEIL. In one volume, 4to. pp. 166, sewed. 1870. 5*s*.

Syed Ahmad.—A SERIES OF ESSAYS ON THE LIFE OF MOHAMMED, and Subjects subsidiary thereto. By SYED AHMAD KHAN BAHADOR, C.S.I., Author of the "Mohammedan Commentary on the Holy Bible," Honorary Member of the Royal Asiatic Society, and Life Honorary Secretary to the Allygurh Scientific Society. 8vo. pp. 532, with 4 Genealogical Tables, 2 Maps, and a Coloured Plate, handsomely bound in cloth. 1870. £1 10*s*.

Wherry.—Commentary on the Quran. See Trübner's Oriental Series, page 5.

ASSAMESE.

Bronson.—A DICTIONARY IN ASSAMESE AND ENGLISH. Compiled by M. BRONSON, American Baptist Missionary. 8vo. calf, pp. viii. and 609. £2 2*s*.

ASSYRIAN (CUNEIFORM, ACCAD, BABYLONIAN).

Budge.—ASSYRIAN TEXTS, Selected and Arranged, with Philological Notes. By E. A. BUDGE, B.A., M.R.A.S., Assyrian Exhibitioner, Christ's College, Cambridge. (New Volume of the Archaic Classics.) Crown 4to. cloth, pp. viii. and 44. 1880. 7*s*. 6*d*.

Budge.—THE HISTORY OF ESARHADDON. See "Trübner's Oriental Series," p. 4.

Catalogue (A), of leading Books on Egypt and Egyptology, and on Assyria and Assyriology, to be had at the affixed prices, of Trübner and Co. pp. 40. 1880. 1*s*.

Clarke.—RESEARCHES IN PRE-HISTORIC AND PROTO-HISTORIC COMPARATIVE PHILOLOGY, MYTHOLOGY, AND ARCHÆOLOGY, in connexion with the Origin of Culture in America and the Accad or Sumerian Families. By HYDE CLARKE. Demy 8vo. sewed, pp. xi. and 74. 1875. 2*s*. 6*d*.

Cooper.—An Archaic Dictionary, Biographical, Historical and Mythological; from the Egyptian and Etruscan Monuments, and Papyri. By W. R. COOPER. London, 1876. 8vo. cloth. 15*s*.

Hincks.—SPECIMEN CHAPTERS OF AN ASSYRIAN GRAMMAR. By the late Rev. E. HINCKS, D.D., Hon. M.R.A.S. 8vo., sewed, pp. 44. 1*s*.

Lenormant (F.)—CHALDEAN MAGIC; its Origin and Development. Translated from the French. With considerable Additions by the Author. London, 1877. 8vo. pp. 440. 12*s*.

Luzzatto.—GRAMMAR OF THE BIBLICAL CHALDAIC LANGUAGE AND THE TALMUD BABYLONICAL IDIOMS. By S. D. LUZZATTO. Translated from the Italian by J. S. GOLDAMMER. Cr. 8vo. cl., pp. 122. 7*s*. 6*d*.

Rawlinson.—NOTES ON THE EARLY HISTORY OF BABYLONIA. By Colonel RAWLINSON, C.B. 8vo. sd., pp. 48. 1*s*.

Rawlinson.—A COMMENTARY ON THE CUNEIFORM INSCRIPTIONS OF BABYLONIA AND ASSYRIA, including Readings of the Inscription on the Nimrud Obelisk, and Brief Notice of the Ancient Kings of Nineveh and Babylon, by Major H. C. RAWLINSON. 8vo. pp. 84, sewed. London, 1850. 2*s*. 6*d*.

Rawlinson.—INSCRIPTION OF TIGLATH PILESER I., KING OF ASSYRIA, B.C. 1150, as translated by Sir H. RAWLINSON, FOX TALBOT, Esq., Dr. HINCKS. and Dr. OPPERT. Published by the Royal Asiatic Society. 8vo. sd., pp. 74. 2*s*.

Rawlinson.—OUTLINES OF ASSYRIAN HISTORY, from the Inscriptions of Nineveh. By Lieut. Col. RAWLINSON, C.B., followed by some Remarks by A. H. LAYARD, Esq., D.C.L. 8vo., pp. xliv., sewed. London, 1852. 1*s*.

Records of the Past: being English Translations of the Assyrian and the Egyptian Monuments. Published under the sanction of the Society of Biblical Archæology. Edited by S. BIRCH. Vols. 1 to 12. 1874 to 1879. £1 11*s*. 6*d*. or 3*s*. 6*d*. each vol.

Renan.—AN ESSAY ON THE AGE AND ANTIQUITY OF THE BOOK OF NABATHÆAN AGRICULTURE. To which is added an Inaugural Lecture on the Position of the Shemitic Nations in the History of Civilization. By M. ERNEST RENAN, Membre de l'Institut. Crown 8vo., pp. xvi. and 148, cloth. 3*s*. 6*d*.

Sayce.—AN ASSYRIAN GRAMMAR FOR COMPARATIVE PURPOSES. By A. H. SAYCE, M.A. 12mo. cloth, pp. xvi. and 188. 1872. 7*s*. 6*d*.

Sayce.—AN ELEMENTARY GRAMMAR and Reading Book of the Assyrian Language, in the Cuneiform Character: containing the most complete Syllabary yet extant, and which will serve also as a Vocabulary of both Accadian and Assyrian. London, 1875. 4to. cloth. 9*s*.

Sayce.—LECTURES upon the Assyrian Language and Syllabary London, 1877. Large 8vo. 9*s*. 6*d*.

Sayce.—BABYLONIAN LITERATURE. Lectures. London, 1877. 8vo. 4*s*.

Smith.—THE ASSYRIAN EPONYM CANON; containing Translations of the Documents of the Comparative Chronology of the Assyrian and Jewish Kingdoms, from the Death of Solomon to Nebuchadnezzar. By E. SEITH. London, 1876. 8vo. 9*s*.

AUSTRALIAN LANGUAGES.

Grey.—HANDBOOK OF AFRICAN, AUSTRALIAN, AND POLYNESIAN PHILOLOGY, as represented in the Library of His Excellency Sir George Grey, K.C.B., Her Majesty's High Commissioner of the Cape Colony. Classed, Annotated, and Edited by Sir GEORGE GREY and Dr. H. I. BLEEK.

Vol. I. Part 1.—South Africa. 8vo. pp. 186. 20*s*.
Vol. I. Part 2.—Africa (North of the Tropic of Capricorn). 8vo. pp. 70. 4*s*.
Vol. I. Part 3.—Madagascar. 8vo. pp. 24 1*s*.
Vol. II. Part 1.—Australia. 8vo. pp. iv. and 44. 3*s*.
Vol. II. Part 2.—Papuan Languages of the Loyalty Islands and New Hebrides, comprising those of the Islands of Nengone, Lifu, Aneitum, Tana, and others. 8vo. pp. 12. 1*s*.
Vol. II. Part 3.—Fiji Islands and Rotuma (with Supplement to Part II., Papuan Languages, and Part I., Australia). 8vo. pp. 34. 2*s*.
Vol. II. Part 4.—New Zealand, the Chatham Islands, and Auckland Islands. 8vo. pp. 76. 7*s*.
Vol. II. Part 4 (*continuation*).—Polynesia and Borneo. 8vo. pp. 77-154. 7*s*.
Vol. III. Part 1.—Manuscripts and Incunables. 8vo. pp. viii. and 24. 2*s*.
Vol. IV. Part 1.—Early Printed Books. England. 8vo. pp. vi. and 266. 12*s*.

Ridley.—KÁMILARÓI, AND OTHER AUSTRALIAN LANGUAGES. By the Rev. WILLIAM RIDLEY, M.A. Second Edition. Revised and enlarged by the Author; with Comparative Tables of Words from twenty Australian Languages, and Songs, Traditions, Laws, and Customs of the Australian Race. Small 4to., cloth, pp. vi. and 172. 1877. 10*s*. 6*d*.

BASQUE.

Van Eys.—OUTLINES OF BASQUE GRAMMAR. By W. J. VAN EYS. Crown 8vo. pp. xii. and 52, cloth. 1883. 3*s*. 6*d*.

BENGALI.

Browne.—A BÁNGÁLI PRIMER, in Roman Character. By J. F. BROWNE, B.C.S. Crown 8vo. pp. 32, cloth. 1881. 2*s*,

Charitabali (The); OR, INSTRUCTIVE BIOGRAPHY BY ISVARACHANDRA VIDYÁSAGARA. With a Vocabulary of all the Words occurring in the Text, by J. F. BLUMHARDT, Bengali Lecturer University College, London; and Teacher of Bengali Cambridge University. 12mo. pp. 120-iv.-48, cloth. 1884. 5*s*.

Mitter.—BENGALI AND ENGLISH DICTIONARY for the Use of Schools. Revised and improved. 8vo. cloth. Calcutta, 1860. 7*s*. 6*d*.

Sykes.—ENGLISH AND BENGALI DICTIONARY for the Use of Schools. Revised by GOPEE KISSEN MITTER. 8vo. cloth. Calcutta, 1874. 7*s*. 6*d*.

Yates.—A BENGÁLÍ GRAMMAR. By the late Rev. W. YATES, D.D. Reprinted, with improvements, from his Introduction to the Bengálí Language. Edited by I. WENGER. Fcap. 8vo. bds, pp. iv. and 150. Calcutta, 1864. 4*s*.

BRAHOE.

Bellew.—FROM THE INDUS TO THE TIGRIS. A Narrative; together with together with a Synoptical Grammar and Vocabulasy of the Brahoe language. See p. 19.

BURMESE.

Hough's GENERAL OUTLINES OF GEOGRAPHY (in Burmese). Re-written and enlarged by Rev. JAS. A. HASWELL. Large 8vo. pp. 368. Rangoon, 1874. 9*s.*

Judson.—A DICTIONARY, English and Burmese, Burmese and English. By A. JUDSON. 2 vols. 8vo. pp. iv. and 968, and viii. and 786. £3 3*s.*

Sloan.—A PRACTICAL METHOD with the Burmese Language. By W. H. SLOAN. Large 8vo. pp. 232. Rangoon, 1876. 12*s.* 6*d.*

CHINESE.

Acheson.—AN INDEX TO DR. WILLIAMS'S "SYLLABIC DICTIONARY OF THE CHINESE LANGUAGE." Arranged according to Sir THOMAS WADE'S System of Orthography. Royal 8vo. pp. viii. and 124. Half bound. Hongkong. 1879. 18*s.*

Baldwin.—A MANUAL OF THE FOOCHOW DIALECT. By Rev. C. C. BALDWIN, of the American Board Mission. 8vo. pp. viii.–256. 18*s.*

Balfour.—TAOIST TEXTS. See page 34.

Balfour.—THE DIVINE CLASSIC OF NAN-HUA. Being the Works of Chuang-Tsze, Taoist Philosopher. With an Excursus, and copious Annotations in English and Chinese. By H. BALFOUR, F.R.G.S. Demy 8vo. pp. xxxviii. and 426, cloth. 1881. 14*s.*

Balfour.—WAIFS AND STRAYS FROM THE FAR EAST; being a Series of Disconnected Essays on Matters relating to China. By F. H. Balfour. 8vo. pp. 224, cloth. 1876. 10*s.* 6*d.*

Beal.—THE BUDDHIST TRIPITAKA, as it is known in China and Japan. A Catalogue and Compendious Report. By SAMUEL BEAL, B.A. Folio, sewed, pp. 117. 7*s.* 6*d.*

Beal.—THE DHAMMAPADA. See "Trübner's Oriental Series," page 3.

Beal.—Buddhist Literature. See p. 32.

Bretschneider.—See page 21.

Chalmers.—THE SPECULATIONS ON METAPHYSICS, POLITY, AND MORALITY OF "THE OLD PHILOSOPHER" LAU TSZE. Translated from the Chinese, with an Introduction by John Chalmers, M.A. Fcap. 8vo. cloth, xx. and 62. 4*s.* 6*d.*

Chalmers.—THE ORIGIN OF THE CHINESE; an Attempt to Trace the connection of the Chinese with Western Nations, in their Religion, Superstitions, Arts Language, and Traditions. By JOHN CHALMERS, A.M. Foolscap 8vo. cloth, pp. 78. 5*s.*

Chalmers.—A CONCISE KHANG-HSI CHINESE DICTIONARY. By the Rev. J. CHALMERS, LL.D., Canton. Three Vols. Royal 8vo. bound in Chinese style, pp. 1000. £1 10*s.*

Chalmers. — THE STRUCTURE OF CHINESE CHARACTERS, UNDER 300 Primary Forms; after the Shwoh-wan, 100 A.D., and the Phonetic Shwoh-wan 1833. By JOHN CHALMERS, M.A., LL.D. 8vo. pp. x-199, with a plate, cloth. 1882. 12*s.* 6*d.*

China Review; OR, NOTES AND QUERIES ON THE FAR EAST. Published bi-monthly. Edited by E. J. EITEL. 4to. Subscription, £1 10*s.* per volume.

Dennys.—A HANDBOOK OF THE CANTON VERNACULAR OF THE CHINESE LANGUAGE. Being a Series of Introductory Lessons, for Domestic and Business Purposes. By N. B. DENNYS, M.R.A.S., Ph.D. 8vo. cloth, pp. 4, 195, and 31. £1 10*s.*

Dennys.—The Folk-Lore of China, and its Affinities with that of the Aryan and Semitic Races. By N. B. Dennys, Ph.D., F.R.G.S., M.R.A.S., author of "A Handbook of the Canton Vernacular," etc. 8vo. cloth, pp. 168. 10*s.* 6*d.*

Douglas.—Chinese Language and Literature. Two Lectures delivered at the Royal Institution, by R. K. Douglas, of the British Museum, and Professor of Chinese at King's College. Cr. 8vo. cl. pp. 118. 1875. 5*s.*

Douglas.—Chinese-English Dictionary of the Vernacular or Spoken Language of Amoy, with the principal variations of the Chang-Chew and Chin-Chew Dialects. By the Rev. Carstairs Douglas, M.A., LL.D., Glasg., Missionary of the Presbyterian Church in England. 1 vol. High quarto, cloth, double columns, pp. 632. 1873. £3 3*s.*

Douglas.—The Life of Jenghiz Khan. Translated from the Chinese, with an Introduction, by Robert Kennaway Douglas, of the British Museum, and Professor of Chinese, King's College, London. Cr. 8vo. cloth, pp. xxxvi.–106. 1877. 5*s.*

Edkins.—A Grammar of Colloquial Chinese, as exhibited in the Shanghai Dialect. By J. Edkins, B.A. Second edition, corrected. 8vo. half-calf, pp. viii. and 225. Shanghai, 1868. 21*s.*

Edkins.—A Vocabulary of the Shanghai Dialect. By J. Edkins. 8vo. half-calf, pp. vi. and 151. Shanghai, 1869. 21*s.*

Edkins.—Religion in China. A Brief Account of the Three Religions of the Chinese. By Joseph Edkins, D.D. Post 8vo. cloth. 7*s.* 6*d.*

Edkins.—A Grammar of the Chinese Colloquial Language, commonly called the Mandarin Dialect. By Joseph Edkins. Second edition. 8vo. half-calf, pp. viii. and 279. Shanghai, 1864. £1 10*s.*

Edkins.—Introduction to the Study of the Chinese Characters. By J. Edkins, D.D., Peking, China. Roy. 8vo. pp. 340, paper boards. 18*s.*

Edkins.—China's Place in Philology. An attempt to show that the Languages of Europe and Asia have a common origin. By the Rev. Joseph Edkins. Crown 8vo, pp. xxiii.—403, cloth. 10*s.* 6*d.*

Edkins.—Chinese Buddhism. See "Trübner's Oriental Series," p. 4.

Edkins.—Progressive Lessons in the Chinese Spoken Language, with Lists of Common Words and Phrases, and an Appendix containing the Laws of Tones in the Pekin Dialect. Fourth Edition, 8vo. Shanghai, 1881. 14*s.*

Eitel.—A Chinese Dictionary in the Cantonese Dialect. By Ernest John Eitel, Ph.D. Tubing. Will be completed in four parts. Parts I. to IV. 8vo. sewed, 12*s.* 6*d.* each.

Eitel.—Handbook for the Student of Chinese Buddhism. By the Rev. E. J. Eitel, of the London Missionary Society. Cr. 8vo. pp. viii., 224, cl. 18*s*

Eitel.—Feng-Shui: or, The Rudiments of Natural Science in China. By Rev. E. J. Eitel, M.A., Ph.D. Demy 8vo. sewed, pp. vi. and 84. 6*s.*

Faber.—A systematical Digest of the Doctrines of Confucius, according to the Analects, Great Learning, and Doctrine of the Mean, with an Introduction on the Authorities upon Confucius and Confucianism. By Ernst Faber, Rhenish Missionary. Translated from the German by P. G. von Möllendorff. 8vo. sewed, pp. viii. and 131. 1875. 12*s.* 6*d.*

Faber.—Introduction to the Science of Chinese Religion. A Critique of Max Müller and other Authors. By E. Faber. 8vo. paper, pp. xii. and 154. Hong Kong, 1880. 7*s.* 6*d.*

Faber.—THE MIND OF MENCIUS. See "Trübner's Oriental Series,' page 4.

Ferguson.—CHINESE RESEARCHES. First Part: Chinese Chronology and Cycles. By T. FERGUSON. Crown 8vo. pp. vii. and 274, sewed. 1880 10*s*. 6*d*.

Giles.—A DICTIONARY OF COLLOQUIAL IDIOMS IN THE MANDARIN DIALECT. By HERBERT A. GILES. 4to. pp. 65. £1 8*s*.

Giles.—THE SAN TZU CHING; or, Three Character Classic; and the Ch'Jen Tsu Wen; or, Thousand Character Essay. Metrically Translated by HERBERT A. GILES. 12mo. pp. 28. 2*s*. 6*d*.

Giles.—SYNOPTICAL STUDIES IN CHINESE CHARACTER. By HERBERT A. GILES. 8vo. pp. 118. 15*s*.

Giles.—CHINESE SKETCHES. By HERBERT A. GILES, of H.B.M.'s China Consular Service. 8vo. cl., pp. 204. 10*s*. 6*d*.

Giles.—A GLOSSARY OF REFERENCE ON SUBJECTS CONNECTED WITH THE Far East. By H. A. GILES, of H.M. China Consular Service. 8vo. sewed, pp. v.-183. 7*s*. 6*d*.

Giles.—CHINESE WITHOUT A TEACHER. Being a Collection of Easy and Useful Sentences in the Mandarn Dialect. With a Vocabulary. By HERBERT A. GILES. 12mo. pp. 60. 6*s*. 6*d*.

Hernisz.—A GUIDE TO CONVERSATION IN THE ENGLISH AND CHINESE LANGUAGES, for the use of Americans and Chinese in California and elsewhere. By STANISLAS HERNISZ. Square 8vo. pp. 274, sewed. 10*s*. 6*d*.

The Chinese characters contained in this work are from the collections of Chinese groups engraved on steel, and cast into moveable types, by Mr. Marcellin Legrand, engraver of the Imperial Printing Office at Paris. They are used by most of the missions to China.

Kidd.—CATALOGUE OF THE CHINESE LIBRARY OF THE ROYAL ASIATIC SOCIETY. By the Rev. S. KIDD. 8vo. pp. 58, sewed. 1*s*.

Legge.—THE CHINESE CLASSICS. With a Translation, Critical and Exegetical Notes, Prolegomena, and Copious Indexes. By JAMES LEGGE, D.D., of the London Missionary Society. In seven vols.

Vol. I. containing Confucian Analects, the Great Learning, and the Doctrine of the Mean. 8vo. pp. 526, cloth. £2 2*s*.

Vol. II., containing the Works of Mencius. 8vo. pp. 634, cloth. £2 2*s*.

Vol. III. Part I. containing the First Part of the Shoo-King, or the Books of Tang, the Books of Yu, the Books of Hea, the Books of Shang, and the Prolegomena. Royal 8vo. pp. viii. and 280, cloth. £2 2*s*.

Vol. III. Part II. containing the Fifth Part of the Shoo-King, or the Books of Chow, and the Indexes. Royal 8vo. pp. 281—736, cloth. £2 2*s*.

Vol. IV. Part I. containing the First Part of the She-King, or the Lessons from the States; and the Prolegomena. Royal 8vo. cloth, pp. 182–244. £2 2*s*.

Vol. IV. Part II. containing the 2nd, 3rd and 4th Parts of the She-King, or the Minor Odes of the Kingdom, the Greater Odes of the Kingdom, the Sacrificial Odes and Praise-Songs, and the Indexes. Royal 8vo. cloth, pp. 540. £2 2*s*.

Vol. V. Part I. containing Dukes Yin, Hwan, Chwang, Min, He, Wan, Seuen, and Ch'ing; and the Prolegomena. Royal 8vo. cloth, pp. xii., 148 and 410. £2 2*s*.

Vol. V. Part II. Contents:—Dukes Seang, Ch'aon, Ting, and Gal, with Tso's Appendix, and the Indexes. Royal 8vo. cloth, pp. 526. £2 2*s*.

Legge.—The Chinese Classics. Translated into English. With Preliminary Essays and Explanatory Notes. By James Legge, D.D., LL.D. Crown 8vo. cloth. Vol. I. The Life and Teachings of Confucius. pp. vi. and 338. 10*s*. 6*d*. Vol. II. The Life and Works of Mencius. pp. 412. 12*s*. Vol. III. The She King, or The Book of Poetry. pp. viii. and 432. 12*s*.

Legge.—Inaugural Lecture on the Constituting of a Chinese Chair in the University of Oxford. Delivered in the Sheldonian Theatre, Oct. 27th, 1876, by Rev. James Legge, M.A., LL.D., Professor of the Chinese Language and Literature at Oxford. 8vo. pp. 28, sewed. 6*d*.

Legge.—Confucianism in Relation to Christianity. A Paper Read before the Missionary Conference in Shanghai, on May 11, 1877. By Rev. James Legge, D.D., LL.D. 8vo. sewed, pp. 12. 1877. 1*s*. 6*d*.

Legge.—A Letter to Professor Max Müller, chiefly on the Translation into English of the Chinese Terms *Ti* and *Shang Ti*. By J. Legge, Professor of Chinese Language and Literature in the University of Oxford. Crown 8vo. sewed, pp. 30. 1880. 1*s*.

Leland.—Fusang; or, the Discovery of America by Chinese Buddhist Priests in the Fifth Century. By Charles G. Leland. Cr. 8vo. cloth, pp. xix. and 212. 1875. 7*s*. 6*d*.

Leland.—Pidgin-English Sing-Song; or Songs and Stories in the China-English Dialect. With a Vocabulary. By Charles G. Leland. Crown 8vo. pp. viii. and 140, cloth. 1876. 5*s*.

Lobscheid.—English and Chinese Dictionary, with the Punti and Mandarin Pronunciation. By the Rev. W. Lobscheid, Knight of Francis Joseph, C.M.I.R.G.S.A., N.Z.B.S.V., etc. Folio, pp. viii. and 2016. In Four Parts. £8 8*s*.

Lobscheid.—Chinese and English Dictionary, Arranged according to the Radicals. By the Rev. W. Lobscheid, Knight of Francis Joseph, C.M.I.R.G.S.A., N.Z.B.S.V., &c. 1 vol. imp. 8vo. double columns, pp. 600 bound. £2 8*s*.

M'Clatchie.—Confucian Cosmogony. A Translation (with the Chinese Text opposite) of section 49 (Treatise on Cosmogony) of the "Complete Works" of the Philosopher Choo-Foo-Tze, with Explanatory Notes. By the Rev. Thomas M'Clatchie, M.A. Small 4to. pp. xviii. and 162. 1874. £1 1*s*.

Macgowan.—A Manual of the Amoy Colloquial. By Rev. J. Macgowan, of the London Missionary Society. Second Edition. 8vo. half-bound, pp. 206. Amoy, 1880. £1 10*s*.

Macgowan.—English and Chinese Dictionary of the Amoy Dialect. By Rev. J. Macgowan, London Missionary Society. Small 4to. half-bound, pp. 620. Amoy, 1883. £3 3*s*.

Maclay and Baldwin.—An Alphabetic Dictionary of the Chinese Language in the Foochow Dialect. By Rev. R. S. Maclay, D.D., of the Methodist Episcopal Mission, and Rev. C. C. Baldwin, A.M., of the American Board of Mission. 8vo. half-bound, pp. 1132. Foochow, 1871. £4 4*s*.

Mayers.—The Anglo-Chinese Calendar Manual. A Handbook of Reference for the Determination of Chinese Dates during the period from 1860 to 1879. With Comparative Tables of Annual and Mensual Designations, etc. Compiled by W. F. Mayers, Chinese Secretary, H.B.M.'s Legation, Peking. 2nd Edition. Sewed, pp. 28. 7*s*. 6*d*.

Mayers.—THE CHINESE GOVERNMENT. A Manual of Chinese Titles, Categorically arranged, and Explained with an Appendix. By W. F. MAYERS, Chinese Secretary to H.B.M.'s Legation at Peking. Royal 8vo. cloth, pp. viii.-160. 1878. £1 10*s*.

Medhurst.—CHINESE DIALOGUES, QUESTIONS, and FAMILIAR SENTENCES, literally translated into English, with a view to promote commercial intercourse and assist beginners in the Language. By the late W. H. MEDHURST, D.D. A new and enlarged Edition. 8vo. pp. 226. 18*s*.

Möllendorff.—MANUAL OF CHINESE BIBLIOGRAPHY, being a List of Works and Essays relating to China. By P. G. and O. F. VON MÖLLENDORFF, Interpreters to H.I.G.M.'s Consulates at Shanghai and Tientsin. 8vo. pp. viii. and 378. £1 10*s*.

Morrison.—A DICTIONARY OF THE CHINESE LANGUAGE. By the Rev. R. MORRISON, D.D. Two vols. Vol. I. pp. x. and 762; Vol. II. pp. 828, cloth. Shanghae, 1865. £6 6*s*.

Peking Gazette.—Translation of the Peking Gazette for 1872, 1873, 1874, 1875, 1876, 1877, 1878, and 1879. 8vo. cloth. 10*s*. 6*d*. each.

Piry.—LE SAINT EDIT, Etude de Littérature Chinoise. Préparée par A. THEOPHILE PIRY, du Service des Douanes Maritimes de Chine. Chinese Text with French Translation. 4to. cloth, pp. xx. and 320. 21*s*.

Playfair.—CITIES AND TOWNS OF CHINA. 25*s*. See page 27.

Ross.—A MANDARIN PRIMER. Being Easy Lessons for Beginners, Transliterated according to the European mode of using Roman Letters. By Rev. JOHN ROSS, Newchang. 8vo. wrapper, pp. 122. 7*s*. 6*d*.

Rudy.—THE CHINESE MANDARIN LANGUAGE, after Ollendorff's New Method of Learning Languages. By CHARLES RUDY. In 3 Volumes. Vol. I. Grammar. 8vo. pp. 248. £1 1*s*.

Scarborough.—A COLLECTION OF CHINESE PROVERBS. Translated and Arranged by WILLIAM SCARBOROUGH, Wesleyan Missionary, Hankow. With an Introduction, Notes, and Copious Index. Cr. 8vo. pp. xliv. and 278. 10*s*.6*d*

Smith.—A VOCABULARY OF PROPER NAMES IN CHINESE AND ENGLISH. of Places, Persons, Tribes, and Sects, in China, Japan, Corea, Assam, Siam, Burmah, The Straits, and adjacent Countries. By F. PORTER SMITH, M.B., London, Medical Missionary in Central China. 4to. half-bound, pp. vi., 72, and x. 1870. 10*s*. 6*d*.

Stent.—A CHINESE AND ENGLISH VOCABULARY IN THE PEKINESE DIALECT. By G. E. STENT. Second Edition, 8vo. pp. xii.-720, half bound. 1877. £2.

Stent.—A CHINESE AND ENGLISH POCKET DICTIONARY. By G. E. STENT. 16mo. pp. 250. 1874. 15*s*.

Stent.—THE JADE CHAPLET, in Twenty-four Beads. A Collection of Songs, Ballads, etc. (from the Chinese). By GEORGE CARTER STENT, M.N.C.B.R.A.S., Author of "Chinese and English Vocabulary," "Chinese and English Pocket Dictionary," "Chinese Lyrics," "Chinese Legends," etc. Cr. 8o. cloth, pp. 176. 5*s*.

Vaughan.—The Manners and Customs of the Chinese of the Straits Settlements. By J. D. VAUGHAN. Royal 8vo. boards. Singapore, 1879. 7*s*. 6*d*.

Vissering.—ON CHINESE CURRENCY. Coin and Paper Money. With a Facsimile of a Bank Note. By W. Vessering. Royal 8vo. cloth, pp. xv. and 219. Leiden, 1877. 18*s*.

Williams.—A SYLLABIC DICTIONARY OF THE CHINESE LANGUAGE, arranged according to the Wu-Fang Yuen Yin, with the pronunciation of the Characters as heard in Peking, Canton, Amoy, and Shanghai. By S. WELLS WILLIAMS. 4to. cloth, pp. lxxxiv. and 1252. 1874. £5 5*s.*

Wylie.—NOTES ON CHINESE LITERATURE; with introductory Remarks on the Progressive Advancement of the Art; and a list of translations from the Chinese, into various European Languages. By A. WYLIE, Agent of the British and Foreign Bible Society in China. 4to. pp. 296, cloth. Price, £1 16*s.*

COREAN.

Ross—A COREAN PRIMER. Being Lessons in Corean on all Ordinary Subjects. Transliterated on the principles of the Mandarin Primer by the same author. By the Rev. JOHN ROSS, Newchang. Demy 8vo. stitched. pp. 90. 10*s.*

DANISH.

Otté.—HOW TO LEARN DANO-NORWEGIAN. A Manual for Students of Dano-Norwegian, and especially for Travellers in Scandinavia. Based upon the Ollendorffian System of teaching languages, and adapted for Self-Instruction. By E. C. OTTE. Second Edition. Crown 8vo. pp. xx.-338, cloth. 1884. 7*s.* 6*d.* (Key to the Exercises, pp. 84, cloth, price 3*s.*)

Otté.—SIMPLIFIED GRAMMAR OF THE DANISH LANGUAGE. By E. C. OTTE. Crown 8vo. pp. viii.-66, cloth. 1884. 2*s.* 6*d.*

EGYPTIAN (COPTIC, HIEROGLYPHICS).

Birch.—EGYPTIAN TEXTS: I. Text, Transliteration and Translation —II. Text and Transliteration.—III. Text dissected for analysis.—IV. Determinatives, etc. By S. Birch. London, 1877. Large 8vo. 12*s.*

Catalogue (**C**) of leading Books on Egypt and Egyptology on Assyria and Assyriology. To be had at the affixed prices of Trübner and Co. 8vo., pp. 40. 1880. 1*s.*

Chabas.—LES PASTEURS EN EGYPTE.—Mémoire Publié par l'Academie Royale des Sciences à Amsterdam. By F. CHABAS. 4to. sewed, pp. 56. Amsterdam, 1868. 6*s.*

Clarke.—MEMOIR ON THE COMPARATIVE GRAMMAR OF EGYPTIAN, COPTIC, AND UDE. By HYDE CLARKE, Cor. Member American Oriental Society; Mem. German Oriental Society, etc., etc. Demy 8vo. sd., pp. 32. 2*s.*

Egyptologie.—(Forms also the Second Volume of the First Bulletin of the Congrès Provincial des Orientalistes Français.) 8vo. sewed, pp. 604, with Eight Plates. Saint-Etiene, 1880. 8*s.* 6*d.*

Lieblein.—RECHERCHES SUR LA CHRONOLOGIE EGYPTIENNE d'après les listes Généalogiques. By J. LIEBLEIN. Roy. 8vo. sewed, pp. 147, with Nine Plates. Christiana, 1873. 10*s.*

Records of the Past, being English Translations of the Assyrian and the Egyptian Monuments. *Published under the Sanction of the Society of Biblical Archæology.* Edited by Dr. S. Birch.

Vols. I. to XII., 1874–79. 3*s.* 6*d.* each. (Vols. I., III., V., VII., IX., XI., contain Assyrian Texts.)

Renouf.—Elementary Grammar of the Ancient Egyptian Language, in the Hieroglyphic Type. By Le Page Renouf. 4to., cloth. 1875. 12*s.*

ENGLISH (Early and Modern English and Dialects).

Ballad Society (The).—Subscription—Small paper, one guinea, and large paper, three guineas, per annum. List of publications on application.

Boke of Nurture (The). By John Russell, about 1460–1470 Anno Domini. The Boke of Keruynge. By Wynkyn de Worde, Anno Domini 1513. The Boke of Nurture. By Hugh Rhodes, Anno Domini 1577. Edited from the Originals in the British Museum Library, by Frederick J. Furnivall, M.A., Trinity Hall, Cambridge, Member of Council of the Philological and Early English Text Societies. 4to. half-morocco, gilt top, pp. xix. and 146, 28, xxviii. and 56. 1867. 1*l.* 11*s.* 6*d.*

Charnock.—Verba Nominalia; or Words derived from Proper Names. By Richard Stephen Charnock, Ph. Dr. F.S.A., etc. 8vo. pp. 326, cloth. 14*s.*

Charnock.—Ludus Patronymicus; or, the Etymology of Curious Surnames. By Richard Stephen Charnock, Ph.D., F.S.A., F.R.G.S. Crown 8vo., pp. 182, cloth. 7*s.* 6*d.*

Charnock.—A Glossary of the Essex Dialect. By R. S. Charnock. 8vo. cloth, pp. x. and 64, . 1880. 3*s.* 6*d.*

Chaucer Society (The). — Subscription, two guineas per annum. *List of Publications on application.*

Eger and Grime; an Early English Romance. Edited from Bishop Percy's Folio Manuscript, about 1650 a.d. By John W. Hales, M.A., Fellow and late Assistant Tutor of Christ's College, Cambridge, and Frederick J. Furnivall, M.A., of Trinity Hall, Cambridge. 1 vol. 4to., pp. 64, (only 100 copies printed), bound in the Roxburghe style. 10*s.* 6*d.*

Early English Text Society's Publications. Subscription, one guinea per annum.

1. Early English Alliterative Poems. In the West-Midland Dialect of the Fourteenth Century. Edited b R. Morris, Esq., from an unique Cottonian MS. 16*s.*
2. Arthur (about 1440 a.d.). Edited by F. J. Furnivall, Esq., from the Marquis of Bath's unique MS. 4*s.*
3. Ane Compendious and Breue Tractate concernyng ye Office and Dewtie of Kyngis, etc. By William Lauder. (1556 a.d.) Edited by F. Hall, Esq., D.C.L. 4*s.*
4. Sir Gawayne and the Green Knight (about 1320-30 a.d.). Edited by R. Morris, Esq., from an unique Cottonian MS. 10*s.*

5. Of the Orthographie and Congruitie of the Britan Tongue; a treates, noe shorter than necessarie, for the Schooles, be Alexander Hume. Edited for the first time from the unique MS. in the British Museum (about 1617 A.D.), by Henry B. Wheatley, Esq. 4*s*.

6. Lancelot of the Laik. Edited from the unique MS. in the Cambridge University Library (ab. 1500), by the Rev. Walter W. Skeat, M.A. 8*s*.

7. The Story of Genesis and Exodus, an Early English Song, of about 1250 A.D. Edited for the first time from the unique MS. in the Library of Corpus Christi College, Cambridge, by R. Morris, Esq. 8*s*.

8 Morte Arthure; the Alliterative Version. Edited from Robert Thornton's unique MS. (about 1440 A.D.) at Lincoln, by the Rev. George Perry, M.A., Prebendary of Lincoln. 7*s*.

9. Animadversions uppon the Annotacions and Corrections of some Imperfections of Impressiones of Chaucer's Workes, reprinted in 1598; by Francis Thynne. Edited from the unique MS. in the Bridgewater Library. By G. H. Kingsley, Esq., M.D., and F. J. Furnivall, Esq., M.A. 10*s*.

10. Merlin, or the Early History of King Arthur. Edited for the first time from the unique MS. in the Cambridge University Library (about 1450 A.D.), by Henry B. Wheatley, Esq. Part I. 2*s*. 6*d*.

11. The Monarche, and other Poems of Sir David Lyndesay. Edited from the first edition by Johne Skott, in 1552, by Fitzedward Hall, Esq., D.C.L. Part I. 3*s*.

12. The Wright's Chaste Wife, a Merry Tale, by Adam of Cobsam (about 1462 A.D.), from the unique Lambeth MS. 306. Edited for the first time by F. J. Furnivall, Esq., M.A. 1*s*.

13. Seinte Marherete, þe Meiden ant Martyr. Three Texts of ab. 1200, 1310, 1330 A.D. First edited in 1862, by the Rev. Oswald Cockayne, M.A., and now re-issued. 2*s*.

14. Kyng Horn, with fragments of Floriz and Blauncheflur, and the Assumption of the Blessed Virgin. Edited from the MSS. in the Library of the University of Cambridge and the British Museum, by the Rev. J. Rawson Lumby. 3*s*. 6*d*

15. Political, Religious, and Love Poems, from the Lambeth MS. No. 306, and other sources. Edited by F. J. Furnivall, Esq., M.A. 7*s*. 6*d*.

16. A Tretice in English breuely drawe out of þ book of Quintis essencijs in Latyn, þ Hermys þ prophete and king of Egipt after þ flood of Noe, fader of Philosophris, hadde by reuelacioun of an aungil of God to him sente. Edited from the Sloane MS. 73, by F. J. Furnivall, Esq., M.A. 1*s*.

17. Parallel Extracts from 29 Manuscripts of Piers Plowman, with Comments, and a Proposal for the Society's Three-text edition of this Poem. By the Rev. W. Skeat, M.A. 1*s*.

18. Hali Meidenhead, about 1200 A.D. Edited for the first time from the MS. (with a translation) by the Rev. Oswald Cockayne, M.A. 1*s*.

19. The Monarche, and other Poems of Sir David Lyndesay. Part II., the Complaynt of the King's Papingo, and other minor Poems. Edited from the First Edition by F. Hall, Esq., D.C.L. 3*s*. 6*d*.

20. Some Treatises by Richard Rolle de Hampole. Edited from Robert of Thornton's MS. (ab. 1440 A.D.), by Rev. George G. Perry, M.A. 1*s*.

21. MERLIN, OR THE EARLY HISTORY OF KING ARTHUR. Part II. Edited by HENRY B. WHEATLEY, Esq. 4*s*.

22. THE ROMANS OF PARTENAY, OR LUSIGNEN. Edited for the first time from the unique MS. in the Library of Trinity College, Cambridge, by the Rev. W. W. SKEAT. M.A. 6*s*.

23. DAN MICHEL'S AYENBITE OF INWYT, or Remorse of Conscience, in the Kentish dialect, 1340 A.D. Edited from the unique MS. in the British Museum, by RICHARD MORRIS, Esq. 10*s*. 6*d*.

24. HYMNS OF THE VIRGIN AND CHRIST; THE PARLIAMENT OF DEVILS, and Other Religious Poems. Edited from the Lambeth MS. 853, by F. J. FURNIVALL, M.A. 3*s*.

25. THE STACIONS OF ROME, and the Pilgrim's Sea-Voyage and Sea-Sickness, with Clene Maydenhod. Edited from the Vernon and Porkington MSS., etc., by F. J. FURNIVALL, Esq., M.A. 1*s*.

26. RELIGIOUS PIECES IN PROSE AND VERSE. Containing Dan Jon Gaytrigg's Sermon; The Abbaye of S. Spirit; Sayne Jon, and other pieces in the Northern Dialect. Edited from Robert of Thorntone's MS. (ab. 1460 A.D.), by the Rev. G. PERRY, M.A. 2*s*.

27. MANIPULUS VOCABULORUM: a Rhyming Dictionary of the English Language, by PETER LEVINS (1570). Edited, with an Alphabetical Index by HENRY B. WHEATLEY. 12*s*.

28. THE VISION OF WILLIAM CONCERNING PIERS PLOWMAN, together with Vita de Dowel, Dobet et Dobest. 1362 A.D., by WILLIAM LANGLAND. The earliest or Vernon Text; Text A. Edited from the Vernon MS., with full Collations, by Rev. W. W. SKEAT, M.A. 7*s*.

29. OLD ENGLISH HOMILIES AND HOMILETIC TREATISES. (Sawles Warde and the Wohunge of Ure Lauerd: Ureisuns of Ure Louerd and of Ure Lefdi, etc.) of the Twelfth and Thirteenth Centuries. Edited from MSS. in the British Museum, Lambeth, and Bodleian Libraries; with Introduction, Translation, and Notes. By RICHARD MORRIS. *First Series*. Part I. 7*s*.

30. PIERS, THE PLOUGHMAN'S CREDE (about 1394). Edited from the MSS. by the Rev. W. W. SKEAT, M.A. 2*s*.

31. INSTRUCTIONS FOR PARISH PRIESTS. By JOHN MYRC. Edited from Cotton MS. Claudius A. II., by EDWARD PEACOCK, Esq., F.S.A., etc., etc. 4*s*.

32. THE BABEES BOOK, Aristotle's A B C, Urbanitatis, Stans Puer ad Mensam, The Lytille Childrenes Lytil Boke. THE BOKES OF NURTURE of Hugh Rhodes and John Russell, Wynkyn de Worde's Boke of Kervynge, The Booke of Demeanor, The Boke of Curtasye, Seager's Schoole of Vertue, etc., etc. With some French and Latin Poems on like subjects, and some Forewords on Education in Early England. Edited by F. J. FURNIVALL, M.A., Trin. Hall, Cambridge. 15*s*.

33. THE BOOK OF THE KNIGHT DE LA TOUR LANDRY, 1372. A Father's Book for his Daughters, Edited from the Harleian MS. 1764, by THOMAS WRIGHT Esq., M.A., and Mr. WILLIAM ROSSITER. 8*s*.

34. OLD ENGLISH HOMILIES AND HOMILETIC TREATISES. (Sawles Warde, and the Wohunge of Ure Lauerd: Ureisuns of Ure Louerd and of Ure Lefdi, etc.) of the Twelfth and Thirteenth Centuries. Edited from MSS. in the British Museum, Lambeth, and Bodleian Libraries; with Introduction, Translation, and Notes, by RICHARD MORRIS. *First Series*. Part 2. 8*s*.

35. SIR DAVID LYNDESAY'S WORKS. PART 3. The Historie of ane Nobil and Wailzeand Sqvyer, WILLIAM MELDRUM, umqvhyle Laird of Cleische and Bynnis, compylit be Sir DAUID LYNDESAY of the Mont *alias* Lyoun King of Armes. With the Testament of the said Williame Meldrum, Squyer, compylit alswa be Sir Dauid Lyndesay, etc. Edited by F. HALL, D.C.L. 2*s.*

36. MERLIN, OR THE EARLY HISTORY OF KING ARTHUR. A Prose Romance (about 1450–1460 A.D.), edited from the unique MS. in the University Library, Cambridge, by HENRY B. WHEATLEY. With an Essay on Arthurian Localities, by J. S. STUART GLENNIE, Esq. Part III. 1869. 12*s.*

37. SIR DAVID LYNDESAY'S WORKS. Part IV. Ane Satyre of the thrie estaits, in commendation of vertew and vitvperation of vyce. Maid be Sir DAVID LINDESAY, of the Mont, *alias* Lyon King of Armes. At Edinbvrgh. Printed be Robert Charteris, 1602. Cvm privilegio regis. Edited by F. HALL, Esq., D.C.L. 4*s.*

38. THE VISION OF WILLIAM CONCERNING PIERS THE PLOWMAN, together with Vita de Dowel, Dobet, et Dobest, Secundum Wit et Resoun, by WILLIAM LANGLAND (1377 A.D.). The "Crowley" Text; or Text B. Edited from MS. Laud Misc. 581, collated with MS. Rawl. Poet. 38, MS. B. 15. 17. in the Library of Trinity College, Cambridge, MS. Dd. 1. 17. in the Cambridge University Library, the MS. in Oriel College, Oxford, MS. Bodley 814, etc. By the Rev. WALTER W. SKEAT, M.A., late Fellow of Christ's College, Cambridge. 10*s.* 6*d.*

39. THE "GEST HYSTORIALE" OF THE DESTRUCTION OF TROY. An Alliterative Romance, translated from Guido De Colonna's "Hystoria Troiana." Now first edited from the unique MS. in the Hunterian Museum, University of Glasgow, by the Rev. GEO. A. PANTON and DAVID DONALDSON. Part I. 10*s.* 6*d.*

40. ENGLISH GILDS. The Original Ordinances of more than One Hundred Early English Gilds: Together with the olde usages of the cite of Wynchestre; The Ordinances of Worcester; The Office of the Mayor of Bristol; and the Customary of the Manor of Tettenhall-Regis. From Original MSS. of the Fourteenth and Fifteenth Centuries. Edited with Notes by the late TOULMIN SMITH, Esq., F.R.S. of Northern Antiquaries (Copenhagen). With an Introduction and Glossary, etc., by his daughter, LUCY TOULMIN SMITH. And a Preliminary Essay, in Five Parts, ON THE HISTORY AND DEVELOPMENT OF GILDS, by LUJO BRENTANO, Doctor Juris Utriusque et Philosophiæ. 21*s.*

41. THE MINOR POEMS OF WILLIAM LAUDER, Playwright, Poet, and Minister of the Word of God (mainly on the State of Scotland in and about 1568 A.D., that year of Famine and Plague). Edited from the Unique Originals belonging to S. CHRISTIE-MILLER, Esq., of Britwell, by F. J. FURNIVALL, M.A., Trin. Hall, Camb 3*s.*

42. BERNARDUS DE CURA REI FAMULIARIS, with some Early Scotch Prophecies, etc. From a MS., KK 1. 5, in the Cambridge University Library. Edited by J. RAWSON LUMBY, M.A., late Fellow of Magdalen College, Cambridge. 2*s.*

43. RATIS RAVING, and other Moral and Religious Pieces, in Prose and Verse. Edited from the Cambridge University Library MS. KK 1. 5, by J. RAWSON LUMBY, M.A., late Fellow of Magdalen College, Cambridge. 3*s.*

44. JOSEPH OF ARIMATHIE: otherwise called the Romance of the Seint Graal, or Holy Grail: an alliterative poem, written about A.D. 1350, and now first printed from the unique copy in the Vernon MS. at Oxford. With an appendix, containing "The Lyfe of Joseph of Armathy," reprinted from the black-letter copy of Wynkyn de Worde; "De sancto Joseph ab

Arimathia," first-printed by Pynson, A.D. 1516; and "The Lyfe of Joseph of Arimathia," first printed by Pynson, A.D. 1520. Edited, with Notes and Glossarial Indices, by the Rev. Walter W. Skeat, M.A. 5s.

45. King Alfred's West-Saxon Version of Gregory's Pastoral Care. With an English translation, the Latin Text, Notes, and an Introduction Edited by Henry Sweet, Esq., of Balliol College, Oxford. Part I. 10s.

46. Legends of the Holy Rood; Symbols of the Passion and Cross-Poems. In Old English of the Eleventh, Fourteenth, and Fifteenth Centuries. Edited from MSS. in the British Museum and Bodleian Libraries; with Introduction, Translations, and Glossarial Index. By Richard Morris, LL.D. 10s.

47. Sir David Lyndesay's Works. Part V. The Minor Poems of Lyndesay. Edited by J. A. H. Murray, Esq. 3s.

48. The Times' Whistle: or, A Newe Daunce of Seven Satires, and other Poems: Compiled by R. C., Gent. Now first Edited from MS. Y. 8. 3. in the Library of Canterbury Cathedral; with Introduction, Notes, and Glossary, by J. M. Cowper. 6s.

49. An Old English Miscellany, containing a Bestiary, Kentish Sermons, Proverbs of Alfred, Religious Poems of the 13th century. Edited from the MSS. by the Rev. R. Morris, LL.D. 10s.

50. King Alfred's West-Saxon Version of Gregory's Pastoral Care. Edited from 2 MSS., with an English translation. By Henry Sweet, Esq., Balliol College, Oxford. Part II. 10s.

51. Þe Liflade of St. Juliana, from two old English Manuscripts of 1230 A.D. With renderings into Modern English, by the Rev. O. Cockayne and Edmund Brock. Edited by the Rev. O. Cockayne, M.A. Price 2s.

52. Palladius on Husbondrie, from the unique MS., ab. 1420 A.D., ed. Rev. B. Lodge. Part I. 10s.

53. Old English Homilies, Series II., from the unique 13th-century MS. in Trinity Coll. Cambridge, with a photolithograph; three Hymns to the Virgin and God, from a unique 13th-century MS. at Oxford, a photolithograph of the music to two of them, and transcriptions of it in modern notation by Dr. Rimbault, and A. J. Ellis, Esq., F.R.S.; the whole edited by the Rev. Richard Morris, LL.D. 8s.

54. The Vision of Piers Plowman, Text C (completing the three versions of this great poem), with an Autotype; and two unique alliterative Poems: Richard the Redeles (by William, the author of the *Vision*); and The Crowned King; edited by the Rev. W. W. Skeat, M.A. 18s.

55. Generydes, a Romance, edited from the unique MS., ab. 1440 A.D., in Trin. Coll. Cambridge, by W. Aldis Wright, Esq., M.A., Trin. Coll. Cambr. Part I. 3s.

56. The Gest Hystoriale of the Destruction of Troy, translated from Guido de Colonna, in alliterative verse; edited from the unique MS. in the Hunterian Museum, Glasgow, by D. Donaldson, Esq., and the late Rev. G. A. Panton. Part II. 10s. 6d.

57. The Early English Version of the "Cursor Mundi," in four Texts, from MS. Cotton, Vesp. A. iii. in the British Museum; Fairfax MS. 14. in the Bodleian; the Göttingen MS. Theol. 107; MS. R. 3, 8, in Trinity College, Cambridge. Edited by the Rev. R. Morris, LL.D. Part I. with two photo-lithographic facsimiles by Cooke and Fotheringham. 10s. 6d.

58. The Blickling Homilies, edited from the Marquis of Lothian's Anglo-Saxon MS. of 971 A.D., by the Rev. R. Morris, LL.D. (With a Photolithograph). Part I. 8s.

59. The Early English Version of the "Cursor Mundi;" in four Texts, from MS. Cotton Vesp. A. iii. in the British Museum; Fairfax MS. 14. in the Bodleian; the Göttingen MS. Theol. 107; MS. R. 3, 8, in Trinity College, Cambridge. Edited by the Rev. R. Morris, LL.D. Part II. 15*s.*

60. Meditacyuns on the Soper of our Lorde (perhaps by Robert of Brunne). Edited from the MSS. by J. M. Cowper, Esq. 2*s.* 6*d.*

61. The Romance and Prophecies of Thomas of Erceldoune, printed from Five MSS. Edited by Dr. James A. H. Murray. 10*s.* 6*d.*

62. The Early English Version of the "Cursor Mundi," in Four Texts. Edited by the Rev. R. Morris, M.A., LL.D. Part III. 15*s.*

63. The Blickling Homilies. Edited from the Marquis of Lothian's Anglo-Saxon MS. of 971 A.D., by the Rev. R. Morris, LL.D. Part II. 4*s.*

64. Francis Thynne's Emblemes and Epigrams, A.D. 1600, from the Earl of Ellesmere's unique MS. Edited by F. J. Furnivall, M.A. 4*s.*

65. Be Domes Dæge (Bede's De Die Judicii) and other short Anglo-Saxon Pieces. Ed. from the unique MS. by the Rev. J. Rawson Lumby, B.D. 2*s.*

66. The Early English Version of the "Cursor Mundi," in Four Texts. Edited by Rev. R. Morris, M.A., LL.D. Part IV. 10*s.*

67. Notes on Piers Plowman. By the Rev. W. W. Skeat, M.A. Part I. 21*s.*

68. The Early English Version of the "Cursor Mundi," in Four Texts. Edited by Rev. R. Morris, M.A., LL.D. Part V. 25*s.*

69. Adam Davy's Five Dreams about Edward II. The Life of Saint Alexius. Solomon's Book of Wisdom. St. Jerome's 15 Tokens before Doomsday. The Lamentation of Souls. Edited from the Laud MS. 622, in the Bodleian Library, by F. J. Furnivall, M.A. 5*s.*

70. Generydes, a Romance. Edited by W. Aldis Wright, M.A. Part II. 4*s.*

71. The Lay Folk's Mass-Book, 4 Texts. Edited by Rev. Canon Simmons. 25*s.*

72. Palladius on Husbondrie, englisht (ab. 1420 A.D.). Part II. Edited by S. J. Herrtage, B.A. 5*s.*

73. The Blickling Homilies, 971 A.D. Edited by Rev. Dr. R. Morris. Part III. 8*s.*

74. English Works of Wyclif, hitherto unprinted. Edited by F. D. Matthew. 20*s.*

75. Catholicon Anglicum, an early English Dictionary, from Lord Monson's MS., A.D. 1483. Edited with Introduction and Notes by S. J. Herrtage, B.A.; and with a Preface by H. B. Wheatley. 20*s.*

76. Aelfric's Metrical Lives of Saints, in MS. Cott. Jul. E. 7. Edited by Rev. Prof. Skeat, M.A. Part I. 10*s.*

77. Beowulf. The unique MS. Autotyped and Transliterated. Edited by Professor Zupitza, Ph.D. 25*s.*

78. The Fifty Earliest English Wills in the Court of Probate, 1387-1439. Edited by F. J. Furnivall, M.A. 7*s.*

79. King Alfred's Orosius from Lord Tollemache's 9th Century MS. Part I. Edited by H. Sweet, M.A. 13*s.*

Extra Volume. Facsimile of the Epinal Glossary, 8th Century, edited by H. Sweet. 15.

80. The Anglo-Saxon Life of St. Katherine and its Latin Original. Edited by Dr. Einenkel. 12*s.*

Extra Series. Subscriptions—Small paper, one guinea; large paper two guineas, per annum.

1. THE ROMANCE OF WILLIAM OF PALERNE (otherwise known as the Romance of William and the Werwolf). Translated from the French at the command of Sir Humphrey de Bohun, about A.D. 1350, to which is added a fragment of the Alliterative Romance of Alisaunder, translated from the Latin by the same author, about A.D. 1340; the former re-edited from the unique MS. in the Library of King's College, Cambridge, the latter now first edited from the unique MS. in the Bodleian Library, Oxford. By the Rev. WALTER W. SKEAT, M.A. 8vo. sewed, pp. xliv. and 328. 13*s.*

2. ON EARLY ENGLISH PRONUNCIATION, with especial reference to Shakspere and Chaucer; containing an investigation of the Correspondence of Writing with Speech in England, from the Anglo-Saxon period to the present day, preceded by a systematic Notation of all Spoken Sounds by means of the ordinary Printing Types; including a re-arrangement of Prof. F. J. Child's Memoirs on the Language of Chaucer and Gower, and reprints of the rare Tracts by Salesbury on English, 1547, and Welsh, 1567, and by Barcley on French, 1521 By ALEXANDER J. ELLIS, F.R.S. Part I. On the Pronunciation of the XIVth, XVIth, XVIIth, and XVIIIth centuries. 8vo. sewed, pp. viii. and 416. 10*s.*

3. CAXTON'S BOOK OF CURTESYE, printed at Westminster about 1477–8, A.D., and now reprinted, with two MS. copies of the same treatise, from the Oriel MS. 79, and the Balliol MS. 354. Edited by FREDERICK J. FURNIVALL, M.A. 8vo. sewed, pp. xii. and 58. 5*s.*

4. THE LAY OF HAVELOK THE DANE; composed in the reign of Edward I., about A.D. 1280. Formerly edited by Sir F. MADDEN for the Roxburghe Club, and now re-edited from the unique MS. Laud Misc. 108, in the Bodleian Library, Oxford, by the Rev. WALTER W. SKEAT, M.A. 8vo. sewed, pp. lv. and 160. 10*s.*

5. CHAUCER'S TRANSLATION OF BOETHIUS'S "DE CONSOLATIONE PHILOSOPHIE." Edited from the Additional MS. 10,340 in the British Museum. Collated with the Cambridge Univ. Libr. MS. Ii. 3. 21. By RICHARD MORRIS. 8vo. 12*s.*

6 THE ROMANCE OF THE CHEVELERE ASSIGNE. Re-edited from the unique manuscript in the British Museum, with a Preface, Notes, and Glossarial Index, by HENRY H. GIBBS, Esq., M.A. 8vo. sewed, pp. xviii. and 38. 3*s.*

7. ON EARLY ENGLISH PRONUNCIATION, with especial reference to Shakspere and Chaucer. By ALEXANDER J. ELLIS, F.R.S., etc., etc. Part II. On the Pronunciation of the XIIIth and previous centuries, of Anglo-Saxon, Icelandic, Old Norse and Gothic, with Chronological Tables of the Value of Letters and Expression of Sounds in English Writing. 10*s.*

8. QUEENE ELIZABETHES ACHADEMY, by Sir HUMPHREY GILBERT. A Booke of Precedence, The Ordering of a Funerall, etc. Varying Versions of the Good Wife, The Wise Man, etc., Maxims, Lydgate's Order of Fools, A Poem on Heraldry, Occleve on Lords' Men, etc., Edited by F. J. FURNIVALL, M.A., Trin. Hall, Camb. With Essays on Early Italian and German Books of Courtesy, by W. M. ROSSETTI, Esq., and E. OSWALD, Esq. 8vo. 13*s.*

9. THE FRATERNITYE OF VACABONDES, by JOHN AWDELEY (licensed in 1560-1, imprinted then, and in 1565), from the edition of 1575 in the Bodleian Library. A Caueat or Warening for Commen Cursetors vulgarely called Vagabones, by THOMAS HARMAN, ESQUIERE. From the 3rd edition of 1567, belonging to Henry Huth, Esq., collated with the 2nd edition of 1567,

in the Bodleian Library, Oxford, and with the reprint of the 4th edition of 1573. A Sermon in Praise of Thieves and Thievery, by PARSON HABEN OR HYBERDYNE, from the Lansdowne MS. 98, and Cotton Vesp. A. 25. Those parts of the Groundworke of Conny-catching (ed. 1592), that differ from *Harman's Caueat*. Edited by EDWARD VILES & F. J. FURNIVALL. 8vo. 7*s*. 6*d*.

10. THE FYRST BOKE OF THE INTRODUCTION OF KNOWLEDGE, made by Andrew Borde, of Physycke Doctor. A COMPENDYOUS REGYMENT OF A DYETARY OF HELTH made in Mountpyllier, compiled by Andrewe Boorde, of Physycke Doctor. BARNES IN THE DEFENCE OF THE BERDE: a treatyse made, answerynge the treatyse of Doctor Borde upon Berdes. Edited, with a life of Andrew Boorde, and large extracts from his Breuyary, by F. J FURNIVALL, M.A., Trinity Hall, Camb. 8vo. 18*s*.

11. THE BRUCE; or, the Book of the most excellent and noble Prince, Robert de Broyss, King of Scots: compiled by Master John Barbour, Archdeacon of Aberdeen. A.D. 1375. Edited from MS. G 23 in the Library of St. John's College, Cambridge, written A.D. 1487; collated with the MS. in the Advocates' Library at Edinburgh, written A.D. 1489, and with Hart's Edition, printed A.D. 1616; with a Preface, Notes, and Glossarial Index, by the Rev. WALTER W. SKEAT, M.A. Part I 8vo. 12*s*.

12. ENGLAND IN THE REIGN OF KING HENRY THE EIGHTH. A Dialogue between Cardinal Pole and Thomas Lupset, Lecturer in Rhetoric at Oxford. By THOMAS STARKEY, Chaplain to the King. Edited, with Preface, Notes, and Glossary, by J. M. COWPER. And with an Introduction, containing the Life and Letters of Thomas Starkey, by the Rev. J. S. BREWER, M.A. Part II. 12*s*. (*Part I., Starkey' Life and Letters, is in preparation.*

13. A SUPPLICACYON FOR THE BEGGARS. Written about the year 1529, by SIMON FISH. Now re-edited by FREDERICK J. FURNIVALL. With a Supplycacion to our moste Soueraigne Lorde Kynge Henry the Eyght (1544 A.D.), A Supplication of the Poore Commons (1546 A.D.), The Decaye of England by the great multitude of Shepe (1550-3 A.D.). Edited by J. MEADOWS COWPER. 6*s*.

14. ON EARLY ENGLISH PRONUNCIATION, with especial reference to Shakspere and Chaucer. By A. J. ELLIS, F.R.S., F.S.A. Part III. Illustrations of the Pronunciation of the XIVth and XVIth Centuries. Chaucer, Gower, Wycliffe, Spenser, Shakspere, Salesbury, Barcley, Hart, Bullokar, Gill. Pronouncing Vocabulary. 10*s*.

15. ROBERT CROWLEY'S THIRTY-ONE EPIGRAMS, Voyce of the Last Trumpet, Way to Wealth, etc., 1550-1 A.D. Edited by J. M. COWPER, Esq. 12*s*.

16. A TREATISE ON THE ASTROLABE; addressed to his son Lowys, by Geoffrey Chaucer, A.D. 1391. Edited from the earliest MSS. by the Rev. WALTER W. SKEAT, M.A., late Fellow of Christ's College, Cambridge. 10*s*.

17. THE COMPLAYNT OF SCOTLANDE, 1549, A.D., with an Appendix of four Contemporary English Tracts. Edited by J. A. H. MURRAY, Esq. Part I. 10*s*.

18. THE COMPLAYNT OF SCOTLANDE, etc. Part II. 8*s*.

19. OURE LADYES MYROURE, A.D. 1530, edited by the Rev. J. H. BLUNT, M.A., with four full-page photolithographic facsimiles by Cooke and Fotheringham. 24*s*.

20. LONELICH'S HISTORY OF THE HOLY GRAIL (ab. 1450 A.D.), translated from the French Prose of SIRES ROBIERS DE BORRON. Re-edited fron the Unique MS. in Corpus Christi College, Cambridge, by F. J. Furnivall, Esq. M.A. Part I. 8*s*.

21. Barbour's Bruce. Edited from the MSS. and the earliest printed edition by the Rev. W. W. Skeat, M.A. Part II. 4*s*.

22. Henry Brinklow's Complaynt of Roderyck Mors, somtyme a gray Fryre, unto the Parliament Howse of Ingland his naturall Country, for the Redresse of certen wicked Lawes, euel Customs, and cruel Decreys (ab. 1542); and The Lamentacion of a Christian Against the Citie of London, made by Roderigo Mors, A.D. 1545. Edited by J. M. Cowper, Esq. 9*s*.

23. On Early English Pronunciation, with especial reference to Shakspere and Chaucer. By A. J. Ellis, Esq., F.R.S. Part IV. 10*s*.

24. Lonelich's History of the Holy Grail (ab. 1450 A.D.), translated from the French Prose of Sires Robiers de Borron. Re-edited from the Unique MS. in Corpus Christi College, Cambridge, by F. J. Furnivall, Esq., M.A. Part II. 10*s*.

25. The Romance of Guy of Warwick. Edited from the Cambridge University MS. by Prof. J. Zupitza, Ph.D. Part I. 20*s*.

26. The Romance of Guy of Warwick. Edited from the Cambridge University MS. by Prof J. Zupitza, Ph.D. (The 2nd or 15th century version.) Part II. 14*s*.

27. The English Works of John Fisher, Bishop of Rochester (died 1535). Edited by Professor J. E. B. Mayor, M.A. Part I., the Text. 16*s*.

28. Lonelich's History of the Holy Grail. Edited by F. J. Furnivall, M.A. Part III. 10*s*.

29. Barbour's Bruce Edited from the MSS. and the earliest Printed Edition, by the Rev. W. W. Skeat, M.A. Part III. 21*s*.

30. Lonelich's History of the Holy Grail. Edited by F. J. Furnivall, Esq., M.A. Part IV. 15*s*.

31. Alexander and Dindimus. Translated from the Latin about A.D. 1340–50. Re-edited by the Rev. W. W. Skeat, M.A. 6*s*.

32. Starkey's "England in Henry VIII.'s Time." Part I. Starkey's Life and Letters. Edited by S. J. Herrtage, B.A. 8*s*.

33. Gesta Romanorum: the Early English Versions. Edited from the MSS. and Black-letter Editions, by S. J. Herrtage, B.A. 15*s*.

34. Charlemagne Romances: No. I. Sir Ferumbras. Edited from the unique Ashmole MS. by S. J. Herrtage, B.A. 15*s*.

35. Charlemagne Romances: II. The Sege off Malayne, Sir Otuell, etc. Edited by S. J. Herrtage, B.A. 12*s*.

36. Charlemagne Romances: III. Lyf of Charles the Grete, Pt. 1. Edited by S. J. Herrtage, B.A. 16*s*.

37. Charlemagne Romances: IV. Lyf of Charles the Grete, Pt. 2. Edited by S. J. Herrtage, B.A. 15*s*.

38. Charlemagne Romances: V. The Sowdone of Babylone. Edited by Dr. Hausknecht. 15*s*.

39. Charlemagne Romances: VI. The Taill of Rauf Colyear, Roland, Otuel, etc. Edited by Sydney J. Herrtage, B.A. 15*s*.

40. Charlemagne Romances: VII. Houn of Burdeux. By Lord Berners. Edited by S. L. Lee, B.A. Part I. 15*s*.

41. Charlemagne Romances: VIII. Huon of Burdeux. By Lord Berners. Edited by S. L. Lee, B.A. Part II. 15*s*.

English Dialect Society's Publications. Subscription, 1873 to 1876, 10*s*. 6*d*. per annum; 1877 and following years, 20*s*. per annum.

1873.

1. Series B. Part 1. Reprinted Glossaries, I.–VII. Containing a Glossary of North of England Words, by J. H.; Glossaries, by Mr. MARSHALL; and a West-Riding Glossary, by Dr. WILLAN. 7*s*. 6*d*.
2. Series A. Bibliographical. A List of Books illustrating English Dialects. Part I. Containing a General List of Dictionaries, etc.; and a List of Books relating to some of the Counties of England. 4*s*. 6*d*.
3. Series C. Original Glossaries. Part I. Containing a Glossary of Swaledale Words. By Captain HARLAND. 4*s*.

1874.

4. Series D. The History of English Sounds. By H. SWEET, Esq. 4*s*. 6*d*.
5. Series B. Part II. Reprinted Glossaries. VIII.–XIV. Containing seven Provincial English Glossaries, from various sources. 7*s*.
6. Series B. Part III. Reprinted Glossaries. XV.–XVII. Ray's Collection of English Words not generally used, from the edition of 1691; together with Thoresby's Letter to Ray, 1703. Re-arranged and newly edited by Rev. WALTER W. SKEAT. 8*s*.

6*. Subscribers to the English Dialect Society for 1874 also receive a copy of 'A Dictionary of the Sussex Dialect.' By the Rev. W. D PARISH.

1875.

7. Series D. Part II. The Dialect of West Somerset. By F. T. ELWORTHY, Esq. 3*s*. 6*d*.
8. Series A. Part II. A List of Books Relating to some of the Counties of England. Part II. 6*s*.
9. Series C. A Glossary of Words used in the Neighbourhood of Whitby. By F. K. ROBINSON. Part I. A—P. 7*s*. 6*d*.
10. Series C. A Glossary of the Dialect of Lancashire. By J. H. NODAL and G. MILNER. Part I. A—E. 3*s*. 6*d*.

1876.

11. On the Survival of Early English Words in our Present Dialects. By Dr. R. MORRIS. 6*d*.
12. Series C. Original Glossaries. Part III. Containing Five Original Provincial English Glossaries. 7*s*.
13. Series C. A Glossary of Words used in the Neighbourhood of Whitby. By F. K. Robinson. Part II. P—Z. 6*s* 6*d*.
14. A Glossary of Mid-Yorkshire Words, with a Grammar. By C. CLOUGH ROBINSON. 9*s*.

1877.

15. A GLOSSARY OF WORDS used in the Wapentakes of Manley and Corringham, Lincolnshire. By EDWARD PEACOCK, F.S.A. 9*s*. 6*d*.
16. A Glossary of Holderness Words. By F. ROSS, R. STEAD, and T. HOLDERNESS. With a Map of the District. 7*s*. 6*d*.
17. On the Dialects of Eleven Southern and South-Western Counties, with a new Classification of the English Dialects By Prince LOUIS LUCIEN. BONAPARTE. With Two Maps. 1*s*.

18. Bibliographical List. Part III. completing the Work, and containing a List of Books on Scottish Dialects, Anglo-Irish Dialect, Cant and Slang, and Americanisms, with additions to the English List and Index. Edited by J. H. NODAL. 4*s.* 6*d.*
19. An Outline of the Grammar of West Somerset. By F. T. ELWORTHY, ESQ. 5*s.*

1878.

20. A Glossary of Cumberland Words and Phrases. By WILLIAM DICKINSON, F.L.S. 6*s.*
21. Tusser's Five Hundred Pointes of Good Husbandrie. Edited with Introduction, Notes and Glossary, by W. PAINE and SIDNEY J. HERRTAGE, B.A. 12*s.* 6*d.*
22. A Dictionary of English Plant Names. By JAMES BRITTEN, F.L.S., and ROBERT HOLLAND. Part I. (A to F). 8*s.* 6*d.*

1879.

23. Five Reprinted Glossaries, including Wiltshire, East Anglian, Suffolk, and East Yorkshire Words, and Words from Bishop Kennett's Parochial Antiquities. Edited by the Rev. Professor SKEAT, M.A. 7*s.*
24. Supplement to the Cumberland Glossary (No. 20). By W. DICKINSON, F.L.S. 1*s.*
25. Specimens of English Dialects. First Volume. I. Devonshire; Exmoor Scolding and Courtship. Edited, with Notes and Glossary, by F. T. ELWORTHY. II. Westmoreland: Wm. de Worfat's Bran New Wark. Edited by Rev. Prof. SKEAT. 8*s.* 6*d.*
26. A Dictionary of English Plant Names. By J. BRITTEN and R. HOLLAND. Part II. (G to O). 1880. 8*s.* 6*d.*

1880.

27. Glossary of Words in use in Cornwall. I. West Cornwall. By Miss M. A. COURTNEY. II. East Cornwall. By THOMAS Q. COUCH. With Map. 6*s.*
28. Glossary of Words and Phrases in use in Antrim and Down. By WILLIAM HUGH PATTERSON, M.R.I.A. 7*s.*
29. An Early English Hymn to the Virgin. By F. J. FURNIVALL, M.A., and A. J. ELLIS, F.R.S. 6*d.*
30. Old Country and Farming Words. Gleaned from Agricultural Books. By JAMES BRITTEN, F.L.S. 10*s.* 6*d.*

1881.

31. The Dialect of Leicestershire. By the Rev. A. B. EVANS, D.D., and SEBASTIAN EVANS, LL.D. 10*s.* 6*d.*
32. Five Original Glossaries. Isle of Wight, Oxfordshire, Cumberland, North Lincolnshire and Radnorshire. By various Authors. 7*s.* 6*d.*
33. George Eliot's Use of Dialect. By W. E. A. AXON. (Forming No. 4 of "Miscellanies.") 6*d.*
34. Turner's Names of Herbes, A.D. 1548. Edited (with Index and Indentification of Names) by JAMES BRITTEN, F.L.S. 6*s.* 6*d.*

1882.

35. Glossary of the Lancashire Dialect. By J. H. NODAL and GEO. MILNER. Part II. (F to Z). 6*s.*
36. West Worcester Words. By MRS. CHAMBERLAIN. 8vo. sewed. 4*s.* 6*d.*

37. Fitzherbert's Book of Husbandry, A.D. 1534. Edited with Introduction, Notes, and Glossarial Index. By the REV. PROFESSOR SKEAT. 8vo. sewed. 8*s.* 6*d.*

38. Devonshire Plant Names. By the REV. HILDERIC FRIEND. 8vo. sewed. 5*s.*

1883.

39. A Glossary of the Dialect of Aldmondbury and Huddersfield. By the Rev. A. EASHER, M.A., and the Rev. THOS. LEES, M.A. 8vo. sewed. 8*s.* 6*d.*

40. HAMPSHIRE WORDS AND PHRASES. Compiled and Edited by the Rev. Sir WILLIAM H. COPE, Bart. 6*s.*

41. NATHANIEL BAILEY'S ENGLISH DIALECT WORDS OF THE 18TH CENTURY. Edited by W. E. A. AXON. 9*s.*

41.* THE TREATYSE OF FYSSHINGE WITH AN ANGLE. By JULIANA BARNES. An earlier form (circa 1450) edited with Glossary, by THOMAS SATCHELL, and by him presented to the subscribers for 1883.

Furnivall.—EDUCATION IN EARLY ENGLAND. Some Notes used as Forewords to a Collection of Treatises on "Manners and Meals in the Olden Time," for the Early English Text Society. By FREDERICK J. FURNIVALL, M.A., Trinity Hall, Cambridge, Member of Council of the Philological and Early English Text Societies. 8vo. sewed. pp. 74. 1*s.*

Gould.—GOOD ENGLISH; or, Popular Errors in Language. By E. S. GOULD. Revised Edition. Crown 8vo. cloth, pp. xii. and 214. 1880. 6*s.*

Hall.—ON ENGLISH ADJECTIVES IN -ABLE, with Special Reference to RELIABLE. By FITZEDWARD HALL, C.E., M.A., Hon.D.C.L. Oxon.; formerly Professor of Sanskrit Language and Literature, and of Indian Jurisprudence, in King's College, London. Crown 8vo. cloth, pp. viii. and 238. 7*s.* 6*d.*

Hall.—MODERN ENGLISH. By FITZEDWARD HALL, M.A., Hon. D.C.L., Oxon. Cr. 8vo. cloth, pp. xvi. and 394. 10*s.* 6*d.*

Jackson.—SHROPSHIRE WORD-BOOK; A Glossary of Archaic and Provincial Words, etc., used in the County. By GEORGINA F. JACKSON. 8vo. pp. xcvi. and 524. 1881. 31*s.* 6*d.*

Koch.—A HISTORICAL GRAMMAR OF THE ENGLISH LANGUAGE. By C. F. KOCH. Translated into English Edited, Enlarged, and Annotated by the Rev. R. MORRIS, LL.D., M.A. [*Nearly ready.*

Manipulus Vocabulorum.—A Rhyming Dictionary of the English Language. By Peter Levins (1570) Edited, with an Alphabetical Index, by HENRY B. WHEATLEY. 8vo. pp. xvi. and 370, cloth. 14*s.*

Manning.—AN INQUIRY INTO THE CHARACTER AND ORIGIN OF THE POSSESSIVE AUGMENT in English and in Cognate Dialects. By the late JAMES MANNING, Q.A.S., Recorder of Oxford. 8vo. pp. iv. and 90. 2*s.*

Palmer.—LEAVES FROM A WORD HUNTER'S NOTE BOOK. Being some Contributions to English Etymology. By the Rev. A. SMYTHE PALMER, B.A., sometime Scholar in the University of Dublin. Cr. 8vo. cl. pp. xii.-316. 7*s.* 6*d.*

Percy.—BISHOP PERCY'S FOLIO MANUSCRIPTS—BALLADS AND ROMANCES. Edited by John W. Hales, M.A., Fellow and late Assistant Tutor of Christ's College, Cambridge; and Frederick J. Furnivall, M.A., of Trinity Hall, Cambridge; assisted by Professor Child, of Harvard University, Cambridge, U.S.A., W. Chappell, Esq., etc. In 3 volumes. Vol. I., pp. 610; Vol. 2, pp. 681.; Vol. 3, pp. 640. Demy 8vo. half-bound, £4 4*s.* Extra demy 8vo. half-bound, on Whatman's ribbed paper, £6 6*s.* Extra royal 8vo., paper covers, on Whatman's best ribbed paper, £10 10*s.* Large 4to., paper covers, on Whatman's best ribbed paper, £12.

Philological Society. Transactions of the, contains several valuable Papers on Early English. For contents see page 16.

Stratmann.—A DICTIONARY OF THE OLD ENGLISH LANGUAGE. Compiled from the writings of the XIIIth, XIVth, and XVth centuries. By FRANCIS HENRY STRATMANN. 3rd Edition. 4to. with Supplement. In wrapper. £1 16*s.*

Stratmann.—AN OLD ENGLISH POEM OF THE OWL AND THE NIGHTINGALE Edited by FRANCIS HENRY STRATMANN. 8vo. cloth, pp. 60. 3*s.*

Sweet.—A HISTORY OF ENGLISH SOUNDS, from the Earliest Period, including an Investigation of the General Laws of Sound Change, and full Word Lists. By HENRY SWEET. Demy 8vo. cloth, pp. iv. and 164. 4*s.* 6*d.*

Turner.—THE ENGLISH LANGUAGE. A Concise History of the English Language, with a Glossary showing the Derivation and Pronunciation of the English Words. By R. TURNER. In German and English on opposite pages. 18mo. sewed, pp. viii. and 80. 1884. 1*s.* 6*d.*

Vere.—STUDIES IN ENGLISH; or, Glimpses of the Inner Life of our Language. By M. SCHELE DE VERE, LL.D., Professor of Modern Languages in the University of Virginia. 8vo. cloth, pp. vi. and 365. 12*s.* 6*d.*

Wedgwood.—A DICTIONARY OF ENGLISH ETYMOLOGY. By HENSLEIGH WEDGWOOD. Third revised Edition. With an Introduction on the Formation of Language. Imperial 8vo., double column, pp. lxxii. and 746. 21*s.*

Wright.—FEUDAL MANUALS OF ENGLISH HISTORY. A Series of Popular Sketches of our National History, compiled at different periods, from the Thirteenth Century to the Fifteenth, for the use of the Feudal Gentry and Nobility. (In Old French). Now first edited from the Original Manuscripts. By THOMAS WRIGHT, Esq., M.A. Small 4to. cloth, pp. xxiv. and 184. 1872. 15*s.*

Wright.—ANGLO-SAXON AND OLD-ENGLISH VOCABULARIES, Illustrating the Condition and Manners of our Forefathers, as well as the History of the Forms of Elementary Education, and of the Languages Spoken in this Island from the Tenth Century to the Fifteenth. Edited by THOMAS WRIGHT, Esq., M.A., F.S.A., etc. Second Edition, edited and collated, by RICHARD WULCKER. 2 vols. 8vo. pp. xx.-408, and iv.-486, cloth. 1884. 28*s.*

FRISIAN.

Cummins.—A GRAMMAR OF THE OLD FRIESIC LANGUAGE. By A. H. CUMMINS, A.M. Crown 8vo. cloth, pp. x. and 76. 1881. 3*s.* 6*d.*

Oera Linda Book, from a Manuscript of the Thirteenth Century, with the permission of the Proprietor, C. Over de Linden, of the Helder. The Original Frisian Text, as verified by Dr. J. O. OTTEMA; accompanied by an English Version of Dr. Ottema's Dutch Translation, by WILLIAM R. SANDBACH. 8vo. cl. pp. xxvii. and 223. 5*s.*

GAUDIAN (See under "HOERNLE," page 40.)

OLD GERMAN.

Douse.—GRIMM'S LAW; A STUDY: or, Hints towards an Explanation of the so-called "Lautverschiebung." To which are added some Remarks on the Primitive Indo-European *K*, and several Appendices. By T. LE MARCHANT DOUSE. 8vo. cloth, pp. xvi. and 230. 10*s.* 6*d.*

Kroeger.—THE MINNESINGER OF GERMANY. By A. E. KROEGER. 12mo. cloth, pp. vi. and 284. 7*s.*

CONTENTS.—Chapter I. The Minnesinger and the Minnesong.—II. The Minnelay.—III. The Divine Minnesong.—IV. Walther von der Vogelweide.—V. Ulrich von Lichtenstein.—VI. The Metrical Romances of the Minnesinger and Gottfried von Strassburg's "Tristan and Isolde."

GIPSY.

Leland.—English Gipsy Songs. In Rommany, with Metrical English Translations. By Charles G. Leland, Author of "The English Gipsies," etc.; Prof. E. H. Palmer; and Janet Tuckey. Crown 8vo. cloth, pp. xii. and 276. 7*s*. 6*d*.

Leland.—The English Gipsies and their Language. By Charles G. Leland. Second Edition. Crown 8vo. cloth, pp. 276. 7*s*. 6*d*.

Leland.—The Gypsies.—By C. G. Leland. Crown 8vo. pp. 372, cloth. 1882. 10*s*. 6*d*.

Paspati.—Études sur les Tchinghianés (Gypsies) ou Bohémiens de L'Empire Ottoman. Par Alexandre G. Paspati, M.D. Large 8vo. sewed, pp. xii. and 652. Constantinople, 1871. 28*s*.

GOTHIC.

Skeat.—A Moeso-Gothic Glossary, with an Introduction, an Outline of Moeso-Gothic Grammar, and a List of Anglo-Saxon and Modern English Words etymologically connected with Moeso-Gothic. By the Rev. W. W. Skeat. Small 4to. cloth, pp. xxiv.and 342. 1868. 9*s*.

GREEK (Modern and Classic).

Bizyenos.—ΑΤΘΙΔΕΣ ΑΥΡΑΙ Poems. By M. Bizyenos. With Frontispiece Etched by Prof. A. Legros. Royal 8vo. pp. viii.-312. Printed on hand-made paper, and richly bound. 1884. £1 11*s*. 6*d*.

Buttmann.—A Grammar of the New Testament Greek. By A. Buttmann. Authorized translation by Prof J. H. Thayer, with numerous additions and corrections by the author. Demy 8vo. cloth, pp. xx. and 474. 1873. 14*s*.

Contopoulos.—A Lexicon of Modern Greek-English and English Modern Greek. By N. Contopoulos. In 2 vols. 8vo. cloth. Part I. Modern Greek-English, pp. 460. Part II. English-Modern Greek, pp. 582. £1 7*s*.

Contopoulos.—Handbook of Greek and English Dialogues and Correspondence. Fcap. 8vo. cloth, pp. 238. 1879. 2*s*. 6*d*.

Geldart.—A Guide to Modern Greek. By E. M. Geldart. Post 8vo. cloth, pp. xii. and 274. 1883. 7*s*. 6*d*. Key, cloth, pp. 28. 2*s*. 6*d*.

Geldart.—Simplified Grammar of Modern Greek. By E. M. Geldart, M.A. Crown 8vo. pp. 68, cloth. 1883. 2*s*. 6*d*.

Lascarides.—A Comprehensive Phraseological English-Ancient and Modern Greek Lexicon. Founded upon a manuscript of G. P. Lascarides, Esq., and Compiled by L. Myriantheus, Ph. D. In 2 vols. foolscap 8vo. pp. xii. and 1,338, cloth. 1882. £1 10*s*.

Newman.—Comments on the Text of Æschylus. By F. W. Newman. Demy 8vo. pp. xii. and 144, cloth. 1884. 5*s*.

Sophocles.—Romaic or Modern Greek Grammar. By E. A. Sophocles. 8vo. pp. xxviii. and 196. 10*s*. 6*d*.

GUJARATI.

Minocheherji.—Pahlavi, Gujarâti and English Dictionary. By Jamaspji Dastur Minocheherji Jamasp Asana. 8vo. Vol. I., pp. clxii. and 1 to 168. Vol. II., pp. xxxii and pp. 169 to 440. 1877 and 1879. Cloth. 14*s*. each. (To be completed in 5 vols.)

Shápurjí Edaljí.—A Grammar of the Gujarátí Language. By Shápurjí Edaljí. Cloth, pp. 127. 10*s*. 6*d*.

Shápurjí Edaljí.—A DICTIONARY, GUJRATI AND ENGLISH. By SHÁPURJÍ EDALJÍ. Second Edition. Crown 8vo. cloth, pp. xxiv. and 874. 21*s.*

GURMUKHI (PUNJABI).

Adi Granth (The); OR, THE HOLY SCRIPTURES OF THE SIKHS, translated from the original Gurmukī, with Introductory Essays, by Dr. ERNEST TRUMPP, Professor Regius of Oriental Languages at the University of Munich, etc. Roy. 8vo. cloth, pp. 866. £2 12*s.* 6*d.*

Singh.—SAKHEE BOOK; or, The Description of Gooroo Gobind Singh's Religion and Doctrines, translated from Gooroo Mukhi into Hindi, and afterwards into English. By SIRDAR ATTAR SINGH, Chief of Bhadour. With the author's photograph. 8vo. pp. xviii. and 205. 15*s.*

HAWAIIAN.

Andrews.—A DICTIONARY OF THE HAWAIIAN LANGUAGE, to which is appended an English-Hawaiian Vocabulary, and a Chronological Table of Remarkable Events. By LORRIN ANDREWS. 8vo. pp. 560, cloth. £1 11*s.* 6*d.*

HEBREW.

Bickell.—OUTLINES OF HEBREW GRAMMAR. By GUSTAVUS BICKELL, D.D. Revised by the Author; Annotated by the Translator, SAMUEL IVES CURTISS, junior, Ph.D. With a Lithographic Table of Semitic Characters by Dr. J. EUTING. Cr. 8vo. sd., pp. xiv. and 140. 1877. 3*s.* 6*d.*

Collins.—A GRAMMAR AND LEXICON OF THE HEBREW LANGUAGE, entitled Sefer Hassoham. By RABBI MOSEH BEN YITSHAK, of England. Edited from a MS. in the Bodleian Library of Oxford, and collated with a MS. in the Imperial Library of St. Petersburg, with Additions and Corrections. By G. W. COLLINS, M.A., Corpus Christi College, Camb., Hon. Hebrew Lecturer, Keble College, Oxford. Part I. 4to. pp. 112, wrapper. 1884. 7*s.* 6*d.*

Gesenius.—HEBREW AND ENGLISH LEXICON OF THE OLD TESTAMENT, including the Biblical Chaldee, from the Latin. By EDWARD ROBINSON. Fifth Edition. 8vo. cloth, pp. xii. and 1160. £1 16*s.*

Gesenius.—HEBREW GRAMMAR. Translated from the Seventeenth Edition. By Dr. T. J. CONANT. With Grammatical Exercises, and a Chrestomathy by the Translator. 8vo. cloth, pp. xvi.–364. £1.

Hebrew Literature Society (Publications of). Subscription £1 1*s.* per Series. 1872-3. *First Series.*

Vol. I. Miscellany of Hebrew Literature. Demy 8vo. cloth, pp. viii. and 228. 10*s.*

Vol. II. The Commentary of Ibn Ezra on Isaiah. Edited from MSS., and Translated with Notes, Introductions, and Indexes, by M. FRIEDLÄNDER, Ph.D. Vol. I. Translation of the Commentary. Demy 8vo. cloth, pp. xxviii. and 332. 10*s.* 6*d.*

Vol. III. The Commentary of Ibn Ezra. Vol. II. The Anglican Version of the Book of the Prophet Isaiah amended according to the Commentary of Ibn Ezra. Demy 8vo. cloth, pp. 112. 4*s.* 6*d.*

1877. *Second Series.*

Vol. I. Miscellany of Hebrew Literature. Vol. II. Edited by the Rev. A. LÖWY. Demy 8vo. cloth, pp. vi. and 276. 10*s.* 6*d.*

Vol. II. The Commentary of Ibn Ezra. Vol. III. Demy 8vo. cloth, pp. 172. 7*s.*

Vol. III. Ibn Ezra Literature. Vol. IV. Essays on the Writings of Abraham Ibn Ezra. By M. FRIEDLÄNDE, Ph.D. Demy 8vo. cloth, pp. x.–252 and 78. 12*s.* 6*d.*

1881. *Third Series.*

Vol. I. The Guide of the Perplexed of Maimonides. Translated from the original text and annotated by M. Friedländer, Ph.D. Demy, 8vo. pp. lxxx. —370, cloth. £1 5*s*.

Herson.—Talmudic Miscellany. See Trübner's Oriental Series, page 4.

Land.—The Principles of Hebrew Grammar. By J. P. N. Land, Professor of Logic and Metaphysic in the University of Leyden. Translated from the Dutch by Reginald Lane Poole, Balliol College, Oxford. Part I. Sounds. Part II. Words. Crown 8vo. pp. xx. and 220, cloth. 7*s*. 6*d*.

Mathews.—Abraham ben Ezra's Unedited Commentary on the Canticles, the Hebrew Text after two MS., with English Translation by H. J. Mathews, B.A., Exeter College, Oxford. 8vo. cl. limp, pp. x., 34, 24. 2*s*. 6*d*.

Nutt.—Two Treatises on Verbs containing Feeble and Double Letters by R. Jehuda Hayug of Fez, translated into Hebrew from the original Arabic by R. Moses Gikatilia, of Cordova; with the Treatise on Punctuation by the same Author, translated by Aben Ezra. Edited from Bodleian MSS. with an English Translation by J. W. Nutt, M.A. Demy 8vo. sewed, pp. 312. 1870. 7*s*. 6*d*.

Semitic (Songs of the). In English Verse. By G. E. W. Cr. 8vo. cloth, pp. 140. 5*s*.

Spiers.—The School System of the Talmud, and an Address delivered delivered at the Beth Hamidrash on the occasion of the Conclusion of the Talmudical Treatise, Baba Metsia. By the Rev. B. Spiers. Cloth 8vo. pp. 48. 1882. 2*s*. 6*d*.

Weber.—System der altsynagogalen Palästinischen Theologie. By Dr. Ferd. Weber. 8vo. sewed. Leipzig, 1880. 7*s*.

HINDI.

Ballantyne.—Elements of Hindí and Braj Bhákà Grammar. By the late James R. Ballantyne, LL.D. Second edition, revised and corrected Crown 8vo., pp. 44, cloth. 5*s*.

Bate.—A Dictionary of the Hindee Language. Compiled by J. D. Bate. 8vo. cloth, pp. 806. £2 12*s*. 6*d*.

Beames.—Notes on the Bhojpurí Dialect of Hindí, spoken in Western Behar. By John Beames, Esq., B.C.S., Magistrate of Chumparun. 8vo. pp. 26, sewed. 1868. 1*s*. 6*d*.

Browne.—A Hindi Primer. In Roman Character. By J. F. Browne, B.C.S. Crown 8vo. pp. 36, cloth. 1882. 2*s*. 6*d*.

Etherington.—The Student's Grammar of the Hindí Language. By the Rev. W. Etherington, Missionary, Benares. Second edition. Crown 8vo. pp. xiv., 255, and xiii., cloth. 1873. 12*s*.

Hoernle.—Hindi Grammar. See page 42.

Kellogg.—A Grammar of the Hindi Language, in which are treated the Standard Hindî, Braj, and the Eastern Hindî of the Ramayan of Tulsi Das; also the Colloquial Dialects of Marwar, Kumaon, Avadh, Baghelkhand, Bhojpur, etc., with Copious Philological Notes. By the Rev. S. H. Kellogg, M.A. Royal 8vo. cloth, pp. 400. 21*s*.

Mahabharata. Translated into Hindi for Madan Mohun Bhatt, by Krishnachandradharmadhikarin of Benares. (Containing all but the Harivansâ.) 3 vols. 8vo. cloth, pp. 574, 810, and 1106. £3 3*s*.

Mathuráprasáda Misra.—A TRILINGUAL DICTIONARY, being a Comprehensive Lexicon in English, Urdú, and Hindí, exhibiting the Syllabication, Pronunciation, and Etymology of English Words, with their Explanation in English, and in Urdú and Hindí in the Roman Character. By MATHURAPRASADA MISRA, Second Master, Queen's College, Benares. 8vo. cloth, pp. xv. and 1330, Benares, 1865. £2 2s.

HINDUSTANI.

Ballantyne.—HINDUSTANI SELECTIONS IN THE NASKHI AND DEVANAGARI Character. With a Vocabulary of the Words. Prepared for the use of the Scottish Naval and Military Academy, by JAMES R. BALLANTYNE. Royal 8vo. cloth, pp. 74. 3s. 6d.

Craven.—The Popular Dictionary in English and Hindustani and Hindustani and English, with a Number of Useful Tables. By the Rev. T. CRAVEN, M.A. Fcap. 8vo. pp. 214, cloth. 1882. 3s. 6d.

Dowson.—A GRAMMAR OF THE URDU OR HINDUSTANI LANGUAGE. By JOHN DOWSON, M.R.A.S. 12mo. cloth, pp. xvi. and 264. 10s. 6d.

Dowson.—A HINDUSTANI EXERCISE BOOK. Containing a Series of Passages and Extracts adapted for Translation into Hindustani. By JOHN DOWSON, M.R.A.S. Crown 8vo. pp. 100. Limp cloth, 2s. 6d.

Eastwick.—KHIRAD AFROZ (the Illuminator of the Understanding). By Maulaví Hafízu'd-dín. A New Edition of Hindústaní Text, carefully revised, with Notes, Critical and Explanatory. By EDWARD B. EASTWICK, F.R.S., Imperial 8vo. cloth, pp. xiv. and 319. Re-issne, 1867. 18s.

Fallon.—A NEW HINDUSTANI-ENGLISH DICTIONARY. With Illustrations from Hindustani Literature and Folk-lore. By S. W. FALLON, Ph.D. Halle. Roy. 8vo. cloth, pp. xxviii. and 1216 and x. Benares, 1879. £5 5s.

Fallon.—ENGLISH-HINDUSTANI DICTIONARY. With Illustrations from English Literature and Colloquial English Translated into Hindustani. By S. W. FALLON. Royal 8vo. pp. iv.-674, sewed. £2 2s.

Fallon.—A HINDUSTANI-ENGLISH LAW AND COMMERCIAL DICTIONARY. By S. W. FALLON. 8vo. cloth, pp. ii. and 284. Benares, 1879. £1 1s.

Ikhwánu-s Safá; or, BROTHERS OF PURITY. Describing the Contention between Men and Beasts as to the Superiority of the Human Race. Translated from the Hindustání by Professor J. DOWSON, Staff College, Sandhurst. Crown 8vo. pp. viii. and 156, cloth. 7s.

Khirad-Afroz (The Illuminator of the Understanding). By Maulaví Hafízu'd-dín. A new edition of the Hindústání Text, carefully revised, with Notes, Critical and Explanatory. By E. B. EASTWICK, M.P., F.R.S. 8vo. cloth, pp. xiv. and 321. 18s.

Lutaifi Hindee (The); OR, HINDOOSTANEE JEST-BOOK, containing a Choice Collection of Humorous Stories in the Arabic and Roman Characters; to which is added a Hindoostanee Poem by MEER MOOHUMMUD TUQUEE. 2nd edition, revised by W. C. Smyth. 8vo. pp. xvi. and 160. 1840. 10s. 6d.; reduced to 5s.

Mathuráprasáda Misra.—A TRILINGUAL DICTIONARY, being a comprehensive Lexicon in English, Urdú, and Hindí, exhibiting the Syllabication, Pronunciation, and Etymology of English Words, with their Explanation in English, and in Urdú and Hindí in the Roman Character. By MATHURÁ-PRASÁDA MISRA, Second Master, Queen's College, Benares. 8vo. pp. xv. and 1330, cloth. Benares, 1865. £2 2s.

Palmer.—HINDUSTANI GRAMMAR. See page 48.

HUNGARIAN.

Singer.—SIMPLIFIED GRAMMAR OF THE HUNGARIAN LANGUAGE. By I. SINGER, of Buda-Pesth. Crown 8vo. cloth, pp. vi. and 88. 1884. 4*s.* 6*d.*

ICELANDIC.

Anderson.—NORSE MYTHOLOGY, or the Religion of our Forefathers. Containing all the Myths of the Eddas carefully systematized and interpreted, with an Introduction, Vocabulary and Index. By R. B. ANDERSON, Prof. of Scandinavian Languages in the University of Wisconsin. Crown 8vo. cloth. Chicago, 1879. 12*s.* 6*d.*

Anderson and Bjarnason.—VIKING TALES OF THE NORTH. The Sagas of Thorstein, Viking's Son, and Fridthjof the Bold. Translated from the Icelandic by R. B. Anderson, M.A., and J. Bjarnason. Also, Tegner's Fridthjof's Saga. Translated into English by G. Stephens. Crown 8vo. cloth, pp. xviii. and 370. Chicago, 1877. 10*s.*

Cleasby.—AN ICELANDIC-ENGLISH DICTIONARY. Based on the MS. Collections of the late Richard Cleasby. Enlarged and completed by G. VIGFÚSSON. With an Introduction, and Life of Richard Cleasby, by G. WEBBE DASENT, D.C.L. 4to. £3 7*s.*

Cleasby.—APPENDIX TO AN ICELANDIC-ENGLISH DICTIONARY. *See* Skeat.

Edda Saemundar Hinns Froda—The Edda of Saemund the Learned. From the Old Norse or Icelandic. By BENJAMIN THORPE. Part I. with a Mythological Index. 12mo. pp. 152, cloth, 3*s.* 6*d.* Part II. with Index of Persons and Places. 12mo. pp. viii. and 172, cloth. 1866. 4*s.*; or in 1 Vol. complete, 7*s.* 6*d.*

Publications of the Icelandic Literary Society of Copenhagen. For Numbers 1 to 54, see "Record," No. 111, p. 14.

55. SKÍRNER TÍDINDI. Hins Islenzka Bókmentafèlags, 1878. 8vo. pp. 176. Kaupmannahöfn, 1878. Price 5*s.*

56. UM SIDBÓTINA Á ISLANDI eptir Þorkel Bjarnason, prest á Reynivöllum. Utgefid af Hinu Islenzka Bokmentafélagi. 8vo. pp. 177. Reykjavik, 1878. Price 7*s.* 6*d.*

57. BISKUPA SÖGUR, gefnar út af Hinu Íslenzka Bókmentafélagi. Annat Bindi III. 1878. 8vo. pp. 509 to 804. Kaupmannahöfn. Price 10*s.*

58. SKÝRSLUR OG REIKNÍNGAR Hins Islenzka Bókmentafèlags, 1877 to 1878. 8vo. pp. 28. Kaupmannahöfn, 1878. Price 2*s.*

59. FRJETTIR FRA ISLANDI, 1877, eptir V. Briem. 8vo. pp. 50. Reykjavik, 1878. Price 2*s.* 6*d.*

60. ALÞÍNGISSTADUR HINN FORNI VID ÖXARA, med Uppdrattum eptir Sigurd Gudmundsson. 8vo. pp. 66, with Map. Kaupmannahöfn, 1878. Price 6*s.*

Skeat.—A LIST OF ENGLISH WORDS, the Etymology of which is illustrated by Comparison with Icelandic. Prepared in the form of an Appendix to Cleasby and Vigfusson's Icelandic-English Dictionary. By the Rev. WALTER W. SKEAT, M.A., English Lecturer and late Fellow of Christ's College, Cambridge; and M.A. of Exeter College, Oxford; one of the Vice-Presidents of the Cambridge Philological Society; and Member of the Council of the Philological Society of London. 1876. Demy 4to. sewed. 2*s.*

JAPANESE.

Aston.—A Grammar of the Japanese Written Language. By W. G. Aston, M.A., Assistant Japanese Secretary, H.B.M.'s Legation, Yedo, Japan. Second edition, Enlarged and Improved. Royal 8vo. pp. 306. 28*s*.

Aston.—A Short Grammar of the Japanese Spoken Language. By W. G. Aston, M.A., H. B. M.'s Legation, Yedo, Japan. Third edition. 12mo. cloth, pp. 96. 12*s*.

Black.—Young Japan, Yokohama and Yedo. A Narrative of the Settlement and the City, from the Signing of the Treaties in 1858 to the close of the Year 1879. With a Glance at the Progress of Japan during a period of Twenty-one Years. By J. R. Black. Two Vols., demy 8vo. pp. xviii. and 418; xiv. and 522, cloth. 1881. £2 2*s*.

Chamberlain.—Classical Poetry of the Japanese. See "Trübner's Oriental Series," page 4.

Hepburn.—A Japanese and English Dictionary. With an English and Japanese Index. By J. C. Hepburn, M.D., LL.D. Second edition. Imperial 8vo. cloth, pp. xxxii., 632 and 201. £8 8*s*.

Hepburn.—Japanese-English and English-Japanese Dictionary. By J. C. Hepburn, M.D., LL.D. Abridged by the Author from his larger work. Small 4to. cloth, pp. vi. and 206. 1873. 18*s*.

Hoffmann, J. J.—A Japanese Grammar. Second Edition. Large 8vo. cloth, pp. viii. and 368, with two plates. £1 1*s*.

Hoffmann.—Shopping Dialogues, in Japanese, Dutch, and English. By Professor J. Hoffmann. Oblong 8vo. pp. xiii. and 44, sewed. 5*s*.

Hoffmann (Prof. Dr. J. J.)—Japanese-English Dictionary.—Published by order of the Dutch Government. Elaborated and Edited by Dr. L. Serrurier. Vols. 1 and 2. Royal 8vo. Brill, 1881. 12*s*. 6*d*.

Imbrie.—Handbook of English-Japanese Etymology. By W. Imbrie. 8vo. pp. xxiv. and 208, cloth. Tōkiyō, 1880. £1 1*s*.

Metchnikoff.—L'Empire Jǎponais, texte et dessins, par L. Metchnikoff. 4to. pp. viii. and 694. Illustrated with maps, coloured plates and woodcuts. cloth. 1881. £1 10*s*.

Pfoundes.—Tu So Mimi Bokuro. See page 31.

Satow.—An English Japanese Dictionary of the Spoken Language. By Ernest Mason Satow, Japanese Secretary to H.M. Legation at Yedo, and Ishibashi Masakata, of the Imperial Japanese Foreign Office. Second edition. Imp. 32mo., pp. xvi. and 416, cloth. 12*s*. 6*d*.

Suyematz.—Genji Monogatari. The most celebrated of the Classical Japanese Romances. Translated by K. Suyematz. Crown 8vo. pp. xvi. and 254, cloth. 1882. 7*s*. 6*d*.

KANARESE.

Garrett.—A Manual English and Kanarese Dictionary, containing about Twenty-three Thousand Words. By J. Garrett. 8vo. pp. 908, cloth. Bangalore, 1872. 18*s*.

KAYATHI.

Grierson.—A Handbook to the Kayathi Character. By G. A. Grierson, B.C.S., late Subdivisional Officer, Madhubani, Darbhanga. With Thirty Plates in Facsimile, with Translations. 4to. cloth, pp. vi. and 4. Calcutta, 1881. 18*s*.

KELTIC (CORNISH, GAELIC, WELSH, IRISH).

Bottrell.—TRADITIONS AND HEARTHSIDE STORIES OF WEST CORNWALL. By W. BOTTRELL (an old Celt). Demy 12mo. pp. vi. 292, cloth. 1870. Scarce.

Bottrell.—TRADITIONS AND HEARTHSIDE STORIES OF WEST CORNWALL. By WILLIAM BOTTRELL. With Illustrations by Mr. JOSEPH BLIGHT. Second Series. Crown 8vo. cloth, pp. iv. and 300. 6*s*.

English and Welsh Languages.—THE INFLUENCE OF THE ENGLISH and Welsh Languages upon each other, exhibited in the Vocabularies of the two Tongues. Intended to suggest the importance to Philologers, Antiquaries, Ethnographers, and others, of giving due attention to the Celtic Branch of the Indo-Germanic Family of Languages. Square 8vo. sewed, pp. 30. 1869. 1*s*.

Mackay.—THE GAELIC ETYMOLOGY OF THE LANGUAGES OF WESTERN Europe, and more especially of the English and Lowland Scotch, and of their Slang, Cant, and Colloquial Dialects. By CHARLES MACKAY, LL.D. Royal 8vo. cloth, pp. xxxii. and 604. 42*s*.

Rhys.—LECTURES ON WELSH PHILOLOGY. By JOHN RHYS, M.A., Professor of Celtic at Oxford. Second edition, revised and enlarged. Crown 8vo. cloth, pp. viii. and 466. 15*s*.

Spurrell.—A GRAMMAR OF THE WELSH LANGUAGE. By WILLIAM SPURRELL. 3rd Edition. Fcap. cloth, pp. viii.-206. 1870. 3*s*.

Spurrell.—A WELSH DICTIONARY. English-Welsh and Welsh-English. With Preliminary Observations on the Elementary Sounds of the English Language, a copious Vocabulary of the Roots of English Words, a list of Scripture Proper Names and English Synonyms and Explanations. By WILLIAM SPURRELL. Third Edition. Fcap. cloth, pp. xxv. and 732. 8*s*. 6*d*.

Stokes.—GOIDELICA—Old and Early-Middle Irish Glosses: Prose and Verse. Edited by WHITLEY STOKES. Second edition. Medium 8vo. cloth, pp. 192. 1872. 18*s*.

Stokes.—TOGAIL TROI; The Destruction of Troy. Transcribed from the fascimile of the book of Leinster, and Translated with a Glossarial Index of the Rare words. By W. STOKES. 8vo. pp. xv.-188, boards. 1882. 18*s*. A limited edition only, privately printed, Calcutta.

Stokes.—THE BRETON GLOSSES AT ORLEANS. By W. STOKES. 8vo. pp. x.-78, boards. 1880. 10*s*. 6*d*. A limited edition only, privately printed, Calcutta.

Stokes.—THREE MIDDLE-IRISH HOMILIES on the Lives of Saints Patrick, Brigit, and Columba. By W. STOKES. 8vo. pp. xii.-140, boards. 1877. 10*s*. 6*d*. A limited edition only privately printed, Calcutta.

Stokes.—BEUNANS MERIASEK. The Life of Saint Meriasek, Bishop and Confessor. A Cornish Drama. Edited, with a Translation and Notes, by WHITLEY STOKES. Medium 8vo. cloth, pp. xvi.-280, and Facsimile. 1872. 15*s*.

Wright's Celt, Roman, and Saxon.

KONKANI.

Maffei.—A KONKANI GRAMMAR. By ANGELUS F. X. MAFFEI. 8vo. pp. xiv. and 438, cloth. Mangalore, 1882. 18*s*.

Maffei.—AN ENGLISH-KONKANI AND KONKANI-ENGLISH DICTIONARY. 8vo. pp. xii. and 546; xii. and 158. Two parts in one. Half bound. £1 10*s*.

LIBYAN.

Newman.—LIBYAN VOCABULARY. An Essay towards Reproducing the Ancient Numidian Language, out of Four Modern Languages. By F. W. Newman, Emeritus Professor of University College, London; formerly Fellow of Balliol College; and now M.R.A.S. Crown 8vo. pp. vi. and 204, cloth. 1882. 10*s*. 6*d*.

MAHRATTA.

Æsop's Fables.—Originally Translated into Marathi by Sadashiva Kashinath Chhatre. Revised from the 1st ed. 8vo. cloth. Bombay, 1877. 5*s*. 6*d*.

Ballantyne.—A GRAMMAR OF THE MAHRATTA LANGUAGE. For the use of the East India College at Haileybury. By JAMES R. BALLANTYNE, of the Scottish Naval and Military Academy. 4to. cloth, pp. 56. 5*s*.

Bellairs.—A GRAMMAR OF THE MARATHI LANGUAGE. By H. S. K. BELLAIRS, M.A., and LAXMAN Y. ASHKEDKAR, B.A. 12mo. cloth, pp. 90. 5*s*.

Molesworth.—A DICTIONARY, MÁRATHI and ENGLISH. Compiled by J. T. MOLESWORTH, assisted by GEORGE and THOMAS CANDY. Second Edition, revised and enlarged. By J. T. MOLESWORTH. Royal 4to. pp. xxx. and 922, boards. Bombay, 1857. £3 3*s*.

Molesworth.—A COMPENDIUM OF MOLESWORTH'S MARATHI AND ENGLISH DICTIONARY. By BABA PADMANJI. Second Edition. Revised and Enlarged. Demy 8vo. cloth, pp. xx. and 624. 21*s*.

Navalkar.—THE STUDENT'S MARÁTHI GRAMMAR. By G. R. NAVALKAR. New Edition. 8vo. cloth, pp. xvi. and 342. Bombay, 1879. 18*s*.

Tukarama.—A COMPLETE COLLECTION of the Poems of Tukáráma (the Poet of the Maháráshtra). In Marathi. Edited by VISHNU PARASHURAM SHASTRI PANDIT, under the supervision of Sankar Pandurang Pandit, M.A. With a complete Index to the Poems and a Glossary of difficult Words. To which is prefixed a Life of the Poet in English, by Janárdan Sakhárám Gádgil. 2 vols. in large 8vo. cloth, pp. xxxii. and 742, and pp. 728, 18 and 72. Bombay 1873. £1 11*s*. 6*d*. each vol.

MALAGASY.

Parker.—A CONCISE GRAMMAR OF THE MALAGASY LANGUAGE. By G. W. PARKER. Crown 8vo. pp. 66, with an Appendix, cloth. 1883. 5*s*.

Van der Tuuk.—OUTLINES OF A GRAMMAR OF THE MALAGASY LANGUAGE By H. N. VAN DER TUUK. 8vo., pp. 28, sewed. 1*s*.

MALAY.

Dennys.—A HANDBOOK OF MALAY COLLOQUIAL, as spoken in Singapore, Being a Series of Introductory Lessons for Domestic and Business Purposes. By N. B. DENNYS, Ph.D., F.R.G.S., M.R.A.S., etc., Author of "The Folklore of China," "Handbook of Cantonese," etc., etc. 8vo. cloth, pp. 204. 1878. £1 1*s*.

Maxwell.—A MANUAL OF THE MALAY LANGUAGE. With an Introductory Sketch of the Sanskrit Element in Malay. By W. E. MAXWELL, Assistant Resident, Perak, Malay Peninsula. Crown 8vo. cloth, pp. viii–184. 1882. 7*s*. 6*d*.

Swettenham.—VOCABULARY OF THE ENGLISH AND MALAY LANGUAGES. With Notes. By F. A. SWETTENHAM. 2 Vols. Vol. I. English-Malay Vocabulary and Dialogues. Vol. II. Malay-English Vocabulary. Small 8vo. boards. Singapore, 1881. £1.

Van der Tuuk.—SHORT ACCOUNT OF THE MALAY MANUSCRIPTS BELONGING TO THE ROYAL ASIATIC SOCIETY. By H. N. VAN DER TUUK. 8vo., pp. 52. 2*s*. 6*d*.

MALAYALIM.

Gundert.—A MALAYALAM AND ENGLISH DICTIONARY. By Rev. H. GUNDERT, D. Ph. Royal 8vo. pp. viii. and 1116. £2 10*s*.

MAORI.

Grey.—MAORI MEMENTOS: being a Series of Addresses presented by the Native People to His Excellency Sir George Grey, K.C.B., F.R.S. With Introductory Remarks and Explanatory Notes; to which is added a small Collection of Laments, etc. By CH. OLIVER B. DAVIS. 8vo. pp. iv. and 228, cloth. 12*s.*

Williams.—FIRST LESSONS IN THE MAORI LANGUAGE. With a Short Vocabulary. By W. L. WILLIAMS, B.A. Fcap. 8vo. pp. 98, cloth. 5*s.*

PALI.

D'Alwis.—A DESCRIPTIVE CATALOGUE of Sanskrit, Pali, and Sinhalese Literary Works of Ceylon. By JAMES D'ALWIS, M.R.A.S., etc., Vol. I. (all published), pp. xxxii. and 244. 1870. 8*s.* 6*d.*

Beal.—DHAMMAPADA. See "Trübner's Oriental Series," page 3.

Bigandet.—GAUDAMA. See "Trübner's Oriental Series," page 4.

Buddhist Birth Stories. See "Trübner's Oriental Series," page 4.

Bühler.—THREE NEW EDICTS OF AŚOKA. By G. BÜHLER. 16mo. sewed, with Two Facsimiles. 2*s.* 6*d.*

Childers.—A PALI-ENGLISH DICTIONARY, with Sanskrit Equivalents, and with numerous Quotations, Extracts, and References. Compiled by the late Prof. R. C. CHILDERS, late of the Ceylon Civil Service. Imperial 8vo. Double Columns. Complete in 1 Vol., pp. xxii. and 622, cloth. 1875. £3 3*s.*

The first Pali Dictionary ever published.

Childers.—THE MAHÂPARINIBBÂNASUTTA OF THE SUTTA-PITAKA. The Pali Text. Edited by the late Professor R. C. CHILDERS. 8vo. cloth, pp. 72. 5*s.*

Childers.—ON SANDHI IN PALI. By the late Prof. R. C. CHILDERS. 8vo. sewed, pp. 22. 1*s.*

Coomára Swamy.—SUTTA NÍPÁTA; or, the Dialogues and Discourses of Gotama Buddha. Translated from the Pali, with Introduction and Notes. By Sir M. COOMARA SWAMY. Cr. 8vo. cloth, pp. xxxvi. and 160. 1874. 6*s.*

Coomára Swamy.—THE DATHÁVANSA; or, the History of the Tooth-Relic of Gotama Buddha. English Translation only. With Notes. Demy 8vo. cloth, pp. 100. 1874. 6*s.*

Coomára Swamy.—THE DATHÁVANSA; or, the History of the Tooth-Relic of Gotama Buddha. The Pali Text and its Translation into English, with Notes. By Sir M. COOMARA SWAMY, Mudeliár. Demy 8vo. cloth, pp. 174. 1874. 10*s.* 6*d.*

Davids.—See BUDDHIST BIRTH STORIES, "Trübner's Oriental Series," page 4.

Davids.—SÌGIRI, THE LION ROCK, NEAR PULASTIPURA, AND THE 39TH CHAPTER OF THE MAHÂVAMSA. By T. W. RHYS DAVIDS. 8vo. pp. 30. 1*s.* 6*d.*

Dickson.—THE PÂTIMOKKHA, being the Buddhist Office of the Confession of Priests. The Pali Text, with a Translation, and Notes, by J. F. DICKSON. 8vo. sd., pp. 69. 2*s.*

Fausböll.—JÁTAKA. See under JÁTAKA.

Fausböll.—THE DASARATHA-JÁTAKA, being the Buddhist Story of King Râma. The original Pâli Text, with a Translation and Notes by V. FAUSBÖLL. 8vo. sewed, pp. iv. and 48. 2*s.* 6*d.*

Fausböll.—FIVE JÁTAKAS, containing a Fairy Tale, a Comical Story, and Three Fables. In the original Páli Text, accompanied with a Translation and Notes. By V. FAUSBÖLL. 8vo. sewed, pp. viii. and 72. 6*s.*

Fausböll.—TEN JÁTAKAS. The Original Páli Text, with a Translation and Notes. By V. FAUSBÖLL. 8vo. sewed, pp. xiii. and 128. 7*s.* 6*d.*

Fryer.—VUTTODAYA. (Exposition of Metre.) By SANGHARAKKHITA THERA. A Pali Text, Edited, with Translation and Notes, by Major G. E. FRYER. 8vo. pp. 44. 2*s*. 6*d*.

Haas.—CATALOGUE OF SANSKRIT AND PALI BOOKS IN THE LIBRARY OF THE BRITISH MUSEUM. By Dr. ERNST HAAS. Printed by Permission of the Trustees of the British Museum. 4to. cloth, pp. 200. £1 1*s*.

Jataka (The); together with its Commentary. Being Tales of the Anterior Birth of Gotama Buddha. For the first time Edited in the original Pali by V. FAUSBOLL. Demy 8vo. cloth. Vol. I. pp. 512. 1877. 28*s*. Vol. II., pp. 452. 1879. 28*s*. Vol. III. pp. viii.-544. 1883. 28*s*. For Translation see under "Buddhist Birth Stories," page 4.

The "Jataka" is a collection of legends in Pali, relating the history of Buddha's transmigration before he was born as Gotama. The great antiquity of this work is authenticated by its forming part of the sacred canon of the Southern Buddhists, which was finally settled at the last Council in 246 B.C. The collection has long been known as a storehouse of ancient fables, and as the most original attainable source to which almost the whole of this kind of literature, from the Panchatantra and Pilpay's fables down to the nursery stories of the present day, is traceable; and it has been considered desirable, in the interest of Buddhistic studies as well as for more general literary purposes, that an edition and translation of the complete work should be prepared. The present publication is intended to supply this want.—*Athenæum*.

Mahawansa (The)—THE MAHAWANSA. From the Thirty-Seventh Chapter. Revised and edited, under orders of the Ceylon Government, by H. SUMANGALA, and DON ANDRIS DE SILVA BATUWANTUDAWA. Vol. I. Pali Text in Sinhalese character, pp. xxxii. and 436. Vol. II. Sinhalese Translation, pp. lii. and 378 half bound. Colombo, 1877. £2 2*s*.

Mason.—THE PALI TEXT OF KACHCHAYANO'S GRAMMAR, WITH ENGLISH ANNOTATIONS. By FRANCIS MASON, D.D. I. The Text Aphorisms, 1 to 673. II. The English Annotations, including the various Readings of six independent Burmese Manuscripts, the Singalese Text on Verbs, and the Cambodian Text on Syntax. To which is added a Concordance of the Aphorisms. In Two Parts. 8vo. sewed, pp. 208, 75, and 28. Toongoo, 1871. £1 11*s*. 6*d*.

Minayeff.—GRAMMAIRE PALIE. Esquisse d'une Phonétique et d'une Morphologie de la Langue Palie. Traduite du Russe par St. Guyard. By J. MINAYEFF. 8vo. pp. 128. Paris, 1874. 8*s*.

Müller.—SIMPLIFIED GRAMMAR OF THE PALI LANGUAGE. By E. MÜLLER, Crown 8vo. cloth, pp. xvi. and 144. 1884. 7*s*. 6*d*.

Olcott.—BUDDHIST CATECHISM.

Senart.—KACCÂYANA ET LA LITTÉRATURE GRAMMATICALE DU PÂLI. Ire Partie. Grammaire Palie de Kaccâyana, Sutras et Commentaire, publiés avec une traduction et des notes par E. SENART. 8vo. pp. 338. Paris, 1871. 12*s*.

PAZAND.

Maino-i-Khard (The Book of the).—The Pazand and Sanskrit Texts (in Roman characters) as arranged by Neriosengh Dhaval, in the fifteenth century. With an English translation, a Glossary of the Pazand texts, containing the Sanskrit, Rosian, and Pahlavi equivalents, a sketch of Pazand Grammar, and an Introduction. By E. W. WEST. 8vo. sewed, pp. 484. 1871. 16*s*.

PEGUAN.

Haswell.—GRAMMATICAL NOTES AND VOCABULARY OF THE PEGUAN LANGUAGE. To which are added a few pages of Phrases, etc. By Rev. J. M. HASWELL. 8vo. pp. xvi. and 160. 15*s*.

PEHLEWI.

Dinkard (The).—The Original Pehlwi Text, the same transliterated in Zend Characters. Translations of the Text in the Gujrati and English Languages; a Commentary and Glossary of Select Terms. By Peshotun Dustoor Behramjee Sunjana. Vols. I. and II. 8vo. cloth. £2 2*s*.

Haug.—An Old Pahlavi-Pazand Glossary. Ed., with Alphabetical Index, by Destur Hoshangji Jamaspji Asa, High Priest of the Parsis in Malwa. Rev. and Enl., with Intro. Essay on the Pahlavi Language, by M. Haug, Ph.D. Pub. by order of Gov. of Bombay. 8vo. pp. xvi. 152, 268, sd. 1870. 28*s*.

Haug.—A Lecture on an Original Speech of Zoroaster (Yasna 45), with remarks on his age. By Martin Haug, Ph.D. 8vo. pp. 28, sewed. Bombay, 1865. 2*s*.

Haug.—The Parsis. See "Trübner's Oriental Series," page 3.

Haug.—An Old Zand-Pahlavi Glossary. Edited in the Original Characters, with a Transliteration in Roman Letters, an English Translation, and an Alphabetical Index. By Destur Hoshengji Jamaspji, High-priest of the Parsis in Malwa, India. Rev. with Notes and Intro. by Martin Haug, Ph.D. Publ. by order of Gov. of Bombay. 8vo. sewed, pp. lvi. and 132. 15*s*.

Haug.—The Book of Arda Viraf. The Pahlavi text prepared by Destur Hoshangji Jamaspji Asa. Revised and collated with further MSS., with an English translation and Introduction, and an Appendix containing the Texts and Translations of the Gosht-i Fryano and Hadokht Nask. By Martin Haug, Ph.D., Professor of Sanskrit and Comparative Philology at the University of Munich. Assisted by E. W. West, Ph.D. Published by order of the Bombay Government. 8vo. sewed, pp. lxxx., v., and 316. £1 5*s*.

Minocheherji.—Pahlavi, Gujarâti and English Dictionary. By Jamaspji Dastur Minocherji, Jamasp Asana. 8vo. Vol. I. pp. clxii. and 1 to 168, and Vol. II. pp. xxxii. and pp. 169 to 440. 1877 and 1879. Cloth. 14*s*. each. (To be completed in 5 vols.)

Sunjana.—A Grammar of the Pahlvi Language, with Quotations and Examples from Original Works and a Glossary of Words bearing affinity with the Semitic Languages. By Peshotun Dustoor Behramjee Sunjana, Principal of Sir Jamsetjee Jejeeboy Zurthosi Madressa. 8vo. cl., pp. 18–457. 25*s*.

Thomas.—Early Sassanian Inscriptions, Seals and Coins, illustrating the Early History of the Sassanian Dynasty, containing Proclamations of Ardeshir Babek, Sapor I., and his Successors. With a Critical Examination and Explanation of the Celebrated Inscription in the Hâjîâbad Cave, demonstrating that Sapor, the Conqueror of Valerian, was a Professing Christian. By Edward Thomas, F.R.S. Illustrated. 8vo. cloth, pp. 148. 7*s*. 6*d*.

Thomas.—Comments on Recent Pehlvi Decipherments. With an Incidental Sketch of the Derivation of Aryan Alphabets, and Contributions to the Early History and Geography of Tabaristân. Illustrated by Coins. By Edward Thomas, F.R.S. 8vo. pp. 56, and 2 plates, cloth, sewed. 3*s*. 6*d*.

West.—Glossary and Index of the Pahlavi Texts of the Book of Arda Viraf, The Tale of Gosht-I Fryano, The Hadókht Nask, and to some extracts from the Din-Kard and Nirangistan; prepared from Destur Hoshangji Asa's Glossary to the Arda Viraf Namak, and from the Original Texts, with Notes on Pahlavi Grammar. By E. W. West, Ph.D. Revised by Martin Haug, Ph.D. Published by order of the Government of Bombay. 8vo. sewed, pp. viii. and 352. 25*s*.

PENNSYLVANIA DUTCH.

Haldeman. — PENNSYLVANIA DUTCH: a Dialect of South Germany with an Infusion of English. By S. S. HALDEMAN, A.M., Professor of Comparative Philology in the University of Pennsylvania, Philadelphia. 8vo. pp. viii. and 70, cloth. 1872. 3*s*. 6*d*.

PERSIAN.

Ballantyne.—PRINCIPLES OF PERSIAN CALIGRAPHY, illustrated by Lithographic Plates of the TA"LIK characters, the one usually employed in writing the Persian and the Hindūstānī. Second edition. Prepared for the use of the Scottish Naval and Military Academy, by JAMES R. BALLANTYNE. 4to. cloth, pp. 14, 6 plates. 2*s*. 6*d*.

Blochmann.—THE PROSODY OF THE PERSIANS, according to Saifi, Jami, and other Writers. By H. BLOCHMANN, M.A. Assistant Professor, Calcutta Madrasah. 8vo. sewed, pp. 166. 10*s*. 6*d*.

Blochmann.—A TREATISE ON THE RUBA'I entitled Risalah i Taranah. By AGHA AHMAD 'ALI. With an Introduction and Explanatory Notes, by H. BLOCHMANN, M.A. 8vo. sewed, pp. 11 and 17. 2*s*. 6*d*.

Blochmann.—THE PERSIAN METRES BY SAIFI, and a Treatise on Persian Rhyme by Jami. Edited in Persian, by H. BLOCHMANN, M.A. 8vo. sewed pp. 62. 3*s*. 6*d*.

Catalogue of Arabic and Persian Books, Printed in the East. Constantly for sale by Trübner and Co. 16mo. sewed, pp. 46. 1*s*.

Eastwick.—THE GULISTAN. See "Trübner's Oriental Series," page 4.

Finn.—PERSIAN FOR TRAVELLERS. By A. FINN, H.B.M. Consul at RESHT. Part I. Rudiments of Grammar. Part II. English-Persian Vocabulary. Oblong 32mo, pp. xxii.—232, cloth. 1884. 5*s*.

Griffith.—YUSUF AND ZULAIKHA. See "Trübner's Oriental Series," p. 5.

Háfiz of Shíráz.—SELECTIONS FROM HIS POEMS. Translated from the Persian by HERMAN BICKNELL. With Preface by A. S. BICKNELL. Demy 4to., pp. xx. and 384, printed on fine stout plate-paper, with appropriate Oriental Bordering in gold and colour, and Illustrations by J. R. HERBERT, R.A. £2 2*s*.

Haggard and Le Strange.—THE VAZIR OF LANKURAN. A Persian Play. A Text-Book of Modern Colloquial Persian, for the use of European Travellers, Residents in Persia, and Students in India. Edited, with a Grammatical Introduction, a Translation, copious Notes, and a Vocabulary giving the Pronunciation of all the words. By W. H. HAGGARD and GUY LE STRANGE. Crown 8vo. pp. xl.-176 and 56 (Persian Text), cloth. 1882. 10*s*. 6*d*.

Mírkhónd.—THE HISTORY OF THE ATÁBEKS OF SYRIA AND PERSIA. By MUHAMMED BEN KHÁWENDSHÁH BEN MAHMUD, commonly called MÍRKHÓND. Now first Edited from the Collation of Sixteen MSS., by W. H. MORLEY, Barrister-at-law, M.R.A.S. To which is added a Series of Facsimiles of the Coins struck by the Atábeks, arranged and described by W. S. W. Vaux, M.A., M.R.A.S. Roy. 8vo. cloth, 7 Plates, pp. 118. 1848. 7*s*. 6*d*.

Morley.—A Descriptive Catalogue of the Historical Manuscripts in the Arabic and Persian Languages preserved in the Library of the Royal Asiatic Society of Great Britain and Ireland. By WILLIAM H. MORLEY, M.R.A.S. 8vo. pp. viii. and 160, sewed. London, 1854. 2*s*. 6*d*.

Palmer.—THE SONG OF THE REED; and other Pieces. By E. H. PALMER, M.A., Cambridge. Crown 8vo. cloth, pp. 208, 5*s*.

Among the Contents will be found translations from Hafiz, from Omer el Kheiyám, and from other Persian as well as Arabic poets.

Palmer.—A CONCISE PERSIAN-ENGLISH DICTIONARY By E. H. PALMER, M.A., Professor of Arabic in the University of Cambridge. Second Edition. Royal 16mo. pp. viii. and 364, cloth. 1883. 10*s.* 6*d.*

Palmer.—A CONCISE ENGLISH-PERSIAN DICTIONARY. Together with a Simplified Grammar of the Persian Language. By the late E. H. PALMER, M.A., Lord Almoner's Reader and Professor of Arabic, Cambridge. Completed and Edited from the MS. left imperfect at his death. By G. LE STRANGE. Royal 16mo. pp. xii. and 546, cloth. 1883. 10*s.* 6*d.*

Palmer.—PERSIAN GRAMMAR. See page 48.

Redhouse.—THE MESNEVI. See "Trübner's Oriental Series," page 4.

Rieu.—CATALOGUE OF THE PERSIAN MANUSCRIPTS IN THE BRITISH MUSEUM. By CHARLES RIEU, Ph.D., Keeper of the Oriental MSS. Vol. I. 4to. cloth, pp. 432. 1879. £1 5*s.* Vol. II. 4to. cloth, pp. viii. and 446. 1881. 25*s*

Whinfield.—GULSHAN-I-RAZ; The Mystic Rose Garden of Sa'd ud din Mahmud Shabistani. The Persian Text, with an English Translation and Notes, chiefly from the Commentary of Muhammed Bin Yahya Lahiji. By E. H. WHINFIELD, M.A., late of H.M.B.C.S. 4to. pp. xvi., 94, 60, cloth. 1880. 10*s.* 6*d*

Whinfield.—THE QUATRAINS OF OMAR KHAYYÁM. Translated into English Verse by E. H. WHINFIELD, M.A., late of Bengal Civil Service. Post 8vo. cloth, pp. 96. 1881. 5*s.*

PIDGIN-ENGLISH.

Leland.—PIDGIN-ENGLISH SING-SONG; or Songs and Stories in the China-English Dialect. With a Vocabulary. By CHARLES G. LELAND. Fcap. 8vo. cl., pp. viii. and 140. 1876. 5*s.*

POLISH.

Morfill.—A SIMPLIFIED GRAMMAR OF THE POLISH LANGUAGE. By W. R. MORFILL, M.A. Crown 8vo. pp. viii.—64, cloth. 1884. 3*s.* 6*d.*

PRAKRIT.

Cowell.—A SHORT INTRODUCTION TO THE ORDINARY PRAKRIT OF THE SANSKRIT DRAMAS. With a List of Common Irregular Prakrit Words. By Prof. E. B. COWELL. Cr. 8vo. limp cloth, pp. 40. 1875. 3*s.* 6*d.*

Cowell.—PRAKRITA-PRAKASA; or, The Prakrit Grammar of Vararuchi, with the Commentary (Manorama) of Bhamaha; the first complete Edition of the Original Text, with various Readings from a collation of Six MSS. in the Bodleian Library at Oxford, and the Libraries of the Royal Asiatic Society and the East India House; with Copious Notes, an English Translation, and Index of Prakrit Words, to which is prefixed an Easy Introduction to Prakrit Grammar. By EDWARD BYLES COWELL, of Magdalen Hall, Oxford, Professor of Sanskrit at Cambridge. New Edition, with New Preface, Additions, and Corrections. Second Issue. 8vo. cloth, pp. xxxi. and 204. 1868. 14*s.*

PUKSHTO (PAKKHTO, PASHTO).

Bellew.—A GRAMMAR OF THE PUKKHTO OR PUKSHTO LANGUAGE, on a New and Improved System. Combining Brevity with Utility, and Illustrated by Exercises and Dialogues. By H. W. BELLEW, Assistant Surgeon, Bengal Army. Super-royal 8vo., pp. xii. and 156, cloth. 21*s.*

Bellew.—A DICTIONARY OF THE PUKKHTO, OR PUKSHTO LANGUAGE, on a New and Improved System. With a reversed Part, or English and Pukkhto, By H. W. BELLEW, Assistant Surgeon, Bengal Army. Super Royal 8vo. pp. xii. and 356, cloth. 42*s*.

Plowden.—TRANSLATION OF THE KALID-I-AFGHANI, the Text Book for the Pakkhto Examination, with Notes, Historical, Geographical, Grammatical, and Explanatory. By TREVOR CHICHELE PLOWDEN, Captain H.M. Bengal Infantry, and Assistant Commissioner, Panjab. Small 4to. cloth, pp. xx. and 395 and ix. With Map. *Lahore*, 1875. £2 10*s*.

Thorburn.—BANNÚ; or, Our Afghan Frontier. By S. S. THORBURN, I.C.S., Settlement Officer of the Bannú District. 8vo. cloth, pp. x. and 480. 1876. 18*s*.

pp. 171 to 230: Popular Stories, Ballads and Riddles, and pp. 231 to 413: Pashto Proverbs Translated into English. pp. 414 to 473: Pashto Proverbs in Pashto.

Trumpp.—GRAMMAR OF THE PAŠTO, or Language of the Afghans, compared with the Irānian and North-Indian Idioms. By Dr. ERNEST TRUMPP 8vo. sewed, pp. xvi. and 412. 21*s*.

ROUMANIAN.

Torceanu.—SIMPLIFIED GRAMMAR OF THE ROUMANIAN LANGUAGE. By R. TORCEANU. Crown 8vo. pp. viii.-72, cloth. 1883. 5*s*.

RUSSIAN.

Riola.—A GRADUATED RUSSIAN READER, with a Vocabulary of all the Russian Words contained in it. By H. RIOLA. Crown 8vo. pp. viii. and 314. 1879. 10*s*. 6*d*.

Riola.—HOW TO LEARN RUSSIAN. A Manual for Students of Russian, based upon the Ollendorfian system of teaching languages, and adapted for self instruction. By HENRY RIOLA, Teacher of the Russian Language. With a Preface by W. R. S. RALSTON, M.A. Second Edition. Crown 8vo. cloth, pp. 576. 1884. 12*s*.

Key to the above. Crown 8vo. cloth, pp. 126. 1878. 5*s*.

Thompson.—DIALOGUES, RUSSIAN AND ENGLISH. Compiled by A. R. THOMPSON. Crown 8vo. cloth, pp. iv.-132. 1882. 5*s*.

SAMARITAN.

Nutt.—A SKETCH OF SAMARITAN HISTORY, DOGMA, AND LITERATURE. Published as an Introduction to "Fragments of a Samaritan Targum. By J. W. NUTT, M.A. Demy 8vo. cloth, pp. viii. and 172. 1874. 5*s*.

Nutt.—FRAGMENTS OF A SAMARITAN TARGUM. Edited from a Bodleian MS. With an Introduction, containing a Sketch of Samaritan History, Dogma, and Literature. By J. W. NUTT, M.A. Demy 8vo. cloth, pp. viii., 172, and 84. With Plate. 1874. 15*s*.

SAMOAN.

Pratt.—A GRAMMAR AND DICTIONARY of the Samoan Language. By Rev. GEORGE PRATT, Forty Years a Missionary of the London Missionary Society in Samoa. Second Edition. Edited by Rev. S. J. Whitmee, F.R.G.S. Crown 8vo. cloth, pp. viii. and 380. 1878. 18*s*.

SANSKRIT.

Aitareya Brahmanam of the Rig Veda. 2 vols. See under HAUG.

D'Alwis.—A DESCRIPTIVE CATALOGUE OF SANSKRIT, PALI, AND SINHALESE LITERARY WORKS OF CEYLON. By JAMES D'ALWIS, M.R.A.S., Advocate of the Supreme Court, &c., &c. In Three Volumes. Vol. I., pp. xxxii. and 244, sewed. 1870. 8*s*. 6*d*.

Apastambíya Dharma Sutram.—APHORISMS OF THE SACRED LAWS OF THE HINDUS, by APASTAMBA. Edited, with a Translation and Notes, by G. Bühler. By order of the Government of Bombay. 2 parts. 8vo. cloth, 1868-71. £1 4*s*. 6*d*.

Arnold.—LIGHT OF ASIA. See page 31.

Arnold.—INDIAN POETRY. See "Trübner's Oriental Series," page 4.

Arnold.—THE ILIAD AND ODYSSEY OF INDIA. By EDWIN ARNOLD, M.A., C.S.I., F.R.G.S., etc. Fcap. 8vo. sd., pp. 24. 1*s*.

Apte.—THE STUDENT'S GUIDE TO SANSKRIT COMPOSITION. Being a Treatise on Sanskrit Syntax for the use of School and Colleges. 8vo. boards. Poona, 1881. 6*s*.

Apte.—THE STUDENT'S ENGLISH-SANSKRIT DICTIONARY. Roy. 8vo. pp. xii. and 526, cloth. Poona, 1884. 16*s*.

Atharva Veda Prátiçákhya.—See under WHITNEY.

Auctores Sanscriti. Vol. I. The Jaiminîya-Nyâya-Mâlâ-Vistara. Edited for the Sanskrit Text Society under the supervision of THEODOR GOLDSTÜCKER. Parts I. to VII., pp. 582, large 4to. sewed. 10*s*. each part. Complete in one vol., cloth, £3 13*s*. 6*d*. Vol. II. The Institute of Gautama. Edited with an Index of Words, by A. F. STENZLER, Ph.D., Professor of Oriental Languages in the University of Breslau. 8vo. cloth, pp. iv. 78. 1876. 4*s*. 6*d*. Vol. III. Vaitâna Sûtra. The Ritual of the Atharva Veda. Edited with Critical Notes and Indices, by DR. RICHARD GARBE. 8vo. sewed, pp. 119. 1878. 5*s*. Vols. IV. and V. Vardhamana's Ganaratnamahodadhi, with the Author's Commentary. Edited, with Critical Notes and Indices, by J. EGGLING, Ph.D. 8vo. wrapper. Part I., pp. xii. and 240. 1879. 6*s*. Part II., pp. 240. 1881. 6*s*.

Avery.—CONTRIBUTIONS TO THE HISTORY OF VERB-INFLECTION IN SANSKRIT. By J. AVERY. (Reprinted from the Journal of the American Oriental Society, vol. x.) 8vo. paper, pp. 106. 4*s*.

Ballantyne.—SANKHYA APHORISMS OF KAPILA. See "Trübner's Oriental Series," page 6.

Ballantyne.—FIRST LESSONS IN SANSKRIT GRAMMAR; together with an Introduction to the Hitopadésa. Fourth edition. By JAMES R. BALLANTYNE, LL.D., Librarian of the India Office. 8vo. pp. viii. and 110, cloth. 1884. 3*s*. 6*d*.

Benfey.—A PRACTICAL GRAMMAR OF THE SANSKRIT LANGUAGE, for the use of Early Students. By THEODOR BENFEY, Professor of Sanskrit in the University of Göttingen. Second, revised and enlarged, edition. Royal 8vo. pp. viii. and 296, cloth. 10*s*. 6*d*.

Benfey.—A GRAMMAR OF THE LANGUAGE OF THE VEDAS. By Dr. THEODOR BENFEY. In 1 vol. 8vo., of about 650 pages. [*In preparation.*

Benfey.—VEDICA UND VERWANDTES. By THEOD. BENFEY. Crown 8vo. paper, pp. 178. Strassburg, 1877. 7*s*. 6*d*.

Benfey.—VEDICA UND LINGUISTICA.—By TH. BENFEY. Crown 8vo. pp. 254. 10*s*. 6*d*.

Bibliotheca Indica.—A Collection of Oriental Works published by the Asiatic Society of Bengal. Old Series. Fasc. 1 to 235. New Series. Fasc. 1 to 408. (Special List of Contents to be had on application.) Each Fasc. in 8vo., 2*s*.; in 4to., 4*s*.

Bibliotheca Sanskrita.—See Trübner.

Bombay Sanskrit Series. Edited under the superintendence of G. Bühler, Ph. D., Professor of Oriental Languages, Elphinstone College, and F. Kielhorn, Ph. D., Superintendent of Sanskṛit Studies, Deccan College. 1868–84.

1. Panchatantra iv. and v. Edited, with Notes, by G. Bühler, Ph. D. Pp. 84, 16. 3*s.*
2. Nágojíbhaṭṭa's Paribhásheṇduśekhara. Edited and explained by F. Kielhorn, Ph. D. Part I., the Sanskṛit Text and Various Readings. pp. 116. 4*s.*
3. Panchatantra ii. and iii. Edited, with Notes, by G. Bühler, Ph. D. Pp. 86, 14, 2. 3*s.*
4. Panchatantra i. Edited, with Notes, by F. Kielhorn, Ph.D. Pp. 114, 53. 3*s.*
5. Kálidása's Raghuvamṣa. With the Commentary of Mallinátha. Edited, with Notes, by Shankar P. Paṇḍit, M.A. Part I. Cantos I.–VI. 4*s.*
6. Kálidása's Málavikágnimitra. Edited, with Notes, by Shankar P. Paṇḍit, M.A. 4*s.* 6*d.*
7. Nágojíbhaṭṭa's Paribhásheṇduśekhara Edited and explained by F. Kielhorn, Ph.D. Part II. Translation and Notes. (Paribhâshâs, i.–xxxvii.) pp. 184. 4*s.*
8. Kálidása's Raghuvaṁṣa. With the Commentary of Mallinátha. Edited, with Notes, by Shankar P. Paṇḍit, M.A. Part II. Cantos VII.–XIII. 4*s.*
9. Nágojíbhaṭṭa's Paribhásheṇduṣekhara. Edited and explained by F. Kielhorn. Part II. Translation and Notes. (Paribhâshâs xxxviii.–lxix.) 4*s.*
10. Dandin's Dasakumaracharita. Edited with critical and explanatory Notes by G. Bühler. Part I. 3*s.*
11. Bhartrihari's Nitisataka and Vairagyasataka, with Extracts from Two Sanskrit Commentaries. Edited, with Notes, by Kasinath T. Telang. 4*s.* 6*d.*
12. Nagojibhatta's Paribháshendusekhara. Edited and explained by F. Kielhorn. Part II. Translation and Notes. (Paribhâshâs lxx.–cxxii.) 4*s.*
13. Kalidasa's Raghuvaṁṣa, with the Commentary of Mallinátha. Edited, with Notes, by Shankar P. Paṇḍit. Part III. Cantos XIV.–XIX. 4*s.*
14. Vikramânkadevacharita. Edited, with an Introduction, by G. Bühler. 3*s.*
15. Bhavabhûti's Mâlatî-Mâdhava. With the Commentary of Jagaddhara, edited by Ramkrishna Gopal Bhandarkar. 14*s.*
16. The Vikramorvasîyam. A Drama in Five Acts. By Kâlidâsa. Edited with English Notes by Shankar P. Pandit, M.A. pp. xii. and 129 (Sanskrit Text) and 148 (Notes). 1879. 6*s.*
17. Hemachdra's Desînâmâlâ, with a glossary by Dr. Pischel and Dr. Bühler. Part I. 10*s.*

18—22 and 26. Patanjah's Vyakaranamahabhāthya. By Dr. Kielhorn. Part I—IV. Vol. I. II. Part II. Each part 6*s.*

23. The Vâsishthadharmasastram. Aphorisms on the Sacred Law of the Aryas, as taught in School of Vasishtha. Edited by Rev. A. A. Fuhrur. 8vo. sewed. 1883. 2*s.* 6*d.*
24. Kadambari. Edited by Peter Peterson. 8vo. sewed. 1883. 15*s.*
25. Kirtikaumudi. Sri Somesvaradeva, and edited by Abaji Vishnu Kathavati. 8vo. sewed. 1883. 3*s.* 6*d.*

27. MUDRARAKSHASA. By VISAKHADATTA. With the commentary of Dhundhiraj. Edited with critical and explanatory notes by K. T. Telang. 8vo. sewed. 1884. 6*s*.

Borooah.—A COMPANION TO THE SANSKRIT-READING UNDERGRADUATES of the Calcutta University, being a few notes on the Sanskrit Texts selected for examination, and their Commentaries. By ANUNDORAM BOROOAH. 8vo. pp. 64. 3*s*. 6*d*.

Borooah.—A PRACTICAL ENGLISH-SANSKRIT DICTIONARY. By ANUNDORAM BOROOAH, B.A., B.C.S., of the Middle Temple, Barrister-at-Law. Vol. I. A to Falseness. pp. xx.-580-10. Vol. II. Falsification to Oyster, pp. 581 to 1060. With a Supplementary Treatise on Higher Sanskrit Grammar or Gender and Syntax, with copious illustrations from standard Sanskrit Authors and References to Latin and Greek Grammars, pp. vi. and 296. 1879. Vol. III. £1 11*s*. 6*d*. each.

Borooah.—BHAVABHUTI AND HIS PLACE IN SANSKRIT LITERATURE. By ANUNDORAM BOROOAH. 8vo. sewed, pp. 70. 5*s*.

Brhat-Sanhita (The).—See under Kern.

Brown.—SANSKRIT PROSODY AND NUMERICAL SYMBOLS EXPLAINED. By CHARLES PHILIP BROWN, Author of the Telugu Dictionary, Grammar, etc., Professor of Telugu in the University of London. Demy 8vo. pp. 64, cloth. 3*s*. 6*d*.

Burnell.—RIKTANTRAVYĀKARAṆA. A Prātiçākhya of the Samaveda. Edited, with an Introduction, Translation of the Sutras, and Indexes, by A. C. BURNELL, Ph.D. Vol. I. Post 8vo. boards, pp. lviii. and 84. 10*s*. 6*d*.

Burnell.—A CLASSIFIED INDEX to the Sanskrit MSS. in the Palace at Tanjore. Prepared for the Madras Government. By A. C. BURNELL, Ph.D. In 4to. Part I. pp. iv. and 80, stitched, stiff wrapper. Vedic and Technical Literature. Part II. pp. iv. and 80. Philosophy and Law. 1879. Part III. Drama, Epics, Purānas and Tantras, Indices, 1880. 10*s*. each part.

Burnell.—CATALOGUE OF A COLLECTION OF SANSKRIT MANUSCRIPTS. By A. C. BURNELL, M.R.A.S., Madras Civil Service. PART 1. *Vedic Manuscripts.* Fcap. 8vo. pp. 64, sewed. 1870. 2*s*.

Burnell.—DAYADAÇAÇLOKI. TEN SLOKAS IN SANSKRIT, with English Translation. By A. C. BURNELL. 8vo. pp. 11. 2s.

Burnell.—ON THE AINDRA SCHOOL OF SANSKRIT GRAMMARIANS. Their Place in the Sanskrit and Subordinate Literatures. By A. C. BURNELL. 8vo. pp. 120. 10*s*. 6*d*.

Burnell.—THE SÂMAVIDHÂNABRÂHMAṆA (being the Third Brâhmaṇa) of the Sâma Veda. Edited, together with the Commentary of Sâyaṇa, an English Translation, Introduction, and Index of Words, by A. C. BURNELL. Volume I.—Text and Commentary, with Introduction. 8vo. pp. xxxviii. and 104. 12*s*. 6*d*.

Burnell.—THE ARSHEYABRAHMANA (being the fourth Brāhmana) OF THE SAMA VEDA. The Sanskrit Text. Edited, together with Extracts from the Commentary of Sayana, etc. An Introduction and Index of Words. By A. C. BURNELL, Ph.D. 8vo, pp. 51 and 109. 10*s*. 6*d*.

Burnell.—THE DEVATĀDHYĀYABRĀHMAṆA (being the Fifth Brāhmaṇa) of the Sama Veda. The Sanskrit Text edited, with the Commentary of Sāyaṇa, an Index of Words, etc., by A. C. BURNELL, M.R.A.S. 8vo. and Trans., pp. 34. 5*s*.

Burnell.—THE JAIMINĪYA TEXT OF THE ARSHEYABRĀHMAṆA OF THE Sāma Veda. Edited in Sanskrit by A. C. BURNELL, Ph. D. 8vo. sewed, pp. 56. 7*s*. 6*d*.

Burnell. — THE SAMHITOPANISHADBRĀHMAṆA (Being the Seventh Brāhmaṇa) of the Sāma Veda. The Sanskrit Text. With a Commentary, an Index of Words, etc. Edited by A. C. BURNELL, Ph.D. 8vo. stiff boards, pp. 86. 7*s*. 6*d*.

Burnell.—THE VAMÇABRÂHMANA (being the Eighth Brâhmaṇa) of the Sâma Veda. Edited, together with the Commentary of Sâyaṇa, a Preface and Index of Words, by A. C. BURNELL, M.R.A.S., etc. 8vo. sewed, pp. xliii., 12, and xii., with 2 coloured plates. 10*s.* 6*d.*

Burnell.—The Ordinances of Manu. See "Trübner's Oriental Series," page 6.

Catalogue OF SANSKRIT WORKS PRINTED IN INDIA, offered for Sale at the affixed nett prices by TRÜBNER & Co. 16mo. pp. 52. 1*s.*

Chintamon.—A COMMENTARY ON THE TEXT OF THE BHAGAVAD-GÍTÁ; or, the Discourse between Krishna and Arjuna of Divine Matters. A Sanscrit Philosophical Poem. With a few Introductory Papers. By HURRYCHUND CHINTAMON, Political Agent to H. H. the Guicowar Mulhar Rao Maharajah of Baroda. Post 8vo. cloth, pp. 118. 6*s.*

Clark.—MEGHADUTA, THE CLOUD MESSENGER. Poem of Kalidasa. Translated by the late REV. THOMAS CLARK, M.A. Fcap. 8vo. pp. 64, wrapper. 1882. 1*s.*

Colebrooke.—The Life and Miscellaneous Essays of Henry Thomas Colebrooke. The Biography by his son, Sir T. E. COLEBROOKE, Bart., M.P. The Essays edited by Professor Cowell. In 3 vols.

Vol. I. The Life. With Portrait and Map. Demy 8vo. cloth, pp. xii. and 492. 14*s.*

Vols. II. and III. The Essays. A New Edition, with Notes by E. B. Cowell, Professor of Sanskrit at Cambridge. Demy 8vo. cloth, pp. xvi. and 544, and x. and 520. 1873. 28*s.*

Cowell and Eggeling.—CATALOGUE OF BUDDHIST SANSKRIT MANUSCRIPTS in the Possession of the Royal Asiatic Society (Hodgson Collection). By Professors E. B. COWELL and J. EGGELING. 8vo. sd., pp. 56. 2*s.* 6*d.*

Cowell.—SARVA DARSANA SAMGRAHA. See "Trübner's Oriental Series," page 5.

Da Cunha.—THE SAHYADRI KHAṆḌA OF THE SKANDA PURANA; a Mythological, Historical and Geographical Account of Western India. First edition of the Sanskrit Text, with various readings. By J. GERSON DA CUNHA, M.R.C.S. and L.M. Eng., L.R.C.P. Edinb., etc. 8vo. bds. pp. 580. £1 1*s.*

Davies.—HINDU PHILOSOPHY. See "Trübner's Oriental Series," page 4.

Davies.—BHAGAVAD GITA. See "Trübner's Oriental Series," page 5.

Dutt.—KINGS OF KÁSHMÍRA: being a Translation of the Sanskrita Work Rajataranggini of Kahlana Pandita. By J. CH. DUTT. 12mo. paper, pp. v. 302, and xxiii. 4*s.*

Gautama.—THE INSTITUTES OF GAUTAMA. *See Auctores Sanscriti.*

Goldstücker.—A DICTIONARY, SANSKRIT AND ENGLISH, extended and improved from the Second Edition of the Dictionary of Professor H. H. WILSON, with his sanction and concurrence. Together with a Supplement, Grammatical Appendices, and an Index, serving as a Sanskrit-English Vocabulary. By THEODOR GOLDSTÜCKER. Parts I. to VI. 4to. pp. 400. 1856-1863. 6*s.* each

Goldstücker.—PANINI: His Place in Sanskrit Literature. An Investigation of some Literary and Chronological Questions which may be settled by a study of his Work. A separate impression of the Preface to the Facsimile of MS. No. 17 in the Library of Her Majesty's Home Government for India, which contains a portion of the MANAVA-KALPA-SUTRA, with the Commentary of KUMARILA-SWAMIN. By THEODOR GOLDSTÜCKER. Imperial 8vo. pp 268, cloth. £2 2*s.*

Gough.—PHILOSOPHY OF THE UPANISHADS. See Trübner's Oriental Series, page 6.

Griffith.—SCENES FROM THE RAMAYANA, MEGHADUTA, ETC. Translated by RALPH T. H. GRIFFITH, M.A., Principal of the Benares College. Second Edition. Crown 8vo. pp. xviii., 244, cloth. 6*s.*

CONTENTS.—Preface—Ayodhya—Ravan Doomed—The Birth of Rama—The Heir apparent—Manthara's Guile—Dasaratha's Oath—The Step-mother—Mother and Son—The Triumph of Love—Farewell?—The Hermit's Son—The Trial of Truth—The Forest—The Rape of Sita—Rama's Despair—The Messenger Cloud—Khumbakarna—The Suppliant Dove—True Glory—Feed the Poor—The Wise Scholar.

Griffith.—THE RÁMÁYAN OF VÁLMÍKI. Translated into English verse. By RALPH T. H. GRIFFITH, M.A., Principal of the Benares College. 5 vols.

Vol. I., containing Books I. and II. Demy 8vo. pp. xxxii. 440, cloth' 1870. 18*s.* Out of print.

Vol. II., containing Book II., with additional Notes and Index of Names. Demy 8vo. pp. 504, cloth. 18*s.* Out of print.

Vol. III. Demy 8vo. pp. v. and 371, cloth. 1872. 15*s.*

Vol. IV. Demy 8vo. pp. viii. and 432. 1873. 18*s.*

Vol. V. Demy 8vo. pp. 368, cloth. 1875. 15*s.*

Griffith.—KÁLIDÁSA'S BIRTH OF THE WAR GOD. See "Trübner's Oriental Series," page 3.

Haas.—Catalogue of Sanskrit and Pali Books in the Library of the British Museum. By Dr. ERNST HAAS. Printed by Permission of the British Museum. 4to. cloth, pp. 200. £1 1*s.*

Haug.—THE AITAREYA BRAHMANAM OF THE RIG VEDA: containing the Earliest Speculations of the Brahmans on the meaning of the Sacrificial Prayers, and on the Origin, Performance, and Sense of the Rites of the Vedic Religion. Edited, Translated, and Explained by MARTIN HAUG, Ph.D.. 2 vols. Cr. 8vo. Map of the Sacrificial Compound at the Soma Sacrifice, pp. 312 and 544. £2 2*s.*

Hunter.—CATALOGUE OF SANSKRIT MANUSCRIPTS (Buddhist) Collected in Nepâl by B. H. HODGSON, late Resident at the Court of Nepâl. Compiled from Lists in Calcutta, France, and England. By W. W. HUNTER, C.I.E., LL.D. 8vo. pp. 28, wrapper. 1880. 2*s.*

Jacob.—HINDU PANTHEISM. See "Trübner's Oriental Series," page 4.

Jaiminîya-Nyâya-Mâlâ-Vistara.—See under AUCTORES SANSCRITI.

Kásikâ.—A COMMENTARY ON PÂNINI'S GRAMMATICAL APHORISMS. By PANDIT JAYÂDITYA. Edited by PANDIT BÂLA SÂSTRÎ, Prof. Sansk. Coll., Benares. First part, 8vo. pp. 490. Part II. pp. 474. 16*s.* each part.

Kern.—THE ARYABHATIYA, with the Commentary Bhatadîpikâ of Paramadiçvara, edited by Dr. H. KERN. 4to. pp. xii. and 107. 9*s.*

Kern.—THE BRHAT-SANHITÁ; or, Complete System of Natural Astrology of Varâha-Mihira. Translated from Sanskrit into English by Dr. H. KERN, Professor of Sanskrit at the University of Leyden. Part I. 8vo. pp. 50, stitched. Parts 2 and 3 pp. 51–154. Part 4 pp. 155–210. Part 5 pp. 211–266. Part 6 pp. 267–330. Price 2*s.* each part. [*Will be completed in Nine Parts.*

Kielhorn.—A GRAMMAR OF THE SANSKRIT LANGUAGE. By F. KIELHORN, Ph.D., Superintendent of Sanskrit Studies in Deccan College. Registered under Act xxv. of 1867. Demy 8vo. pp. xvi. 260. cloth. 1870. 10*s.* 6*d.*

Kielhorn.—KÂTYÂYANA AND PATANJALI. Their Relation to each other and to Panini. By F. KIELHORN, Ph. D., Prof. of Orient. Lang. Poona. 8vo. pp. 64. 1876. 3*s.* 6*d.*

Laghu Kaumudí. A Sanskrit Grammar. By Varadarája. With an English Version, Commentary, and References. By JAMES R. BALLANTYNE, LL.D., Principal of the Sanskrit College, Benares. 8vo. pp. xxxvi. and 424, cloth. £1 11*s.* 6*d.*

Lanman.—On Noun-Inflection in the Veda. By R. LANMAN, Associate Prof. of Sanskrit in Johns Hopkins University. 8vo. pp. 276, wrapper. 1880. 10*s.*

Lanman.—A SANSKRIT READER, with Vocabulary and Notes. By C. R. LANMAN, Prof. of Sanskrit in Harvard College. Part I. and II.—Text and Vocabulary. Imp. 8vo. pp. xx.—294, cloth. 1884. 10*s.* 6*d.*

Mahabharata.—TRANSLATED INTO HINDI for Madan Mohun Bhatt, by KRISHNACHANDRADHARMADHIKARIN, of Benares. Containing all but the Harivansa. 3 vols. 8vo. cloth. pp. 574, 810, and 1106. £3 3*s.*

Mahábhárata (in Sanskrit), with the Commentary of Nílakaṇṭha. In Eighteen Books: Book I. Ádi Parvan, fol. 248. II. Sabhá do. fol. 82. III. Vana do. fol. 312. IV. Viráṭa do. fol. 62. V. Udyoga do. fol. 180. VI. Bhíshma do. fol. 189. VII. Droṇa do. fol. 215. VIII. Karṇa do fol. 115. IX. Ṣalya do. fol. 42. X. Sauptika do. fol. 19. XI. Strí do. fol. 19. XII. Ṣánti do.:—*a.* Râjadharma, fol. 128; *b.* Ápadharma, fol. 41; *c.* Mokshadharma, fol. 290. XIII. Anuṣâsana Parvan, fol. 207. XIV. Áṣwamedhika do. fol. 78. XV. Áṣramavâsika do. fol. 26. XVI. Mausala do. fol. 7. XVII. Máháprasthânika do. fol. 3. XVIII. Swargarokaṇa do. fol. 8. Printed with movable types. Oblong folio. Bombay, 1863. £12 12*s.*

Maha-Vira-Charita; or, the Adventures of the Great Hero Rama. An Indian Drama in Seven Acts. Translated into English Prose from the Sanskrit of Bhavabhüti. By JOHN PICKFORD, M.A. Crown 8vo. cloth. 5*s.*

Maino-i-Khard (The Book of the).—The Pazand and Sanskrit Texts (in Roman characters) as arranged by Neriosengh Dhaval, in the fifteenth century. With an English translation, a Glossary of the Pazand texts, containing the Sanskrit, Rosian, and Pahlavi equivalents, a sketch of Pazand Grammar, and an Introduction. By E. W. WEST. 8vo. sewed, pp. 484. 1871. 16*s.*

Manava-Kalpa-Sutra; being a portion of this ancient Work on Vaidik Rites, together with the Commentary of KUMARILA-SWAMIN. A Facsimile of the MS. No. 17, in the Library of Her Majesty's Home Government for India. With a Preface by THEODOR GOLDSTÜCKER. Oblong folio, pp. 268 of letterpress and 121 leaves of facsimiles. Cloth. £4 4*s.*

Mandlik.—THE YÁJÑAVALKYA SMṚITI, Complete in Original, with an English Translation and Notes. With an Introduction on the Sources of, and Appendices containing Notes on various Topics of Hindu Law. By V. N. MANDLIK. 2 vols. in one. Roy. 8vo. pp. Text 177, and Transl. pp. lxxxvii. and 532. Bombay, 1880. £3.

Megha-Duta (The). (Cloud-Messenger.) By Kālidāsa. Translated from the Sanskrit into English verse, with Notes and Illustrations. By the late H. H. WILSON, M.A., F.R.S., Boden Professor of Sanskrit in the University of Oxford, etc., etc. The Vocabulary by FRANCIS JOHNSON, sometime Professor of Oriental Languages at the College of the Honourable the East India Company, Haileybury. New Edition. 4to. cloth, pp. xi. and 180. 10*s.* 6*d.*

Muir.—TRANSLATIONS from Sanskrit Writers. See "Trübner's Oriental Series," page 3.

Muir.—ORIGINAL SANSKRIT TEXTS, on the Origin and History of the People of India, their Religion and Institutions. Collected, Translated, and Illustrated by JOHN MUIR, Esq., D.C.L., LL.D., Ph.D.

Vol. I. Mythical and Legendary Accounts of the Origin of Caste, with an Inquiry into its existence in the Vedic Age. Second Edition, re-written and greatly enlarged. 8vo. pp. xx. 532, cloth. 1868. 21*s*.

Vol. II. The Trans-Himalayan Origin of the Hindus, and their Affinity with the Western Branches of the Aryan Race. Second Edition, revised, with Additions. 8vo. pp. xxxii. and 512, cloth. 1871. 21*s*.

Vol. III. The Vedas: Opinions of their Authors, and of later Indian Writers, on their Origin, Inspiration, and Authority. Second Edition, revised and enlarged. 8vo. pp. xxxii. 312, cloth. 1868. 16*s*.

Vol. IV. Comparison of the Vedic with the later representations of the principal Indian Deities. Second Edition Revised. 8vo. pp. xvi. and 524, cloth. 1873. 21*s*.

Vol. V. Contributions to a Knowledge of the Cosmogony, Mythology, Religious Ideas, Life and Manners of the Indians in the Vedic Age. Third Edition. 8vo. pp. xvi. 492, cloth, 1884. 21*s*.

Nagananda; OR THE JOY OF THE SNAKE-WORLD. A Buddhist Drama in Five Acts. Translated into English Prose, with Explanatory Notes, from the Sanskrit of Sri-Harsha-Deva. By PALMER BOYD, B.A., Sanskrit Scholar of Trinity College, Cambridge. With an Introduction by Professor COWELL. Crown 8vo., pp. xvi. and 100, cloth. 4*s*. 6*d*.

Nalopákhyánam.—STORY OF NALA; an Episode of the Mahá-Bhárata. The Sanskrit Text, with Vocabulary, Analysis, and Introduction. By MONIER WILLIAMS, M.A. The Metrical Translation by the Very Rev. H. H. MILMAN, D.D. 8vo. cl. 15*s*.

Naradiya Dharma Sastram; OR, THE INSTITUTES OF NARADA. Translated for the First Time from the unpublished Sanskrit original. By Dr. JULIUS JOLLY, University, Wurzburg. With a Preface, Notes chiefly critical, an Index of Quotations from Narada in the principal Indian Digests, and a general Index. Crown 8vo., pp. xxxv. 144, cloth. 10*s*. 6*d*.

Oppert.—List of Sanskrit Manuscripts in Private Libraries of Southern India. Compiled, Arranged, and Indexed, by GUSTAV OPPERT, Ph.D. Vol. I. Royal 8vo. cloth, pp. 620. 1880. 21*s*.

Oppert.—ON THE WEAPONS, ARMY ORGANIZATION, AND POLITICAL MAXIMS of the Ancient Hindus. With Special Reference to Gunpowder and Fire Arms. By G. OPPERT. 8vo. sewed, pp. vi. and 162. Madras, 1880. 7*s*. 6*d*.

Patanjali.—THE VYÂKARANA-MAHÂBHÂSHYA OF PATANJALI. Edited by F. KIELHORN, Ph.D., Professor of Oriental Languages, Deccan College. Vol. I., Part I. pp. 200. 8*s*. 6*d*.

Rámáyan of Válmiki.—5 vols. See under GRIFFITH.

Ram Jasan.—A SANSKRIT AND ENGLISH DICTIONARY. Being an Abridgment of Professor Wilson's Dictionary. With an Appendix explaining the use of Affixes in Sanskrit. By Pandit RAM JASAN, Queen's College, Benares. Published under the Patronage of the Government, N.W.P. Royal 8vo. cloth, pp. ii. and 707. 28*s*.

Rig-Veda Sanhita.—A COLLECTION OF ANCIENT HINDU HYMNS. Constituting the First Ashtaka, or Book of the Rig-veda; the oldest authority for the religious and social institutions of the Hindus. Translated from the Original Sanskrit by the late H. H. WILSON, M.A. Second Edition, with a Postscript by Dr. FITZEDWARD HALL. Vol. I. 8vo. cloth, pp. lii. and 348. Price 21*s*.

Rig-Veda Sanhita.—A Collection of Ancient Hindu Hymns, constituting the Fifth to Eighth Ashtakas, or books of the Rig-Veda, the oldest Authority for the Religious and Social Institutions of the Hindus. Translated from the Original Sanskrit by the late HORACE HAYMAN WILSON, M.A., F.R.S., etc. Edited by E. B. COWELL, M.A., Principal of the Calcutta Sanskrit College. Vol. IV. 8vo. cloth, pp. 214. 14*s*.
A few copies of Vols. II. and III. still left. [*Vols. V. and VI. in the Press.*

Rig-Veda-Sanhita: THE SACRED HYMNS OF THE BRAHMANS. Translated and explained by F. MAX MÜLLER, M.A., LL.D., Fellow of All Souls' College, Professor of Comparative Philology at Oxford, Foreign Member of the Institute of France, etc., etc. Vol. I. Hymns to the Maruts, or the Storm-Gods. 8vo. cloth, pp. clii. and 264. 1869. 12*s*. 6*d*.

Rig-Veda.—THE HYMNS OF THE RIG-VEDA in the Samhita and Pada Texts. Reprinted from the Editio Princeps. By F. MAX MÜLLER, M.A., etc. Second edition. With the Two Texts on Parallel Pages. In 2 vols. 8vo. pp. 1700, sewed. 1877 32*s*.

Sabdakalpadruma, the well-known Sanskrit Dictionary of RAJÁH RADHAKANTA DEVA. In Bengali characters. 4to. Parts 1 to 40. (In course of publication.) 3*s*. 6*d*. each part.

Sâma-Vidhâna-Brâhmana. With the Commentary of Sâyana. Edited, with Notes, Translation, and Index, by A. C. BURNELL, M.R.A.S. Vol. I. Text and Commentary. With Introduction. 8vo. cloth, pp. xxxviii. and 104. 12*s*. 6*d*.

Sakuntala.—A SANSKRIT DRAMA IN SEVEN ACTS. Edited by MONIER WILLIAMS, M.A. Second Edition. 8vo. cl. £1 1*s*.

Sakuntala.—KÂLIDÂSA'S ÇAKUNTALÂ. The Bengalí Recension. With Critical Notes. Edited by RICHARD PISCHEL. 8vo. cloth, pp. xi. and 210. 14*s*.

Sarva-Sabda-Sambodhini; OR, THE COMPLETE SANSKRIT DICTIONARY. In Telugu characters. 4to. cloth, pp. 1078. £2 15*s*.

Surya-Siddhanta (Translation of the).—*See* Whitney.

Táittiríya-Pratiçakhya.—See WHITNEY.

Tarkavachaspati.—VACHASPATYA, a Comprehensive Dictionary, in Ten Parts. Compiled by TARANATHA TARKAVACHASPATI, Professor of Grammar and Philosophy in the Government Sanskrit College of Calcutta. An Alphabetically Arranged Dictionary, with a Grammatical Introduction and Copious Citations from the Grammarians and Scholiasts, from the Vedas, etc. Parts I. to XIII. 4to. paper. 1873–6. 18*s*. each Part.

Thibaut.—THE SÚLVASÚTRAS. English Translation, with an Introduction. By G. THIBAUT, Ph.D., Anglo-Sanskrit Professor Benares College. 8vo. cloth, pp. 47, with 4 Plates. 5*s*.

Thibaut.—CONTRIBUTIONS TO THE EXPLANATION OF JYOTISHA-VEDÂNGA By G. THIBAUT, Ph.D. 8vo. pp. 27. 1*s*. 6*d*.

Trübner's Bibliotheca Sanscrita. A Catalogue of Sanskrit Literature, chiefly printed in Europe. To which is added a Catalogue of Sanskrit Works printed in India; and a Catalogue of Pali Books. Constantly for sale by Trübner & Co. Cr. 8vo. sd., pp. 84. 2*s*. 6*d*.

Vardhamana.—See Auctores Sanscriti, page 82.

Vedarthayatna (The); or, an Attempt to Interpret the Vedas. A Marathi and English Translation of the Rig Veda, with the Original Samhitâ and Pada Texts in Sanskrit. Parts I. to XXVIII. 8vo. pp. 1—896. Price 3*s*. 6*d*. each.

Vishnu-Purana (The); a System of Hindu Mythology and Tradition. Translated from the original Sanskrit, and Illustrated by Notes derived chiefly from other Purāṇas. By the late H. H. WILSON, M.A., F.R.S., Boden Professor of Sanskrit in the University of Oxford, etc., etc. Edited by FITZEDWARD HALL. In 6 vols. 8vo. Vol. I. pp. cxl. and 200; Vol. II. pp. 343: Vol. III. pp. 348: Vol. IV. pp. 346, cloth; Vol. V. Part I. pp. 392, cloth. 10*s*. 6*d*. each. Vol. V., Part II, containing the Index, compiled by Fitzedward Hall. 8vo. cloth, pp. 268. 12*s*.

Weber.—ON THE RÂMÂYANA. By Dr. ALBRECHT WEBER, Berlin. Translated from the German by the Rev. D. C. Boyd, M.A. Reprinted from "The Indian Antiquary." Fcap. 8vo. sewed, pp. 130. 5*s*.

Weber.—INDIAN LITERATURE. See "Trübner's Oriental Series," page 3.

Whitney.—ATHARVA VEDA PRÁTIÇÁKHYA; or, Çáunakíyá Caturádhyáyiká (The). Text, Translation, and Notes. By WILLIAM D. WHITNEY, Professor of Sanskrit in Yale College. 8vo. pp. 286, boards. £1 11*s*. 6*d*.

Whitney.—SURYA-SIDDHANTA (Translation of the): A Text-book of Hindu Astronomy, with Notes and an Appendix, containing additional Notes and Tables, Calculations of Eclipses, a Stellar Map, and Indexes. By the Rev. E. BURGESS. Edited by W. D. WHITNEY. 8vo. pp. iv. and 354, boards. £1 11*s*. 6*d*.

Whitney.—TÁITTIRÍYA-PRÁTIÇÁKHYA, with its Commentary, the Tribhâshyaratna: Text, Translation, and Notes. By W. D. WHITNEY, Prof. of Sanskrit in Yale College, New Haven. 8vo. pp. 469. 1871. £1 5*s*.

Whitney.—Index Verborum to the Published Text of the Atharva-Veda. By William Dwight Whitney, Professor in Yale College. (Vol. XII. of the American Oriental Society). Imp. 8vo. pp. 384, wide margin, wrapper. 1881. £1 5*s*.

Whitney.—A SANSKRIT GRAMMAR, including both the Classical Language, and the Older Language, and the Older Dialects, of Veda and Brahmana. 8vo. cloth, pp. viii. and 486. 1879. 12*s*.

Williams.—A DICTIONARY, ENGLISH AND SANSCRIT. By MONIER WILLIAMS, M.A. Published under the Patronage of the Honourable East India Company. 4to. pp. xii. 862, cloth. 1851. £3 3*s*.

Williams.—A SANSKRIT-ENGLISH DICTIONARY, Etymologically and Philologically arranged, with special reference to Greek, Latin, German, Anglo-Saxon, English, and other cognate Indo-European Languages. By MONIER WILLIAMS, M.A., Boden Professor of Sanskrit. 4to. cloth, pp. xxv. and 1186 £4 14*s*. 6*d*.

Williams.—A PRACTICAL GRAMMAR OF THE SANSKRIT LANGUAGE, arranged with reference to the Classical Languages of Europe, for the use of English Students, by MONIER WILLIAMS, M.A. 1877. Fourth Edition, Revised. 8vo. cloth. 15*s*.

Wilson.—Works of the late HORACE HAYMAN WILSON, M.A., F.R.S., Member of the Royal Asiatic Societies of Calcutta and Paris, and of the Oriental Soc. of Germany, etc., and Boden Prof. of Sanskrit in the University of Oxford.

Vols. I. and II. ESSAYS AND LECTURES chiefly on the Religion of the Hindus, by the late H. H. WILSON, M.A., F.R.S., etc. Collected and Edited by Dr. REINHOLD ROST. 2 vols. cloth, pp. xiii. and 399, vi. and 416. 21*s*.

Vols. III, IV. and V. ESSAYS ANALYTICAL, CRITICAL, AND PHILOLOGICAL, ON SUBJECTS CONNECTED WITH SANSKRIT LITERATURE. Collected and Edited by Dr. REINHOLD ROST. 3 vols. 8vo. pp. 408, 406, and 390, cloth. Price 36*s*.

Vols. VI., VII., VIII, IX. and X., Part I. VISHNU PURÁNÁ, A SYSTEM OF HINDU MYTHOLOGY AND TRADITION. Vols. I. to V. Translated from the original Sanskrit, and Illustrated by Notes derived chiefly from other Puráṇás. By the late H. H. WILSON, Edited by FITZEDWARD HALL, M.A., D.C.L., Oxon. 8vo., pp. cxl. and 200; 344; 344; 346, cloth. 2*l.* 12*s.* 6*d.*

Vol. X., Part 2, containing the Index to, and completing the Vishnu Puránâ, compiled by Fitzedward Hall. 8vo. cloth, pp. 268. 12*s.*

Vols. XI. and XII. SELECT SPECIMENS OF THE THEATRE OF THE HINDUS. Translated from the Original Sanskrit. By the late HORACE HAYMAN WILSON, M.A., F.R.S. 3rd corrected Ed. 2 vols. 8vo. pp. lxi. and 384; and iv. and 418, cl. 21*s.*

Wilson.—SELECT SPECIMENS OF THE THEATRE OF THE HINDUS. Translated from the Original Sanskrit. By the late HORACE HAYMAN WILSON, M.A., F.R.S. Third corrected edition. 2 vols. 8vo., pp. lxxi. and 384; iv. and 418, cloth. 21*s.*

CONTENTS.

Vol. I.—Preface—Treatise on the Dramatic System of the Hindus—Dramas translated from the Original Sanskrit—The Mrichchakati, or the Toy Cart—Vikram aand Urvasi, or the Hero and the Nymph—Uttara Ráma Charitra, or continuation of the History of Ráma.

Vol. II.—Dramas translated from the Original Sanskrit—Malâti and Mádhava, or the Stolen Marriage—Mudrá Rakshasa, or the Signet of the Minister—Ratnávalí, or the Necklace—Appendix, containing short accounts of different Dramas.

Wilson.—A DICTIONARY IN SANSKRIT AND ENGLISH. Translated, amended, and enlarged from an original compilation prepared by learned Natives for the College of Fort William by H. H. WILSON. The Third Edition edited by Jagunmohana Tarkalankara and Khettramohana Mookerjee. Published by Gyanendrachandra Rayachoudhuri and Brothers. 4to. pp. 1008. Calcutta, 1874. £3 3*s.*

Wilson (H. H.).—See also Megha Duta, Rig-Veda, and Vishnu-Purânâ.

Yajurveda.—THE WHITE YAJURVEDA IN THE MADHYANDINA RECENSION. With the Commentary of Mahidhara. Complete in 36 parts. Large square 8vo. pp. 571. £4 10*s.*

SHAN.

Cushing.—GRAMMAR OF THE SHAN LANGUAGE. By the Rev. J. N. CUSHING. Large 8vo. pp. xii. and 60, boards. Rangoon, 1871. 9*s.*

Cushing.—Elementary Handbook of the Shan Language. By the Rev. J. N. CUSHING, M.A. Small 4to. boards, pp. x. and 122. 1880. 12*s.* 6*d.*

Cushing.—A Shan and English Dictionary. By J. N. CUSHING, M.A. Demy 8vo. cloth, pp. xvi. and 600. 1881. £1 1*s.* 6*d.*

SINDHI.

Trumpp.—GRAMMAR OF THE SINDHI LANGUAGE. Compared with the Sanskrit-Prakrit and the Cognate Indian Vernaculars. By Dr. ERNEST TRUMPP. Printed by order of Her Majesty's Government for India. Demy 8vo. sewed, pp. xvi. and 590. 15*s.*

SINHALESE.

Aratchy.—Athetha Wakya Deepanya, or a Collection of Sinhalese Proverbs, Maxims, Fables, etc. Translated into English. By A. M. S. Aratchy. 8vo. pp. iv. and 84, sewed. Colombo, 1881. 2*s*. 6*d*.

D'Alwis.—A Descriptive Catalogue of Sanskrit, Pali, and Sinhalese Literary Works of Ceylon. By James D'Alwis, M.R.A.S. Vol. I. (all published) pp. xxxii. and 244, sewed. 1877. 8*s*. 6*d*.

Childers.—Notes on the Sinhalese Language. No. 1. On the Formation of the Plural of Neuter Nouns. By the late Prof. R. C. Childers. Demy 8vo. sd., pp. 16. 1873. 1*s*.

Mahawansa (The)—The Mahawansa. From the Thirty-Seventh Chapter. Revised and edited, under orders of the Ceylon Government, by H. Sumangala, and Don Andris de Silva Batuwantudawa. Vol. I. Pali Text in Sinhalese Character, pp. xxxii. and 436.—Vol. II. Sinhalese Translation, pp. lii. and 378, half-bound. Colombo, 1877. £2 2*s*.

Steele.—An Eastern Love-Story. Kusa Jātakaya, a Buddhistic Legend. Rendered, for the first time, into English Verse (with notes) from the Sinhalese Poem of Alagiyavanna Mohottala, by Thomas Steele, Ceylon Civil Service. Crown 8vo. cloth, pp. xii. and 260. London, 1871. 6*s*.

SUAHILI.

Krapf.—Dictionary of the Suahili Language. By the Rev. Dr. L. Krapf. With an Appendix, containing an outline of a Suahili Grammar. The Preface will contain a most interesting account of Dr. Krapf's philological researches respecting the large family of African Languages extending from the Equator to the Cape of Good Hope, from the year 1843, up to the present time. Royal 8vo. pp. xl.-434, cloth. 1882. 30*s*.

SWEDISH.

Otté.—Simplified Grammar of the Swedish Language. By E. C. Otté. Crown 8vo. pp. xii.—70, cloth. 1884. 2*s*. 6*d*.

SYRIAC.

Kalilah and Dimnah (The Book of). Translated from Arabic into Syriac. Edited by W. Wright, LL.D., Professor of Arabic in the University of Cambridge. 8vo. pp. lxxxii.-408, cloth. 1884. 21*s*.

Phillips.—The Doctrine of Addai the Apostle. Now first Edited in a Complete Form in the Original Syriac, with an English Translation and Notes. By George Phillips, D.D., President of Queen's College, Cambridge. 8vo. pp. 122, cloth. 7*s*. 6*d*.

Stoddard.—Grammar of the Modern Syriac Language, as spoken in Oroomiah, Persia, and in Koordistan. By Rev. D. T. Stoddard, Missionary of the American Board in Persia. Demy 8vo. bds., pp. 190. 10*s*. 6*d*.

TAMIL.

Beschi.—CLAVIS HUMANIORUM LITTERARUM SUBLIMIORIS TAMULICI IDIOMATIS. Auctore R. P. CONSTANTIO JOSEPHO BESCHIO, Soc. Jesu, in Madurensi Regno Missionario. Edited by the Rev. K. IHLEFELD, and printed for A. Burnell, Esq., Tranquebar. 8vo. sewed, pp. 171. 10*s*. 6*d*.

Lazarus.—A TAMIL GRAMMAR, Designed for use in Colleges and Schools. By J. LAZARUS. 12mo. cloth, pp. viii. and 230. London, 1879. 5*s*. 6*d*.

TELUGU.

Arden.—A PROGRESSIVE GRAMMAR OF THE TELUGU LANGUAGE, with Copious Examples and Exercises. In Three Parts. Part I. Introduction.—On the Alphabet and Orthography.—Outline Grammar, and Model Sentences. Part II. A Complete Grammar of the Colloquial Dialect. Part III. On the Grammatical Dialect used in Books. By A. H. ARDEN, M.A., Missionary of the C. M. S. Masulipatam. 8vo. sewed, pp. xiv. and 380. 14*s*.

Arden.—A COMPANION Telugu Reader to Arden's Progressive Telugu Grammar. 8vo. cloth, pp. 130. Madras, 1879. 7*s*. 6*d*.

Carr.—ఆంధ్రలోకోక్తి చంద్రిక. A COLLECTION OF TELUGU PROVERBS, Translated, Illustrated, and Explained; together with some Sanscrit Proverbs printed in the Devanâgarî and Telugu Characters. By Captain M. W. CARR, Madras Staff Corps. One Vol. and Supplemnt, royal 8vo. pp. 488 and 148. 31*s*. 6*d*

TIBETAN.

Csoma de Körös.—A DICTIONARY Tibetan and English (only). By A. CSOMA DE KÖRÖS. 4to. cloth, pp. xxii. and 352. Calcutta, 1834. £2 2*s*.

Csoma de Körös.—A GRAMMAR of the Tibetan Language. By A. CSOMA DE KÖRÖS. 4to. sewed, pp xii. and 204, and 40. 1834. 25*s*.

Jaschke.—A TIBETAN-ENGLISH DICTIONARY. With special reference to the prevailing dialects; to which is added an English-Tibetan Vocabulary. By H. A. JASCHKE, late Moravian Missionary at Kijelang, British Lahoul. Compiled and published under the orders of the Secretary of State for India in Council. Royal 8vo. pp. xxii.-672, cloth. 30*s*.

Jaschke.—TIBETAN GRAMMAR. By H. A. JASCHKE. Crown 8vo. pp. viii. and 104, cloth. 1883. 5*s*.

Lewin.—A MANUAL of Tibetan, being a Guide to the Colloquial Speech of Tibet, in a Series of Progressive Exercises, prepared with the assistance of Yapa Ugyen Gyatsho by Major THOMAS HERBERT LEWIN. Oblong 4to. cloth, pp. xi. and 176. 1879. £1 1*s*.

Schiefner.—Tibetan Tales. See "Trübner's Oriental Series," page 5

TURKI.

Shaw.—A SKETCH OF THE TURKI LANGUAGE. As Spoken in Eastern Turkistan (Kàshghar and Yarkand). By ROBERT BARKLAY SHAW, F.R.G.S., Political Agent. In Two Parts. With Lists of Names of Birds and Plants by J. SCULLY, Surgeon, H.M. Bengal Army. 8vo. sewed, Part I., pp. 130. 1875. 7*s*. 6*d*.

TURKISH.

Arnold.—A SIMPLE TRANSLITERAL GRAMMAR OF THE TURKISH LANGUAGE. Compiled from various sources. With Dialogues and Vocabulary. By EDWIN ARNOLD, M.A., C.S.I., F.R.G.S. Pott 8vo. cloth, pp. 80. 1877. 2*s*. 6*d*.

Gibb.—OTTOMAN POEMS. Translated into English Verse in their Original Forms, with Introduction, Biographical Notices, and Notes. Fcap. 4to. pp. lvi. and 272. With a plate and 4 portraits. Cloth. By E. J. W. GIBB. 1882. £1 1*s*.

Gibb.—THE STORY OF JEWĀD, a Romance, by Ali Aziz Efendi, the Cretan. Translated from the Turkish, by E. J. W. GIBB. 8vo. pp. xii. and 238, cloth. 1884. 7*s*.

Hopkins.—ELEMENTARY GRAMMAR OF THE TURKISH LANGUAGE. With a few Easy Exercises. By F. L. HOPKINS, M.A., Fellow and Tutor of Trinity Hall, Cambridge. Cr. 8vo. cloth, pp. 48. 1877. 3*s*. 6*d*.

Redhouse.—On the History, System, and Varieties of Turkish Poetry, Illustrated by Selections in the Original, and in English Paraphrase. With a notice of the Islamic Doctrine of the Immortality of Woman's Soul in the Future State. By J. W. REDHOUSE, M.R.A.S. Demy 8vo. pp 64. 1879. (Reprinted from the Transactions of the Royal Society of Literature) sewed, 1*s*. 6*d*.; cloth, 2*s*. 6*d*.

Redhouse.—THE TURKISH CAMPAIGNER'S VADE-MECUM OF OTTOMAN COLLOQUIAL LANGUAGE; containing a concise Ottoman Grammar; a carefully selected Vocabulary, alphabetically arranged, in two parts, English and Turkish, and Turkish and English; also a few Familiar Dialogues; the whole in English characters. By J. W. REDHOUSE, F.R.A.S. Third Edition. Oblong 32mo pp. viii.-372, limp cloth. 1882. 6*s*.

Redhouse.—A SIMPLIFIED GRAMMAR OF THE OTTOMAN-TURKISH LANGUAGE. By J. W. REDHOUSE, M.R.A.S. Crown 8vo. pp. xii.-204, cloth. 1884. 10*s*. 6*d*.

UMBRIAN.

Newman.—THE TEXT OF THE IGUVINE INSCRIPTIONS, with interlinear Latin Translation and Notes. By FRANCIS W. NEWMAN, late Professor of Latin at University College, London. 8vo. pp. xvi. and 54, sewed. 1868. 2*s*.

URIYA.

Browne.—AN URIYÁ PRIMER IN ROMAN CHARACTER. By J. F. BROWNE, B.C.S. Crown 8vo. pp. 32, cloth. 1882. 2*s*. 6*d*.

Maltby.—A PRACTICAL HANDBOOK OF THE URIYA OR ODIYA LANGUAGE. By THOMAS J. MALTBY, Madras C.S. 8vo. pp. xiii. and 201. 1874. 10*s*. 6*d*.

1000
5,1,85

STEPHEN AUSTIN AND SONS, PRINTERS, HERTFORD.

Printed in the USA
CPSIA information can be obtained
at www.ICGtesting.com
LVHW081916271023
762373LV00010B/1249